The Geological History of the British Isles

The Geological History of the British Isles

George M. Bennison M.Sc., Ph.D.

and

Alan E. Wright B.Sc., Ph.D.

Department of Geology, University of Birmingham

Edward Arnold (Publishers) Ltd London 

First published 1969

SBN 7131 2206 4

Printed in Great Britain by
William Clowes and Sons, Limited, London and Beccles

Preface

When Professor L. J. Wills wrote, in 1929, *The Physiographic Evolution of Great Britain* he remarked that it was a subject of which 'so many items have been, or still are, open to more than one interpretation'. This is just as true today. The addition of almost four decades of geological literature has not settled all the issues then in dispute and has, in fact, raised many further problems. In this short book, which attempts to sketch the outline of the main geological events in the British area through the last 3,000 million years, much detail has had to be omitted. Doubtless in some places the need for brevity has led not merely to simplification but to over-simplification. This is not, of course, an encyclopaedic work but a simple textbook which it is hoped will meet the requirements of the undergraduate student of geology by providing a framework of stratigraphical knowledge which can be filled in by further reading. Geological literature abounds—and pours forth in an ever-increasing flood. Suggestions for further reading to be found at the end of the book (divided by chapters), are not intended to be comprehensive: an attempt has been made to restrict the suggested reading to an amount which a student might realistically be expected to attempt. In the choice, works of recent date have been selected for the most part, since these give current arguments as well as—frequently—reviewing earlier views and listing the older references to literature.

Any book such as this must draw heavily from many sources, too many to be individually acknowledged, although where figures have been derived directly or modified from published work acknowledgement is given in the captions. Similarly, acknowledgement is given for the source of tables. The continuous diligent work of the Institute of Geological Sciences (formerly The Geological Survey) and the personal research of geologists—professional and amateur alike—provide the data on which this work so heavily leans. Our thanks are accorded to our colleagues, especially Dr. I. Strachan, Dr. F. Moseley and Dr. J. Tarney, and to Dr. D. R. Bowes and Dr. R. F. Cheeney for reading drafts of the MS. and for much helpful criticism. We are also indebted to Dr. G. S. Boulton, Dr. G. R. Coope and Dr. C. H. S. Sands for illuminating discussion and help. Professor T. N. George's help and encouragement are here gratefully acknowledged. We should also like to thank Mr. L.

Vaughan and Miss S. Reed for help in the preparation of the figures, and Miss V. Adams and Mr. K. T. Bennison for help with the preparation of the MS.

We are grateful to the Director of the Institute of Geological Sciences for permission to reproduce a number of Geological Survey Copyright diagrams, to Blackie & Son and Professor L. J. Wills for permission to reproduce maps from that indispensable work *A Paleogeographical Atlas*, and to the many authors and editors of journals who have so willingly given permission for their work to be utilized.

University of Birmingham G.M.B.
1968 A.E.W.

Contents

I General Principles

1 Principles of Stratigraphy

Stratification

Examination of rocks, wherever they outcrop, frequently reveals that they are of a layered nature. In rocks of igneous origin, particularly volcanic rocks, the layering may be due to flow-structure initiated in the molten state or—on a larger scale—it may be due to the accumulation of successive lava flows. In the case of rocks altered by metamorphic processes, layering may be present due to the re-orientation of platy minerals, or indeed to the creation of new minerals 'adapted' to the stress conditions: the development of cleavage in many slates is an example of the former and schistosity an example of the latter. However, in the vast majority of rocks, conspicuous layering is called bedding or stratification and owes its origin to the manner in which the rocks are accumulated, successive layers of strata—or beds—being deposited one on top of another. A relatively uniform and lithologically distinct layer is called a bed; this is the fundamental sedimentary unit.

Stratigraphy is the study of strata, i.e. bedded rocks which have been laid down and accumulated as spreads of sediment. Since sequences of such beds provide evidence of sequences of geological events, from the strata (their nature, distribution and structures) the geological history of an area can be deduced. Used in its broadest sense the term stratigraphy is closely analogous to the term historical geology although it is not precisely synonymous.

Uniformitarianism

The assumption that sedimentary rocks were formerly laid down by the deposition of sediments, formed in just the same way as sediments are being formed at the present day, is part of the philosophical approach to geology known as uniformitarianism. It dates back, essentially, only to the eminent eighteenth century Scottish geologist, James Hutton. The doctrine of uniformitarianism can be best summarized in the phrase that 'the present is the key to the past': it postulates that the geological processes which can be observed today (weathering and erosion, deposition and accumulation of sediments as well as the outpouring and solidification of lavas) have taken place in an analogous manner in past geological ages. Hutton, in fact, recognized the difference in mode of origin of sedimentary and igneous rocks, incidentally dispelling earlier fallacies that the latter had crystallized from water. The uniformitarian principle can be applied to the biological aspects of geology, as well as to processes of erosion, transportation and deposition of material. The assumption can be made that animals which thrive now in particular environments have required comparable conditions of life in past ages. Processes may not have always proceeded at the same rate as at present, a fact which did not escape James Hutton. Clearly, uniformitarianism cannot explain all phenomena of which we have evidence. We find ourselves at a geologically atypical time in the history of the Earth by virtue of the fact that the last glaciation was so very recent (ending a mere 10,000 years ago), when in fact ice ages are quite uncommon phenomena. Further, two kinds of geological process are known to have taken place although they cannot be directly observed to be happening. There are those geological processes which take place in the depths of the crust, such as the regional metamorphism of rocks or the emplacement of granite, the effects of which can only be observed later after some millions of years of erosion. There are also those processes which take place very, very slowly; for example, the drifting apart of the continents, where the rate is too slow to be measured—or observed—although the cumulative effects of a slow process establish that it is taking or has taken place. Despite these limitations of the principle of uniformitarianism, it none the less has established a basis from which scientific geological thought could proceed, expanding, qualifying and diverging.

Palaeogeography

Sedimentation can take place in a wide variety of environments. Sediments accumulate on the floors of seas and oceans and in lakes. They accumulate in river courses, and seasonally are spread much wider in times of flood. Sediment is also carried, and therefore subsequently deposited, by both wind and ice. Existing sediments and sedimentary rocks (sediments laid down in former geological ages) are, together with igneous and metamorphic rocks, constantly subjected to the processes of weathering and erosion. Thus rocks are worn down and broken up, transported and redeposited to form sediments of a new generation, which ultimately consolidate into sedimentary rocks. The study of stratigraphy requires, therefore, a consideration of the processes of weathering and the way in which the relative importance of the mechanical and chemical processes relate to climate. From the nature of sediments comprising strata, indications of the conditions in which they were produced and deposited (known as the sedimentary environment), can be deduced. For example, whether sediments were deposited in shallow water or deep, in fresh water or salt, on a stagnant lake bed or in aerated water, may be reflected in certain characters of the sediment.

Most geological processes are inexorably slow, although some can be observed to be taking place. If careful watch is kept, rivers can be seen to be eroding their banks: that they are carrying material in suspension is apparent from the colour of the water, brown from silt and mud especially after heavy rains. Many Alpine rivers, however, being glacier fed are a milky colour from the transported fine-grained rock-flour. A few geological phenomena are much more obvious: the deposition of volcanic ash can be very rapid so that early in the eruption of Paricutin (in 1943) bulldozers were employed to clear it from the streets, and in the Philippines eruption of September 1965 ash buried palm trees so that only the tops were visible after two or three days. On the other hand, geological processes causing widespread elevation or subsidence of parts of the earth's crust take place too slowly, for the most part, to be observable. It is true, of course, that the land surface may be locally elevated or depressed a few feet by a major earthquake and that in areas of crustal instability where mountain building is taking place earthquakes are frequent. However, the elevation of mountains tens of thousands of feet high has certainly, for the most part, taken place slowly. None the less, such mountains have been elevated and sedimentary rocks originally formed beneath the sea may be found at a great height above sea level, a fact first appreciated in the fifteenth century by Leonardo da Vinci. Even these seemingly permanent features of the earth's surface relief are transitory.

The student of geology must appreciate that in the vastness of geological time even the slowest of geological processes are capable of producing great changes. Changes in the distribution of land and sea, of the physiography of the land, and of climate have taken place and their study is called palaeogeography. Much of the evidence for palaeogeographical changes is stratigraphical. For simplicity of treatment it is easier to consider those geographical changes which are adduced to have taken place and then to consider the evidence for those changes, rather than to attempt the converse—the examination of the detailed evidence of the rocks and the elucidation of the palaeogeography by piecing together these fragments of information. The latter process, of course, has been followed in the first instance as research progresses.

From the lithology, mineralogy and fossil content of strata and from their deformation and present distribution (that is to say, from the 'record of the rocks' in its fullest sense), can be deduced a chronological sequence of events from their deposition and consolidation to their subsequent uplift and erosion, i.e. a historical geology. Stratigraphy is here considered to include palaeogeography and in general to be synonymous with historical geology. Where possible, a palaeogeographical approach is used in the firm belief that this is the method of most readily assimilating stratigraphy, which can otherwise all too easily become lists of successions and a morass of not clearly relatable facts. Further, the treatment within this book is essentially chronological although, due to lack of knowledge resulting from the imperfections and ambiguities of the stratigraphical record, it is necessary in places to deal with the strata on a purely regional basis.

Geological time

Time is of course continuous, but the record of past events is incomplete. Not only is this true of the history of the human race but of the whole of geological history. As in the case of the history of mankind, the last few centuries are particularly well documented, and in general the more modern the period the more complete is our knowledge, so the geological record of relatively later events is much more complete: from relatively young rocks it is often possible to derive evidence of their conditions of deposition and the detailed palaeogeography including the climate of that time. Older rocks are liable to have had a complex history subsequent to their formation and so they may be difficult to interpret. Of the first three-quarters of the earth's history the evidence is slight and the problems of interpretation are highly complex. Some periods of earth history are, however, particularly well known, especially the Carboniferous. It has attracted an exceptional amount of attention by geologists, due to its economic importance, and more scientific papers have been written on it than any comparable period.

Frequently the history of the earth has been compared with a book from which some pages or even whole sections are missing. Certainly, the record is far from complete. Yet, since the crust of the earth solidified and an atmosphere came into existence, it is probable that at some place or another on the earth's surface sedimentation has been taking place. Much of this sedimentation has not been consolidated to form rocks but has been removed, sorted and redeposited—possibly many times. Many of the sediments which have eventually been consolidated to form sedimentary rocks were, in time, removed by erosion to provide the materials for a next generation of sedimentary rocks. The oldest known sediments (approx. 3,000 million years old) are considerably younger than the age of the earth and are unlikely to be 'first generation' sediments derived from the weathering of the first formed crust.

Because of erosion many formations are now absent from large areas where they were originally deposited, and attempts to deduce their former distribution and extent comprise some of the most interesting stratigraphical problems. It follows that there must be cases of sediments deposited, consolidated and subsequently completely removed. In the record of strata there are many breaks of differing degrees of size and importance, the major breaks representing appreciable portions of geological time. Locally conspicuous breaks have been frequently used to divide geological strata into Systems and the corresponding intervals of time into Periods. The problems resulting from this practice of utilizing the 'natural breaks' are more fully discussed later in this chapter.

Historians are fortunate in being able to date many of the more important events in years, as well as knowing the chronological order of those events. For example, not only is it known that the Great Fire of London followed after the Plague, but it is also known that they took place in the years 1664–5 and 1666 respectively. More widespread changes such as the Industrial Revolution would be difficult to date if it was not possible to relate the social changes involved to a calendar, since it was an eighteenth century event in Western Europe but a twentieth century happening in China, while in some developing countries it has not yet taken place. Many historical 'events' similarly are far from contemporaneous. Although the Christian era is well defined historically by the birth of Christ, the spread of Christianity occupied many centuries, not reaching western Scotland from Ireland until the fifth century and extending northwards at a much later date. In all these cases we are referring the events to a 'reference' time scale, or calendar, which enables us to compare the date of an event in different places.

An absolute time scale and isotopic (radiometric) dating

Until the present century workers on problems of historical geology were able to date events only relatively. Of course, as a greater number of events are dated relative to one another a more complete sequence is built up, hence dating becomes more precise. Within a limited area there is no fundamental reason, theoretically, why a perfectly adequate and complete chronology could not be built up—if the geological record itself were complete. The need, as A. B. Shaw has pointed out, to have a record divided into equal increments—what is known as an 'absolute' time scale—springs largely from man's habit of measuring time in this way, in hours, days and years. However, the non-contemporaneity of geological events when considering larger regions makes some scale of reference imperative. Clearly, an absolute time scale is a very convenient one.

Early attempts to date geological events in years by means of calculating thicknesses of strata and estimating rates of sedimentation were conspicuously unreliable, although in the very limited example provided by varved sediments (laid down in glacial lakes and exhibiting rhythmic seasonal laminations) a precise chronology has been established for late- and post-glacial events in Scandinavia. Hutton and later geologists appreciated that geological time was immense to have permitted the slow geological processes to have accomplished the geographical changes deducible, and a seventeenth century estimate of the date of the origin of the earth as 4004 B.C. had long been discredited. It is, perhaps, since science has been able to supply figures—to give the age of the earth and the age of rocks in years—that the general reader has become ready to accept the real immensity of geological time. (To be more exact, ages of geological events are given in millions of years and the method involves an experimental error of several per cent. A. Holmes deprecated the use of the term 'Absolute' age as inferring an accuracy which is unwarranted.)

That rocks can be dated stems from the discovery of radioactivity and from calculations of the rate of 'decay' or transmutation of radioactive minerals.

The principal elements used, together with their end-products, are:

uranium → lead, thorium → lead, rubidium → strontium and
potassium → argon.

These elements occur, often in very small amounts, in the mineral constituents of some rocks. The dating of representative rocks from all the geological systems, citing the mineral used and the method employed, is given in two comprehensive tables in Arthur Holmes' *Physical Geology*, 2nd edn. (1965, pp. 360–1 and 368–9), to which the reader is referred for an authoritative and up-to-date account of 'Dating the Pages of Earth History'. Only the limitations of radioactive dating and the precautions which must be taken when considering the evidence are here dealt with. Igneous rocks particularly lend themselves to the dating technique since as a rule they include radioactive minerals. In the case of intrusive igneous rocks, which necessarily post-date the sediments they intrude, the problem arises as to how much older the sediments are than the intrusions. Volcanic rocks, both lavas and tuffs, may be interbedded with sediments permitting accurate dating of the latter—but they may be erupted on to an erosion surface of unknown age. Some sediments, notably glauconitic sandstones, potassium salts and more rarely uraniferous shales, contain radioactive elements which have enabled them to be dated directly. In general it has been the practice to establish the relative ages of sedimentary rocks and their age relationship to associated igneous rocks. Isotopic dating of the latter thus provides a reference point in the time scale to which the complete chronology can be related. The problems of isotopic dating of rocks of high grade regional metamorphism are discussed in Chapter 3.

The law of the order of superposition

Since sediments accumulate by deposition, a thickness of strata gradually building up layer by layer, it follows quite obviously that the highest layer in a succession of strata must be the youngest. All such strata must be arranged in chronological order since successive layers are deposited on top of existing layers. Any stratum in a sequence can be dated relatively to the other strata by its position in the sequence. This simple but vital fundamental principle of stratigraphy applies to all sedimentary rocks except those which have been subjected to tectonic upheaval of such magnitude that either: (1) the strata have been so intensely folded that the succession is in places inverted, i.e. upside down—when, of course, the strata will still be arranged in the original sequence, although now the highest stratum is the oldest and successively younger beds occur at lower levels; or (2) parts of the crustal strata have been thrust or have slid over the top of more recently deposited, younger, rocks. In this latter case the structural position of the strata is no longer a guide to their relative age.

Volcanic rocks, both lava flows and bedded volcanic tuffs, can be dated relatively by their position in a stratigraphical sequence. Care must be taken to distinguish between volcanic rocks, contemporaneous with the deposition of the strata, from intrusive igneous rocks, particularly sills—concordant intrusions generally parallel to the bedding, which have been intruded at a later date. Igneous intrusions are necessarily later than any of the strata which they can be seen to intrude, or to thermally metamorphose. Their age cannot be more specifically determined than this from their field relationship to other rocks; they may immediately post-date the strata they cut or may be very much younger, and to determine their age it is necessary to resort to isotopic dating.

Breaks in the geological record

Every bedding plane of a sedimentary rock represents a plane which was at one point in time the interface between sediment and water (in the case of water-laid sediments), that is to say, each bedding plane was the surface of the sediment already deposited. Normally, if the sediment settles in relatively quiet, undisturbed water, these bedding planes will be parallel, and the rock is said to be even-bedded. It may be true that the bedding planes do in fact represent very temporary pauses in sedimentation. Certainly, the strata although not necessarily exhibiting any visible breaks in deposition rarely represent a record of continuous deposition. This can be grossly demonstrated by comparing the length of time represented by any geological formation with known rates of sedimentation: clearly much of geological time is not represented by sediments. Gaps in the geological succession proved by incompleteness of the sequence are called diastems.

Whether a rock possesses few bedding planes and is massive, or is well bedded, probably depends less on whether sedimentation was continuous or interrupted than on the nature of the material. Where the particles are rounded, such as sand grains or ooliths, massive sandstones and limestones occur, but argillites composed of ultra-microscopic platelets of material are commonly very well bedded, e.g. shale. The formation of platy or 'flaky' minerals such as mica will, of course, make a rock more fissile.

Intervals in sedimentation, perhaps quite brief but long enough to permit organisms to bore and burrow into the semi-consolidated substrate, or to encrust its surface, are a not uncommon occurrence in fossiliferous strata.

When sedimentation is interrupted for some appreciable interval but no apparent erosion of existing sediments takes place, and with renewed deposition further beds are deposited parallel to the existing ones, then the interruption of sedimentation may be demonstrable only by the incompleteness of the fossil sequence (see Chapter 2) and it is called a non-sequence. Such a break can occur when the interruption to sedimentation was caused by strong currents and not by tectonic uplift followed by erosion. The succeeding sediments are of the same type of material. Naturally non-sequences are unprovable in barren unfossiliferous rocks, for example in rocks of great antiquity, or in rocks so altered that fossils have failed to be preserved. Non-sequences are frequently known to occur in strata which have attracted great attention from palaeontologists and which, as a result, have been subdivided in detail. In lesser-known groups of strata and in older strata generally there must have been many similar breaks in sedimentation still unproved, and many which must remain undetectable through lack of faunal evidence.

If sediments, subsequent to their deposition and consolidation, are uplifted so that they are brought into the zone of subaerial weathering and erosion, then they will be gradually removed. These processes work constantly towards reducing the land areas to a flat plain (or peneplain). On the resumption of sedimentation, which might be of either terrestrial type or marine (as a result of subsidence of the land, permitting a new incursion of the sea), further strata will be laid down. A break in sedimentation provable by an erosion surface is called an unconformity and the younger bed is said to be unconformable on the older. Naturally, such a break may correspond to a relatively short interval of geological time or a vastly long one. Unconformities are numerous in the geological record and, as will be seen, are of value and convenience to stratigraphers in dividing up the stratigraphical record, despite the problems this practice creates. Frequently, the uplift which causes already deposited strata to be elevated (incidentally terminating sedimentation), and permits their erosion, as well as the subsequent sinking, is not widely uniform. The older strata may be tilted or even folded during the period of non-deposition and this results in the later strata having a markedly different dip from the beds below the plane of unconformity. Such a case is known as an angular unconformity—the majority of unconformities are of this kind—and was first described (1787) by James Hutton, the eighteenth century Scot, from near Loch Ranza, Arran, where beds of Upper Old Red Sandstone (Devonian age) rest on eroded, steeply dipping, Dalradian schists. He subsequently observed unconformities near Jedburgh and at Siccar Point, Berwickshire, where the Old Red Sandstone, with a gentle dip, rests on upturned eroded Silurian beds.

In the field an unconformity can, as a rule, be proved because the erosion surface or ravinement is conspicuous. In addition, the post-unconformity sediments frequently commence with a conglomeratic deposit. For example, each of the Lower Tertiary marine cycles of the London Syncline commences with a flint pebble conglomerate. Frequently the pebbles in a conglomerate at or near the base of a sedimentary series are fragments derived from an older series of rocks below the plane of unconformity: in fact, the presence of such material derived from an older formation is, as a rule, indicative of an unconformity. The nature of the pebbles in a conglomerate can indicate something of the nature of the land surface supplying the material. The petrology of the contained pebbles in a conglomerate or breccia provides information of the palaeogeography. The Lower Brockram of the Vale of Eden contains pebbles chiefly of Lower Carboniferous Limestone strongly suggesting that, at its time of deposition, the Lower Palaeozoic slates and volcanic rocks of the Lake District were not exposed to erosion. However, slightly later in Permian times, lenses of breccia known as the Upper Brockram were laid down and, though of a similar character to the Lower Brockram, contain a great variety

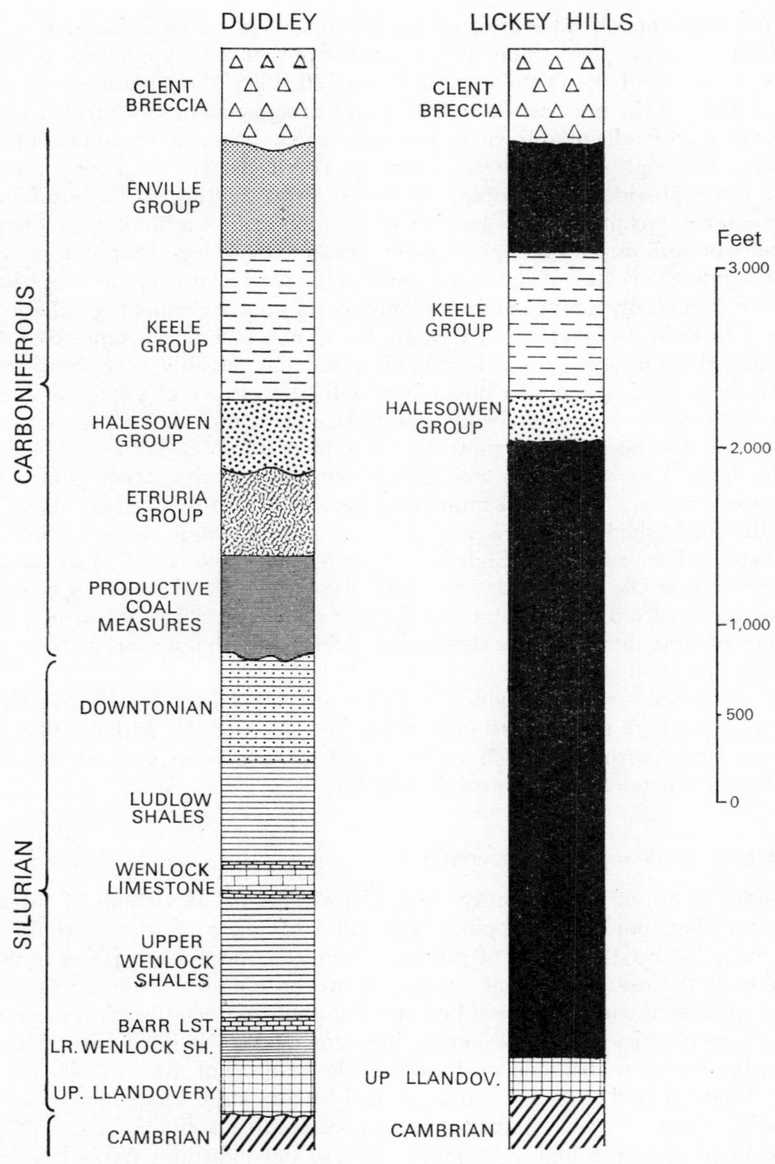

Fig. 1.1 Successions of Palaeozoic rocks in the west Midlands to show the difference in elapsed time represented by unconformities. The succession in the Lickey Hills is less complete than that at Dudley. (Black represents absence of beds at Lickey which are present in the Dudley succession.) From L. J. WILLS, 1950, *Palaeogeography of the Midlands*, 2nd edn. Liverpool Univ. Press, Liverpool; with the author's permission.

of fragments of which the main source was an area of Lower Palaeozoic rocks. Of course this evidence alone is not conclusive that the material was derived from either the Lake District or the Cross Fell Inlier, the nearest outcrops of such rocks at the present. As well as derived fragments of older rocks, a bed may contain fossils of a previous geological age known as derived or remanié fossils. Although such occurrences are not very common, the Lower Brockram again provides an example: the lowest strata (resting on Carboniferous Limestone unconformably) are full of fragments of Carboniferous fossils, chiefly crinoid ossicles conspicuous on weathered surfaces, despite their red staining to the colour of the rock matrix. This particular example of derived fossils is especially significant for not only are they not of Permian age, although found in Permian breccia, but they are found in a rock which represents depositional conditions in which crinoids could not possibly have flourished.

In the case of any angular unconformity the lowest bed of a series of strata is seen to rest on beds of differing ages. Such a phenomenon is known as overstep, and the post-unconformity bed is said to overstep onto successively older beds. Unconformities are often, though not always, accompanied by overstep. Some, while representing long periods of non-deposition cannot be readily established where pre- and post-unconformity strata have similar dip and strike. For example, at Elgin in the north-east of Scotland Permian sandstones rest on Old Red Sandstone, of similar lithology and colour, with little discordance of dip, despite an unconformity representing a very considerable length of time during which thousands of feet of Carboniferous strata were laid down elsewhere.

A geological succession which includes unconformities is an incomplete record. Compare the succession seen in the region of the Lickey Hills, S. Birmingham, with the much more complete succession seen at Dudley (although this too includes breaks) (Fig. 1.1).

Contrasting planes of unconformity

Some planes of unconformity are uniformly flat or, as a result of subsequent tilting, uniformly dipping. This kind of plane of unconformity is generally followed by beds of marine origin and the erosion plane is undoubtedly the result of marine erosion. However, not all marine successions commence with a uniform basal bed on a flat plane of unconformity; in some cases irregularities in the erosion surface remained to be filled with littoral deposits similar in character to those near the margins of the area of deposition. Where a land surface has been eroded by terrestrial agents, peneplanation is seldom if ever complete, so that succeeding strata (of continental facies) will rest on a highly irregular plane of unconformity. Wills has described this type of unconformity as 'buried landscape'. Compare the unconformity of the Torridonian in north-west Scotland—the plane of unconformity varies in height by hundreds of feet when traced along the general strike—with the uniform plane of the sub-Cambrian unconformity in the same area (Fig. 1.2).

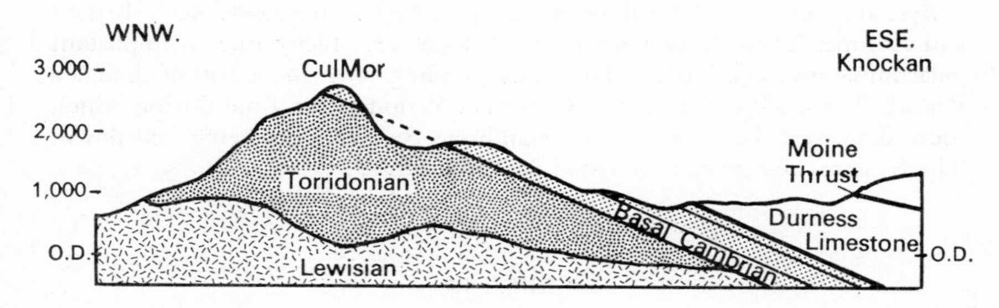

Fig. 1.2 Section showing the comparison between the pre-Cambrian and pre-Torridonian planes of unconformity in the Assynt District of northwest Scotland.

Divisions of the stratigraphical column

Considerable use has been made of the naturally occurring breaks, unconformities, to divide up the stratigraphical column. Within a limited area this is a readily usable method and hence its frequent employment. However, since any unconformity represents a period of time, which in the area under consideration is not represented by any strata, the geological record is necessarily incomplete. Now the earth movements which cause breaks in sedimentation are not contemporaneous over wide areas. Thus if series of strata defined by natural breaks above and below are equated the correlation may be erroneous, deposition commencing in one area sooner than in another; similarly sedimentation may be terminated earlier in one area.

On a large scale problems of definition arise although the major stratigraphical divisions, Systems, have usually been defined where the succession is fairly complete. The Permian, originally defined in Russia by Murchison, occurs in Britain always resting with very marked unconformity on older rocks. The convenience of readily being able to define its base is offset by the fact that it is not always clear how much of the Permian period is unrepresented by strata. On the other hand, in the ancient (Hercynian) massifs of western Europe the major unconformity towards the close of the Palaeozoic Era occurs considerably earlier, and no break occurs at the base of the Permian: there is a transition upwards from beds of late Carboniferous age to beds of similar lithological character of Permian age. Here the succession is complete but the problem now is where to draw the Carboniferous-Permian boundary and how can any such boundary be compared with the Carboniferous-Permian boundary in Britain? Clearly it is desirable that stratigraphical divisions should be contemporaneous, i.e. that the boundaries between such divisions should be time-planes. Although this may be, in fact, an unattainable ideal, the use of fossils in correlation—an essential tool of the stratigrapher, discussed in the next chapter—brings a certainty to correlation that lithological correlations generally lack.

Systems have been defined, originally by reference to successions in Britain and Europe. These have been grouped together, chiefly due to important palaeontological similarities. The corresponding term for a unit of time is a Period. Thus, for example, the Cambrian Period is the time during which rocks defined as belonging to the Cambrian System were being laid down. The Systems recognized are listed in Table 1.1.

TABLE 1.1

Era	*Period or System*		
CAINOZOIC	Quaternary	Holocene[a]	
		Pleistocene	
	Tertiary	Pliocene	
		Miocene	
		Oligocene	
		Eocene	
		Paleocene	
MESOZOIC	Cretaceous		
	Jurassic		
	Triassic		
UPPER PALAEOZOIC	Permian		
	Carboniferous ≡	Pennsylvanian	
		Mississippian	
	Devonian		
LOWER PALAEOZOIC	Silurian		
	Ordovician		
	Cambrian		
PRECAMBRIAN	Eocambrian[b]		
	Proterozoic[b]		
	Archaean[b]		

[a] Considered superfluous by some authors.
[b] Not strictly analogous to the later Systems.

Problems of lithological correlation

The relative ages of beds in a succession can be determined in a limited area according to the Law of Superposition by the relative position of those beds. Further it is possible, by geological mapping, to trace some beds across country and so build up more complete successions. For example:

Locality I	Locality II	Locality III
Bed J	Bed H	Bed D
I	F	C
H	E	B
G	D	A

Clearly the order of succession is A, B, C, D, E, . . . J. Bed G is missing at Locality II but is present at Locality I so that we are able to deduce what appears to be a complete succession. It is true that at other localities (Locality IV, etc.) we may find that the succession is more complete than we had supposed and that additional beds occur, for example, between beds B and C, which we call B_1, B_2, giving in reality a succession A, B, B_1, B_2, C, D, etc. None the less, the original order of succession holds good.

The whole geological succession cannot be seen at any one locality or in any one section (although the Grand Canyon reveals a spectacularly complete partial succession and is unique); neither can the majority of beds be traced directly along their outcrop for great distances. However, some beds, known as marker horizons or index horizons, which are both of wide extent and readily identifiable character, can be recognized at widely separated localities. The Ludlow Bone Bed, recognizable at many outcrops in the Welsh Borders, is an example.

If beds were of constant character difficulties would still arise. These would chiefly be due to post-depositional tectonic structures, as a rule folding and faulting of the strata which, in conjunction with subsequent erosion, create discontinuous outcrops. The Westphalian Coal Measures in Britain and north-western Europe are usually found in synclines or basins (often bounded by major faults), now separated. The establishment of their contemporaneity of deposition over a wide area has been one of the major accomplishments of stratigraphical research, and incidentally one of the most important economically.

However, geological strata are not generally of a constant character laterally. This is especially true of areas of complex geology which have had a tectonic history including several periods of orogenesis. While Gignoux has pointed out that it is the areas of the world with a relatively stable ancient 'shield' core (only occasionally inundated by the sea and with the sediments little folded and disturbed by orogeny, and where strata may be of relatively constant character for hundreds of miles) which are typical, we are not dealing with such a region. North-western Europe has had a complex tectonic and sedimentational history, and Britain is fortunate in its great variety of strata —rocks from almost every period from the Precambrian to late Tertiary occur in a distance of no more than 150 miles. This great variety naturally has attendant problems.

As in present-day seas at one and the same time muds are accumulating in one place while sands are being deposited elsewhere, so geological strata formed contemporaneously show lateral changes in lithology due to environmental differences. In Upper Jurassic times during the 'Corallian' (Oxfordian) epoch a succession of limestones was laid down in the south of England and as far north as Oxford. Contemporaneously muddy sediments were being deposited from about Bedford to the site of the present-day Humber; these beds are called the Ampthill Clay. In north-east Yorkshire, where beds of this age again appear at the surface, they comprise a series of limestones remarkably similar to those of southern England. As a corollary of the above, that beds of different lithological character can be deposited at the same time, it is also true that a bed of particular lithology, and mappable as a recognizable stratigraphic unit, may not have been deposited contemporaneously wherever it occurs. This phenomenon is known as diachronism. The sandy beds which occur towards the end of the Lower Carboniferous Period in the west of England (known locally as the Drybrook Sandstone) are diachronous: sandy deposition commenced much earlier in Carboniferous times further north than in the Bristol district. One way in which this phenomenon may be brought about is by a shifting shoreline. Assuming that coarser material is littoral and that the further from a shoreline the finer is the material being deposited (a generalization which has many exceptions), then an advancing or receding shoreline would cause a lateral shift in the area in which the coarser arenaceous deposits were forming. Wherever this is the case—and it may be fairly general of shallow water sediments—then lithological boundaries which the geologist in the field maps are not 'time-planes'.

Wherever we have an example of an area of deposition which is either shrinking or expanding there will be changes, with time, in the distribution of the deposition of the different lithological types. In an expanding basin of marine sediments the lowest bed will not be of as great extent as the succeeding one, which will spread beyond it and is said to overlap it. Often, but not necessarily, this occurs above an unconformity where marine sediments are re-invading the land and spreading across an erosion surface. Conversely, a shrinking area of deposition will result in successive beds being of lesser extent and this is known as offlap. Examples of both occur where subsidence of the crust is oscillatory and the resulting sedimentation cyclical. An excellent example, the Lower Tertiary beds of the Anglo-Parisian Basin, was described by L. D. Stamp over 30 years ago. Here marine sediments pass laterally into deposits of continental facies and, due to oscillatory subsidence producing alternating expansion and contraction of the area of marine deposits, near the margins of the basin an alternating succession of marine and continental rocks occurs. Overlap and offlap alternate. The overlaps are readily recognizable because the sea advanced over an erosion surface and the lowest bed of each transgression is a flint-pebble conglomerate. The offlaps of marine sedimentation are accompanied, in this case, by lateral and vertical gradation into beds of continental facies. Note that the succession of strata at localities A and B differ very considerably (Fig. 1.3). In such variable successions as this it is very difficult to find a typical succession—or type succession—which can be taken as a standard against which other successions of the same age can be compared. Today this method in stratigraphy, the choice of a type succession, is largely outmoded, although the need remains to find an area where the strata of a particular age are fully developed to permit the division of a System into the Series and Formations which it is hoped to recognize at other localities.

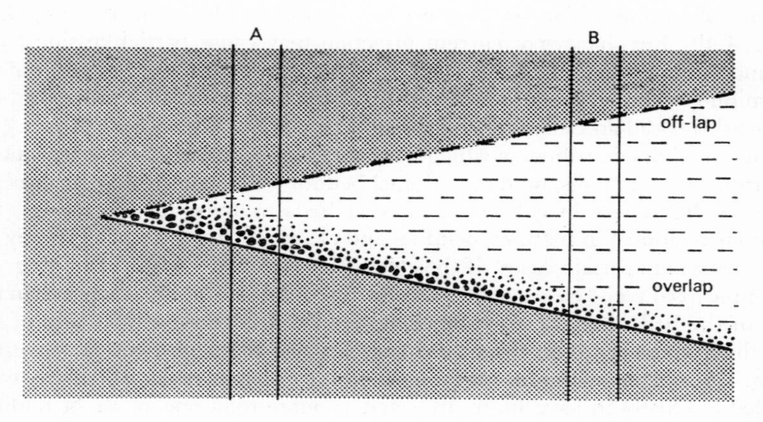

Fig. 1.3 Overlap and offlap of marine sediments (broken lines). Overlap by marine sediments may commence with a basal conglomerate deposited on an erosion surface, and the change from continental to marine facies is abrupt. Note the difference in succession present at two localities A and B.

'Way-upness' of strata

Earth movements may produce structures, essentially overfolds, in which the strata are locally or even regionally inverted. It will be seen in the next chapter that it is often possible to determine whether the beds are the right way up (i.e. in the sequence in which they were deposited), or if they are upside down, by comparing the succession of fossils with a known faunal sequence. However, fossil evidence may be lacking, particularly in rocks which exhibit such structures, for they are often sufficiently metamorphosed to have had all or most indications of life obliterated. Fortunately there are lithological features which provide evidence of 'way-upness' and these may be preserved in rocks of quite a high grade of metamorphism, such as the Moinian and Dalradian rocks of the Scottish Highlands.

Graded bedding occurs when a gradual passage upwards (in a distance of a few millimetres or centimetres) from coarser to finer material is seen. It is especially characteristic of finer grained arenaceous rocks such as greywackes, siltstones and some sandstones. Particles carried by water settle according to

Stokes' Law (with the exception of fine clay particles) so that larger particles settle first, successively finer material settling in order of decreasing size. However, the boundary between the fine material and the succeeding deposit of coarse material is sharp as it represents a renewal of the supply of material (Fig. 1.4a).

Cross-bedding, also referred to as false-bedding or current-bedding (and dune-bedding in the case of aeolian sands), occurs where sand accumulates in sandbanks or deltas at an angle up to 35° (the angle of rest) to the general planes of bedding. Top-set, fore-set and bottom-set beds can be recognized (Fig. 1.4b), but changes of current direction may cause partial erosion truncating the cross-bedding (Fig. 1.4c). Clearly, this feature can be used for determining whether strata are the right way up, and it was first utilized in the Scottish Dalradian Series by Vogt in 1930.

In fine grained sediments which have developed cleavage, but where metamorphism has not obscured the original bedding direction (it may be discernible by differences in colour of different beds or by graded bedding), the angular relationship between bedding and cleavage planes reveals the way up of the strata in certain cases. If the cleavage is associated with the folding, for example axial plane cleavage, then if the dip of the cleavage is steeper than the dip of the beds the latter are the right way up. Conversely, if the dip of the bedding is steeper than the dip of the cleavage then the beds are inverted (Fig. 1.4d). This method must be used with caution as rocks which are overfolded are likely to have been subjected to more than one phase of folding. The bedding-cleavage relation may be superimposed on an already inverted sequence.

Correlation of borehole successions

Experiments carried out by C. and M. Schlumberger in 1927 have since that time been expanded into a system of correlation by geophysical methods widely used in the exploration of oilfield geology for several decades, and latterly applied to a wider range of stratigraphical problems. Its value in correlation has been firmly established and its potential is immense at a time when the search for oil and gas in the North Sea and its neighbouring countries is being intensified: it is a technique employed during the drilling of boreholes. Its great limitation is the difficulty of relating borehole geophysical data to outcrop information.

Instruments lowered down a borehole as it is being drilled provide a continuous record of certain physical properties of the rocks penetrated. The most useful of these characters are resistivity and radioactivity, the latter being divisible into naturally occurring radioactivity (gamma-ray logging) and induced radioactivity (neutron logging). The Schlumberger techniques provide identification—and hence correlation—of strata on a physical-lithological basis and, additionally, give thickness of beds, the presence of water-bearing strata, the quality of coal seams, etc.

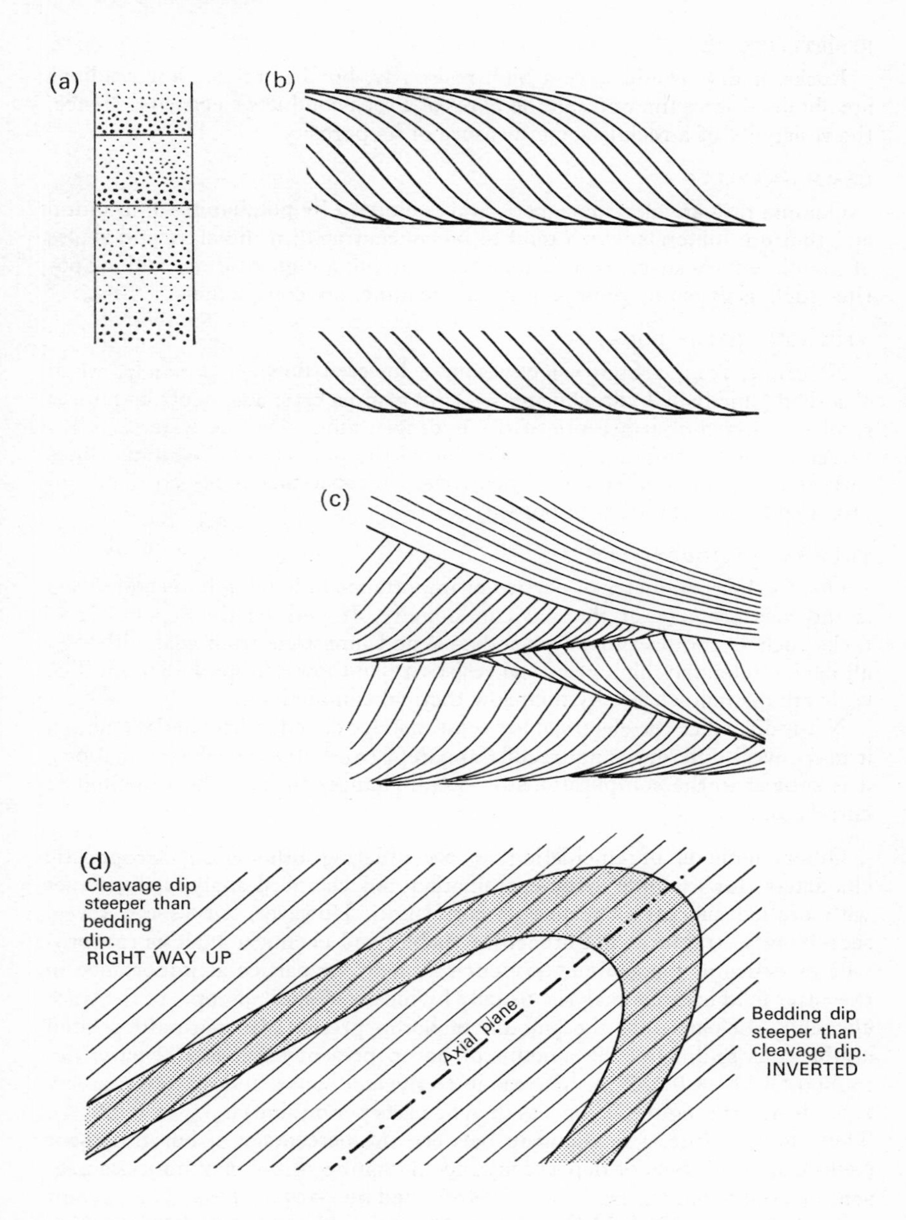

Fig. 1.4 Criteria of 'way-up' of strata: (*a*) graded bedding; (*b*) cross-bedding before and after removal of top-set beds by erosion; (*c*) typical cross-bedding (of a dune sandstone); (*d*) the relationship of axial-plane cleavage to bedding.

RESISTIVITY

Rocks, if dry, would have a high resistivity, but in practice low readings are obtained since the water in the pores of a rock conducts electricity. Hence, the resistivity of a rock is some measure of its porosity.

GAMMA-RAY LOG

Gamma rays are electromagnetic waves emitted by potassium 40, uranium and thorium minerals which tend to be concentrated in the clays and shales of a sedimentary succession. Thus shales, giving a high reading, and evaporites such as gypsum, giving a very low reading, are conspicuous.

NEUTRON-GAMMA LOG

Neutrons, from a high velocity source lowered down a borehole, when absorbed (chiefly by hydrogen atoms), emit gamma rays. The neutron-gamma record is therefore a measure of the hydrogen atoms—hence water or oil—present. It is, by inference, an indication of the porosity of the strata. Since carbon is a good moderator of neutrons, carbonaceous rocks are liable to give a spurious indication of porosity.

THERMAL CONDUCTIVITY

One further parameter of particular importance in logging bore-hole strata is the measurement of thermal conductivity. It permits the separation of rocks such as compact massive sandstone and limestone from coal, although all have moderately high electrical resistivity and low gamma radiation. The coals are distinguishable by their low thermal conductivity.

Naturally, since the electrical log is intimately related to lithologies (though it may reveal differences not readily apparent from lithological examination), it is subject to the complications of lateral changes in facies as a method of correlation.

Other methods of elucidating or accentuating lithologically diagnostic characters (for example x-ray examination and chemical analysis) have met with limited success as means of correlation. However, x-rays can reveal sedimentary structures not otherwise visible, and chemical analysis can provide evidence of the sedimentary environment. Of particular importance in the latter field is rapid analysis, usually by means of spectroscope, of the trace elements—in particular the amount of boron present. In marine-deposited argillaceous sediments the amount of boron present exceeds 109 parts per million (B_2O_3 = 0·035%), whereas in freshwater shales and clays it scarcely exceeds half this figure, being less than 62 parts per million (B_2O_3 = 0·025%). There is, as a rule, a relationship between the percentage of boron present (reflecting conditions of deposition) and the nature of the clay minerals present in a sediment, the latter being determined by x-ray analysis. The amount of work done in this field is still not extensive and at present is of primary value for elucidating the conditions under which rocks were deposited, rather than for purposes of correlation.

2 Fossils and Stratigraphy

Fossilization

Fossils are indications of the former existence of living organisms, both animals and plants. They range from the preservation of the complete animal, exceptionally, to the indirect evidence of the animal, for example its tracks or trails and its coprolites. Usually only the hard skeletal parts or shell of an animal are preserved as fossil, although chitinous skeletons and carapaces of animals are commonly preserved (by the loss of volatiles from the hydrocarbons) as carbonaceous films, as also are plants. Animals lacking a skeleton or shell are infrequently preserved as fossil, and no doubt whole groups of animals have existed on the earth for considerable periods of time without leaving any record. Many living shell-less invertebrate animals are unknown from the geological record, yet it seems unlikely that they have all recently evolved: clearly, it is more probable that they failed to be fossilized. However, soft-bodied animals such as the jelly-fish leave imprints in the wet sand which are occasionally preserved, as are the burrows and casts of worms. While fossils of this kind provide only slight evidence of the nature of the associated animal they may, none the less, be of use to the stratigrapher.

A further condition which must generally be fulfilled, if an organism is to be preserved as a fossil, is that quite soon after death it must be covered by sediment to reduce the chance of bacterial attack and decay. For this reason animals which live in water are much more likely to become fossilized than are terrestrial creatures. A rock may be devoid of fossils ('barren') for three reasons: it may have been deposited in conditions which were unable to support life, life may have existed but failed to be preserved as fossils, or life may have existed and become fossilized yet subsequently destroyed. The contained fossils in a rock may be destroyed by major changes of texture and composition of the parent rock by metamorphic processes (although they are sometimes preserved in rock of quite high metamorphic grade), or they may merely be removed by solution in percolating ground-waters. The mode of preservation of fossils is varied (and is set out in most palaeontology textbooks) and is stratigraphically relevant, for the kind of preservation throws light on the origin of the rock and its subsequent history.

William Smith (1769–1839), the pioneer of English stratigraphy, recognizing that different strata contain different fossils, established that in many cases a bed could be identified by a study of the contained fossils supplementing the lithological characters. It is clear that the recognition of strata by their fossil content is dependent on two factors:

1. throughout geological time there has been, repeatedly, extinction of species and genera, with new ones arising to replace them. This provided a constantly changing pattern of life, in fact due to evolution, although Smith's work and his recognition of the application of fossils to stratigraphy preceded the publication of Darwin's work on evolution by several decades;
2. a single species (or a genus) which dies out does not recur: extinction is complete and final.

The several species which are found together in one bed are called the faunal (or floral) assemblage. In general different outcrops of strata of the same geological age will possess the same faunal assemblage. It is therefore possible to correlate beds of the same age by means of their fossils—as Smith recognized. It is, further, possible to date strata—not of course in years, but relatively by means of their contained fossils—by reference to the established sequences of faunas which have been empirically deduced. This process is of particular importance where the geological structures are of such complexity that relative dating cannot be derived from the law of superposition.

Correlation by means of fossils is complicated by several factors, the most important of which is the lateral (or geographical) variation in a faunal assemblage, which may be relatable to observed lithological differences only in some cases. On a broad scale geographical factors, principally climatic, will affect the composition of a fauna. While it is true that some species, both of present-day and fossil animals, are of almost world-wide distribution, this is exceptional. In seas at the present, reef-building corals flourish in shallow, clear, warm waters of the tropics, but in the colder North Atlantic and in the North Sea, although corals occur, they are either simple (solitary) corals or long-branched compound genera quite different from those of the reefs. In modern seas even in a distance of a few miles changes in fauna occur; on the north-facing coasts of Fife, Scotland, the commonest shell is the cockle *Cardium* but on the south Fife coast it is the mussel *Mytilus*. In the shallow seas off the British coasts occur several species of irregular sea-urchin. *Echinocardium cordatum* is found on clean sandy beaches in the littoral zone, *Spatangus purpureus* in shelly gravels from low-tide level to a depth of 30 fathoms and *S. rachi* in sandy mud at depths of over 100 fathoms. These animals, all contemporary (and rather similar in appearance), although they are marine creatures, all require somewhat different conditions in which to thrive. It follows that in deposits of past geological ages we shall find lateral faunal variations resulting from differences in physical conditions.

Facies

The sum total of the physical conditions under which a deposit is formed is known as the facies. Close to a shore, where the water is very shallow, well-aerated and affected by wave action and currents, occurs the littoral facies. In this region and under these conditions shingle and sand are deposited as a rule. Hence the deposits may alternatively be referred to by a descriptive lithological name, arenaceous facies if the sediments are sandy, or rudaceous facies if the material is of coarser grade. It is noteworthy that most rocks of littoral facies in the geological record are devoid of fossils, presumably since they failed to be preserved.

In shallow seas to a depth of about 600 feet a great variety of animals, many benthonic (living on the sea floor), exist. Many possess shells and clearly have a relatively high probability of becoming fossilized, perhaps of the order of one in a million (see p. 26). Rocks formed from sediments deposited in such conditions frequently contain many fossil shells, mollusca and brachiopoda being very common, and may alternatively be referred to as being of shelly facies.

Sediments laid down in deeper water, usually fine-grained sediments but not necessarily distinguishable on lithological grounds from fine-grained shallow-water deposits, have a quite different faunal content from the shelf-sea deposits.

Strata of the same geological age, if of similar facies, will naturally have very similar assemblages of fossils if not exactly the same. It follows that if they occur in geographically distant areas, the faunal differences may be more conspicuous. On the other hand, rocks formed contemporaneously but of different facies (i.e. deposited under totally different conditions) will be found to have very different faunas. They are analogous to the modern examples cited above.

Due to the non-recurrence of fossil species, rocks formed under closely similar conditions but at different geological periods, for example the reef limestones of Upper Silurian age and those of Lower Carboniferous age, will not contain the same fossil assemblages. In fact, in this example where the two geological series are separated by such a long interval of time, not a single species is common to both, for few species are of this duration. None the less, in both cases, the faunas comprising fossils of the same type, reef-building corals, bryozoans and echinoderms, have a considerable similarity.

In the case of ancient sediments it is often only possible to relate faunal differences to the grosser differences in facies, but there are many good examples of this from the geological record. In the Lower Palaeozoic there are a number of instances of rocks of neritic facies, lithologically consisting of sandstones, siltstones, mudstones and conglomerates, with shelly faunas, having been deposited contemporaneously with fine-grained sediments, shales and greywackes (possibly laid down in deep water) with a solely graptolitic fauna.

The use of fossils in correlation

Some fossils are of very long time range, genera and even species persisting for several geological Periods. In the case of the brachiopod genus *Lingula*, which has existed for over 400 million years (m.y.) from the Ordovician Period until the present day, its constituent species are of long range—*L. squamosa* existing throughout the Carboniferous Period. It follows that species such as this are useless for purposes of correlation. It is essential that, to be of maximum usefulness, fossils should be of short time range. Hence, species of a quickly evolving lineage are most suitable, for where morphological change with time is rapid individual species are of very limited duration. They provide a more accurate 'scale' with which to divide up geological time and therefore enable more precise correlation. In general it is necessary to identify fossils to the specific level before they can be of stratigraphical use. Many genera of molluscs, for example, have existed throughout the Cainozoic Era, some 60 m.y., but their constituent species were sufficiently short-lived to make them of immense value.

To be useful for stratigraphical purposes, fossils must be of animals or plants which were readily distributed and which became geographically widespread. Free-swimming or floating animals such as Goniatites (of Upper Palaeozoic age) and Graptolites (of Ordovician and Silurian age) are obvious choices and have been widely employed. Animals of a sedentary nature such as the Bivalvia (lamellibranchs) and attached forms such as brachiopods, corals and crinoids have also been extensively used in correlation. This is possible because in every case the animals pass through a swimming larval stage which ensures their distribution.

It is desirable that fossils to be used for correlation should be as widespread as possible, so that marine animals greatly restricted by such factors as depth of water and temperature, or terrestrial animals and plants restricted to narrow climatic zones, are of less value. Ideally, to be of maximum value in correlation fossils should be independent of facies. This ideal can never be realized in practice since all animals are adapted to a particular set of ecological conditions—to a greater or lesser extent. Some very highly successful species have only small tolerance to variations from their optimum conditions and, although successful, require very specialized environments. Fossils which are obviously closely tied to a particular environment (facies) are called *facies fossils*. However, it must be remembered that in the broad sense, although not normally included in this somewhat narrowly used term, all fossils are facies fossils and they exist in a particular set of physical conditions which have definable limits. The benthonic faunas of the deep mud-floored ocean and the shallow epi-continental seas have no common elements. It follows that rocks of very different facies will normally have entirely different fossil faunas. It is often difficult to correlate them accurately although it may be possible on wider evidence to establish their approximate contemporaneity.

Type areas and type successions

In practice, for each geological System, both the lithological succession and the sequence of faunas have been worked out in areas where the geological record was believed to be reasonably complete and typical. Initially, type areas were chosen in Britain or Europe. In the case of the Lower Carboniferous, for example, the succession chosen was that exposed in the Avon Gorge, Bristol, where the strata comprise limestones, dolomites, calcite mudstones and, near the top of the succession, sandy limestone. The geographically and geologically similar gorge of the River Meuse at Dinant, Belgium, reveals a comparable succession with similar faunas and could equally be considered as the type succession.

In the limestone successions of the Carboniferous of north-western England there are several important faunal differences and many of the characteristic fossils are absent. The lithologically different strata of this period in other areas, the dominantly argillaceous succession of the mid-Pennine area and the strata of the Hercynian horsts of western Europe, contain entirely different fossils, although here too faunal sequences have been established. The correlation of the different facies of contemporaneous strata presents a major problem in stratigraphy.

Ideally, rocks of deep-water facies should form a more suitable type succession since there are fewer interruptions of sedimentation (unconformities, etc.) and the succession is a more complete record, although the thickness of strata may be less than that of the epi-continental sea areas.

Fossil zones and stages

The earliest use of fossils in correlation was based on the recognition of characteristic fossils in strata of a particular age. This is still the fundamental principle but it can be applied in two ways. The first of these is in the recognition of 'index fossils' by means of which a particular bed may be identified, i.e. in that bed certain species are prevalent. Stemming from this is the widespread practice of learning several hundred fossils and the formations in which they typically occur. In the sense that a characteristic and common fossil may be one of the commonest fossils of a zone, perhaps that species which gives its name to the zone (faunizone), and known as the zone fossil, the practice has more than a merely empirical value to the field geologist who is mapping chiefly lithological boundaries. This second method of using fossils in stratigraphy, the creation of fossil zones, is a logical development from Smith's early use of fossils. A *zone* is defined as *strata* deposited during an interval of time (known as a secule or moment, though these terms are not widely used) throughout which a particular faunal or floral assemblage existed. Hence, a zone is not defined either by lithological changes or by breaks in the depositional record; it is a division of time given in terms of rocks deposited. Although a faunal (or floral) zone is defined by reference to an assemblage of fossils, it is normally named from some characteristic species and this name fossil is known as the zone fossil.

It follows that it may be possible to recognize a zone without actually finding a specimen of the zone fossil itself if other highly characteristic species are found. However, in general a faunal zone or faunizone is recognizable, and definable, because certain species existed together for some time. It is assumed that the constituent species of the fauna have largely overlapping time-ranges, perhaps approximately the same, and that their time-ranges are the same in different areas. Correlation is based on this assumption that a particular faunal assemblage is synchronous in widespread localities.

The term zone, as used here, is a biostratigraphical unit. The word faunizone would be less ambiguous (but floral zone would have to be used where definition was based on plant species, not on animals), since the word zone has unfortunately been used in a wide variety of ways to include non-palaeontological characters: hence terms such as cherty zone and fault zone occur. Even within the field of palaeontology different criteria have been employed; in some cases zones have been defined with reference to the complete fauna but in others definition has been based on the members of only a particular phylum or class. Furthermore, the scale of zonal classification varies greatly from system to system. For example, in the Lower Jurassic where the stratigraphy is known in great detail and fossils are abundant, 20 zones represent the sedimentation of less than 10 m.y., whereas the whole of the Ordovician strata, deposited over a period of perhaps 60 m.y., comprise only 15 zones.

Smaller divisions of time, such as hemera—the corresponding rock unit is an epibole, based on the acme of a single species (approximately the equivalent of a range zone of the American Code of Stratigraphical Nomenclature) —are a useful concept as are other divisions based on the duration of a particular species either locally or on a world-wide basis.

Divisions of strata corresponding to a longer interval of time than a zone, therefore including several zones, are known as stages. The stage has been evolved to augment or to define more precisely in terms of time the divisions of the stratigraphical column called series. The lithologically defined series comprises as a rule several formations, but these are not necessarily all of similar lithology. Due to lateral variations in lithology the series is of only regional application.

TABLE 2.1

BIOSTRATIGRAPHICAL DIVISIONS (Time division)	ERA	PERIOD	STAGE	ZONE
e.g.	Mesozoic	Jurassic	Domerian	Spinatum
LITHOSTRATIGRAPHICAL DIVISIONS			SERIES	FORMATION
e.g.			Lias	Marlstone

There are, therefore, two methods of dividing up the stratigraphical column into its smaller divisions based on (1) lithological units, and (2) palaeontological units. The system can be divided into series which comprise 'groups' of formations or beds, but is also divisible into stages and zones (and in some cases yet finer palaeontological divisions) (see Table 2.1). In problems of correlation, lithological and palaeontological evidence may appear to be conflicting, and neither provides an entirely reliable guide to the contemporaneity of strata in different localities, since neither gives 'absolute' ages.

It has been noted that rocks of the same age can be deposited under different conditions, that is to say they are of different facies; hence contemporaneous strata may be of quite different lithology. Conversely, similarity of lithology is no proof of contemporaneity in many cases. The diachronism of lithologically distinct units, beds, has been discussed in Chapter 1. The variation in the age of a diachronous deposit is presumed from its relation to a biostratigraphical time scale (a sequence of faunas), and if the geographical spread of new species is very rapid, in terms of the vastness of geological time almost instantaneous, no problems arise. However, to be of value in stratigraphical zoning, fossils must be of quite short time range. It is therefore likely that, compared with the time range of a species, the time taken for it to migrate from area to area may be significant. A further complication may arise since most species are to some extent facies controlled: a species living in favourable conditions in an area X is prevented from spreading into an area Y through adverse conditions there. A gradual amelioration of conditions in area Y may eventually permit the spread of the species in question. Here the appearance of this particular species does not indicate that the beds in which it is found are of the same age. In general, a particular fauna will be characteristic of rocks of a particular age, but only so long as conditions for its existence were favourable.

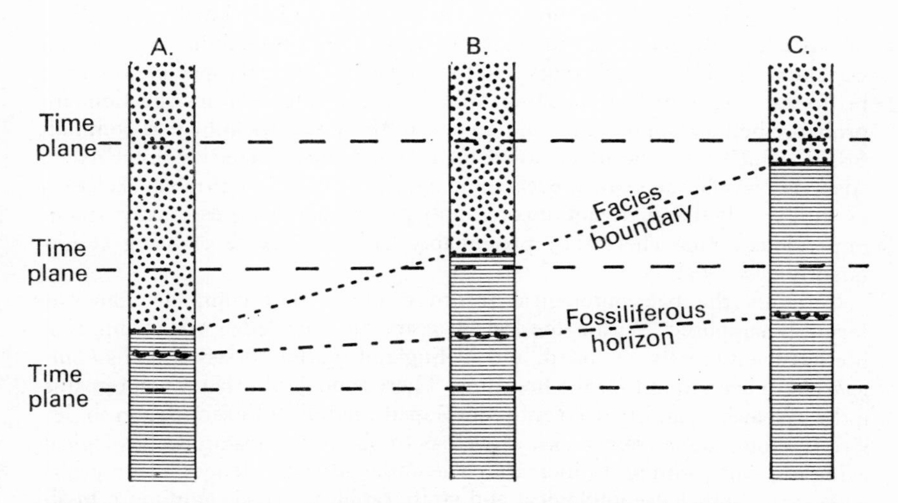

Fig. 2.1 Diachronism of a facies boundary and the migration time of a fossil assemblage. The fossiliferous horizon may be regarded as a time plane if the localities A, B and C are not far distant. As a rule, time-planes cannot be identified.

Ideally, we should be able to construct a system of real or absolute time divisions—chronostratigraphical divisions—to which both lithostratigraphical and biostratigraphical divisions could be related. While isotopic dating does provide a system of dating of this kind, giving dates in millions of years, it is too imprecise to be relevant to this problem. Of course, within a limited area biostratigraphical divisions will approximate to chronostratigraphical divisions. If lithological boundaries fail to coincide with palaeontological divisions it is more probable that the latter can be relied on. In other words, fossils—the basis of the biostratigraphical divisions of the stratigraphical column—provide a useful and usable tool for correlation and for subdivision of strata. Lines indicating contemporaneity, so-called 'time-planes', are normally presented diagrammatically as horizontal, whereas lithological boundaries may slope (through diachronism) and palaeontological boundaries may slope over wide areas (through either migration time or facies control) (Fig. 2.1).

The inadequacy of the fossil record

Over a century ago Charles Darwin expressed disappointment that the fossil record provided less support for the theory of evolution than might have been hoped for. Since that time vast numbers of fossils have been found and many new species and genera described, although the number of examples of fossil lineages which demonstrate evolutionary change in detail is still small. How adequately represented are organisms which formerly existed on earth? A. B. Shaw has examined the statistical probability of organisms being found in the fossil record. It is clear that if one individual in a million, of a particular species, was fossilized there is a high probability, verging on certainty, that if the species survived for only a million years specimens would be found. With marine-dwelling animals the chances of fossilization are probably better than one in a million and in those species known as common fossils, which often occur crowded on bedding planes, considerably better. Also, as Newell has shown, even some species with only chitinous skeletons or with woody tissue are not uncommonly preserved: this is especially true in impervious strata where they presumably had less chance of being subsequently destroyed.

Certainly the palaeontological record is much more complete than was formerly supposed. Within the last 20 years the knowledge of Precambrian life has been greatly extended, and in England marine Triassic fossils (*Lingula*) have been found for the first time. There is no doubt that much remains to be discovered and that recently developed methods of examination to derive the total fauna from rocks, as well as to obtain the maximum geological evidence, will continue to increase palaeontological knowledge. Many problems exist, both palaeontological and stratigraphical, which only more fossil evidence can solve.

It is necessarily debatable how complete our palaeontological knowledge can ultimately be. At least theoretically the fossil record could be fairly complete, since at any given time deposition has been taking place somewhere on the earth's surface, for there are no world-wide unconformities, and not all the facies at a given time would be likely to be inimical to the existence of life and its preservation. However, the observed rates of accumulation of sediments bear little relation to the maximum thickness of strata known for any period of time—since deposits are constantly reworked and redeposited. It follows that the stratigraphical record itself is incomplete and it seems probable that the palaeontological record must also be incomplete. The relatively abrupt rise in the abundance of certain sea-urchins in the Pleistocene (Fig. 2.2) is not due to their sudden success as a group, but is largely attributable to their much greater chance of preservation in very recent sediments. It is an indication of the paucity of the fossil record in rocks no older than the Tertiary.

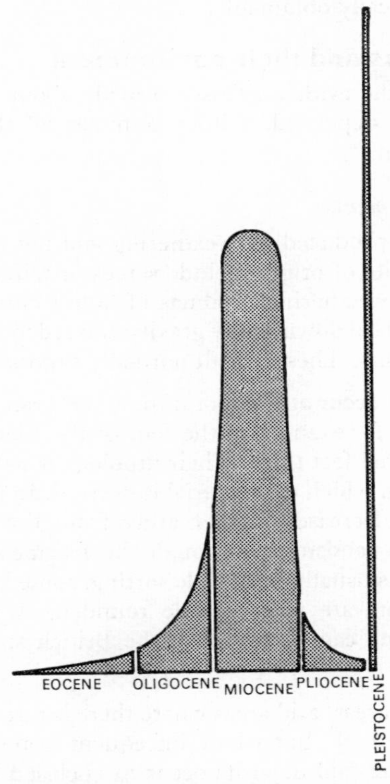

Fig. 2.2 Abundance of families of Clypeastridae in the Caenozoic. The great increase in numbers in the Pleistocene shows the paucity of the fossil record, even in Tertiary age beds. (From the late PROF. F. E. ZEUNER, 1959, *Dating the Past*, Methuen, London. Reproduced with permission.)

If the known palaeontological record is incomplete, what reliance can be placed on the all-important time-ranges of species? Can the evidence from different outcrops be used with confidence that it can provide stratigraphical correlation? Again, Shaw has shown statistically that the known range of a species may be closely approximate to its true time-range (although the latter, in any case, cannot be determined empirically), and he also shows that it is likely that the range at a particular locality may be sufficiently closely related to the total known range. (For all practical purposes this known range for the total available sample can be called the range of the species.) Statistically there is no doubt that the available part of the stratigraphic and palaeontological record is a valid sample. However, apart from the real imperfections of the geological record, there are gaps which must remain unfillable; there are vast areas of formations not exposed due to the cover of later rocks, remote areas as yet unexplored and areas well known but where outcrops and natural exposures are rare. For these reasons the data available to the geologist fall far short of that theoretically obtainable.

Sedimentary rocks and their environment

Before reviewing the evidence fossils provide about the conditions under which sediments are deposited, a brief summary of the main sedimentary environments is given.

Non-marine environments

1. Residual deposits produced by weathering and not transported any great distance from their site of origin include screes or talus slopes. Bauxitic and lateritic deposits are weathering products of hotter climates. Gravels, sometimes re-sorted or carried downhill by gravity and redeposited, occur as well as purely residual deposits. These are all normally unfossiliferous.

2. Piedmont deposits occur at the foot of mountain ranges due to the coalescence of alluvial fans, for example at the foot of the Sierra Nevada, and they may be several hundred feet thick. Their lithology is naturally closely related to the mountains from which the material is derived. In general the size of the fragmental material decreases further away from the mountains, breccias passing laterally into sandstone, although the fragments show a great size range. While there has usually been little sorting, some layers are coarser than others. The fragments are usually little rounded. A number of instances of deposits of this kind can be found in the British stratigraphical succession.

3. Deposits in desert areas, arid areas where there is rare precipitation and the rate of evaporation is high, but where infrequent heavy rain is known, may be of two kinds. Water-laid deposits occur in enclosed basins called bolsons and resemble piedmont deposits. Saline lakes called salinas or playas are subject to intense evaporation, and beds of calcium carbonate and sulphate occur interbedded with even-bedded silts and clays. Deposits of this kind are to be found in the British Permo-Triassic rocks. Wind-deposited sands occur as dunes, often extending many miles, and they have characteristic cross-bedded structure. Grains are rounded and may be well sorted. Most modern desert

sands are not red but many ancient aeolian deposits are, due to both cementation and coating of the grains with ferric oxide. Fossils are virtually unknown in desert deposits.

4. Loess deposits are very fine grained wind-borne material which has been carried for great distances and deposited beyond the margins of the desert areas, where it is presumed to have originated. It is normally unbedded, has vertical joints and tends to stand in near-vertical faces—for example in road cuttings. Rarely, terrestrial-dwelling animals occur as fossils.

5. Glacial deposits, known as drift, are divisible into two: glacial till, deposited by glaciers and ice sheets, and water-laid deposits. The latter comprise outwash sands and gravels deposited by meltwater (the outwash sediments), together with clays, sands and peats laid down in glacial lakes. Till is normally unstratified and is characterized by the complete lack of sorting of the material which ranges from fragments many feet across to particles of clay grade size. This results in a deposit with a clay or sandy clay matrix with pebbles and boulders, which is known as boulder clay. The fragments are chiefly angular, not being rounded during transport, and their petrology (and orientation) provides evidence of the derivation of the till, and hence directions of ice-flow.

Glacial deposits are not fossiliferous excepting the lake deposits, where diatoms and radiolaria sometimes occur in the clays, and in the peats a multitude of insect remains and plant pollen are to be found.

Apart from the extensive Pleistocene glacial deposits, there are only a few examples from the geological record, suggesting that ice ages are rather infrequently occurring phenomena. The Carboniferous (Gondwana) Glaciation of the southern hemisphere and India is well documented, and there are a number of tillites (ancient tills) of Precambrian age. Several occurrences are reported from Britain.

6. Volcanic deposits in the form of ash (= tuff), and the coarser material called agglomerate, often fall onto a land surface and are as a rule bedded. They are readily recognizable both by the angularity of the fragments and by the petrology of the fragments and grains. Although plants and occasionally animals are overcome by pyroclastic deposits, these are normally unfossiliferous except where they fall into water. There they are interbedded with water-laid sediments and both sediments and tuffs may be fossiliferous.

7. Fluviatile deposits are formed by rivers. Rivers deposit shingle, gravel, sand and silt in their beds, although these deposits are constantly re-worked and transported further downstream. Deposits on the inside of meanders build up into shingle banks but are usually subsequently eroded as the river changes its course. In time of flood a river, especially in the lower part of its course, spreads over the floodplain covering it with vast sheets of slowly moving water from which fine material, silt and clay, are deposited as a result of the decreased velocity. Floodplain deposits are liable to include lenticular gravel beds, coarser channel deposits interbedded with finer intervening

material and they show evidence of contemporaneous erosion. Included fossils are chiefly freshwater invertebrates and terrestrial animals and plants. Fluviatile deposits often pass laterally into estuarine and deltaic deposits, described below.

8. Lacustrine deposits. Lakes provide a temporary base level for the rivers which run into them. On entering a lake a river's velocity is checked and its load of coarser material deposited, often building out as delta with fore-set beds deposited at the angle of rest of the sand (up to 35°). Finer material is carried across the delta to be deposited in deeper water, forming bottom-set beds across which the delta gradually advances. Changes in direction of transport of materials and currents naturally produce in most cases complex patterns of cross-bedding. In large lakes the central deposits will be fine grained and with little depositional dip. If not aerated by 'stirring' processes of waves and currents, sulphides may be deposited in anaerobic conditions. Coarser material may accumulate as beach deposits. The most important precipitate in lakes is calcium carbonate which may form tufa or oolitic limestone. Fossils are not uncommon in lake deposits and shelly deposits with fresh-water molluscs and crustaceans occur as well as algal limestones and diatomaceous earth, the latter built up of siliceous plant skeletons.

Marine environments

Seas and oceans cover 70% of the earth's surface. Our knowledge of the shallower-water deposits is extensive but despite recent advances in the scientific study of the sea floor, both directly by using corers and indirectly by using geophysical methods, our knowledge of deep-sea sediments is still relatively small.

1. Neritic Zone. The neritic zone extends from low-tide level to the edge of the continental shelf, where a quite marked steepening of slope of the sea bed occurs. This outer edge of the continental shelf lies at a depth of about 600 feet, although recent data show that it varies from place to place. It is not known whether the variation in the depth at which this 'break' in slope occurs is due to crustal warping or to varying depth in the effects of wave action. Recent work shows that the edges of the continents at the continental shelf are a kind of submarine 'escarpment' and are probably of a fundamental structural nature.

The continental shelf slopes away from the shore very gradually, at 1 in 100 or less, and there is little relief except where it is cut by the heads of submarine canyons.

The littoral subzone, often regarded as part of the neritic zone, includes the shore and tidally affected areas. In some areas special phenomena such as lagoons or coral reefs are present. The deposits are typically boulders, pebbles (= shingle) and sand derived by marine erosion from the cliffs, shelly sands formed from broken up shells, and coral sands in areas where reefs occur. Fine-grained material may also be deposited in the littoral subzone, including carbonaceous muds. Fossils are rare on the whole in ancient littoral deposits, especially in coarser-grained deposits. The distribution of littoral deposits plays an especially important part in determining palaeography.

The deposits of the shallow or epeiric seas of the neritic zone are of sand, shelly sand, gravel, silt, mud and clay. While, in general, coarser material may occur near the shore and progressively finer material in deeper water in a graded sequence, exceptions to this are numerous. In some areas sands and gravels are known to occur towards the seaward edge of the shelf. The picture is complicated by the presence of glacial deposits which occur on the continental shelf in northern latitudes. The lack of grading, resulting in a haphazard distribution of coarse and fine material, may also be due to relatively recent changes in sea level, attributable to the recent glaciation. In general, fineness of grain of deposits cannot—in any geological period—be taken as an indication of the depth of the water in which they were deposited. Since the neritic zone lies within the influence of waves and bottom currents, the sediments are repeatedly moved.

The life of the neritic zone is very varied. It is most abundant to a depth of about 100 feet but a greater number of species occurs in deeper water. As well as swimmers (nekton) and floating creatures (plankton) there is a vast fauna and flora dwelling on the sea floor (benthos). These faunas are very diverse, restricted by physical factors such as temperature, depth of water and the nature of the substrate. Despite this, it is usually possible to correlate contemporaneous neritic deposits by the presence of some widely occurring species.

Ancient neritic deposits are characterized by their lithological variety; sandstones (sometimes green or yellowish due to glauconite), siltstones, shales, mudstones, ironstones and limestones occur. Commonly these sediments are very fossiliferous. Numerous breaks in the stratigraphical record occur, since oscillations of sea level are liable to interrupt sedimentation, although neritic deposits may be quite widespread with little lateral variation.

2. Bathyal Zone. The bathyal zone corresponds to the upper part of the continental slope, extending downwards from the edge of the continental shelf 600 ± feet) to a depth of perhaps 6,000 feet where it merges into the abyssal zone. The slope is very much steeper than that of the continental shelf, being of the order of 1 in 10 in its upper part but flattening out at greater depth. Little light penetrates to the bathyl zone and the water temperature is low, 10°C or less. Wave action does not reach to these depths, but currents of several knots are now known to exist, despite lack of evidence of scouring of the sea bed or movement of already deposited material. The exception to this is the phenomenon of turbidity currents. Flows of high density, due to material carried in suspension, attain high velocities (up to 60 m.p.h.) carrying quite coarse material down the continental slope to great depths. In ancient deposits of this kind characteristic sedimentary structures occur. Apart from this kind of deposition, the rate of deposition in the bathyl zone is very slow. It consists chiefly of terrigenous muds (derived from the land) and carried in

suspension from which it is slowly settled out. As a result, such sediments are relatively thin but are laterally very consistent. Little is known of the benthonic fauna but plankton rains down constantly, and in areas far from the shore, where there is little deposition of terrigenous material, the skeletons of planktonic organisms form the chief deposits. Since aragonite skeletons or tests are readily dissolved and calcite is ultimately soluble, in the deeper parts of the ocean only the remains of siliceous organisms are found.

3. Abyssal Zone. The abyssal zone may be considered to extend downwards from approximately 6,000 feet to the greatest depths of the oceans which, in the 'troughs', exceed 30,000 feet. Only siliceous skeletons such as radiolaria escape solution and much of the abyssal zone is floored by red clay. Its origin is obscure but must be in part windborne dust and much may be of volcanic origin. Abyssal deposits are rarely found as rocks, although some radiolarian cherts may be of very deep water origin. Rocks which must have originated in the bathyal zone, deep-water marine shales with only planktonic fauna and greywackes with sparse fauna, but characterized by sedimentary structures known collectively as sole-marks or bottom-structures, are well known however.

4. Rather special conditions exist in some seas which are connected to other seas or the ocean by a shallow threshold. If the water does not circulate freely, anaerobic lifeless conditions become established and the concentration of lime may be high. In the geological record there are many examples of marine sediments which possess an 'impoverished' fauna, the variety of forms of life being restricted by adverse conditions.

Intermediate environments

1. Estuarine. Today the term estuary is applied to the wide mouth of a river affected by tides which constantly redistribute the detrital material. The fauna of such areas is essentially marine although it may differ considerably from that of more open shores. The term estuarine has been used in a wider sense in its application to ancient sediments, covering water-laid sediments which, on faunal evidence, either are not marine or are of doubtful provenance. Such sediments contain freshwater fossils, drifted terrestrial remains, both animal and plant, or no fossils at all. They may be presumed to be estuarine since they represent a lateral or a vertical transition from undisputable marine sediments to typical fluviatile or terrestrial deposits.

2. Deltaic. Many rivers on entering the littoral zone of the sea deposit their load, partly because of the lessened velocity of the water and partly due to the flocculation of the sediment in the presence of sodium ions. Sediment may be built out on to the continental shelf as a delta in the absence of strong coastal currents. Vast spreads of deltaic origin are common in the geological record although, as a rule, it is not possible to deduce the configuration of an actual delta: such spreads of sediment must represent a sufficiently long period of time for the geography to have varied considerably, at least in detail. Typically

arenaceous and cross-bedded, deltaic deposits show considerable variations in thickness laterally. They contain few fossils except in interbedded shales.

The colour of rocks

Arenaceous rocks (arenites) are normally light coloured, since quartz forms a very large percentage of the rock, and variations in colour depend largely upon the nature of the cementing material. Colour banded sandstones and striped siltstones are common, light and darker layers alternating. In some sandstones the darker bands are carbonaceous.

Argillaceous rocks (lutites = argillites) range in colour from grey to black, although they may weather to a reddish or brown hue due to oxidation of contained iron. The black coloration may be due either to the presence of carbonaceous material of organic origin or to finely divided pyrites. However, black shales can be formed in a variety of environments from non-marine to geosynclinal deeps.

Red beds are sediments ranging in grade from conglomerate or breccia to shale and are pigmented by haematite. In sandstones this may coat the grains as well as cement them together. On the other hand, hydrated oxides of iron give rise to yellow or brown rocks, while ferrous silicate, chlorite and carbonate are responsible for the green or grey colours. Red coloration was at one time interpreted as indicating deposition under desert conditions, since some red sandstones are clearly of aeolian origin and may have a high percentage of recognizable felspar grains, although red sediments are rare in present day deserts. It is now known that red beds, while possessing this dominating character of redness, are not all of the same type or origin. However, certain conclusions can be drawn, that either the parent material from which the sediments were derived was red, or that parent material was weathered in a tropical or subtropical climate. Further, it is evident that the red beds did not encounter a reducing environment.

Faunal evidence of sedimentary environment

In general a great deal of information about the conditions of deposition of a sedimentary rock can be deduced from the fossils it contains, supplementing the evidence of lithology, etc. In rocks of Cainozoic age some of the species found as fossil are still in existence, providing direct evidence of environmental conditions—assuming that the organisms have not appreciably changed their physical requirements. In the case of older rocks, in which all the fossils are of extinct species, analogies can be made. For example, it may be assumed that Jurassic rocks containing reef-building scleractinian corals were deposited in clear, warm shallow seas comparable with those of the present day in which coral reefs flourish. In fact present-day reefs are made up of entirely different species and genera of scleractinians and of alcyonaria which predominate. In yet older reefs, those of Palaeozoic age, the corals found are of entirely different kinds from modern corals, comprising tetracorals and tabulate corals. Probably these, too, flourished in warm clear seas (there is confirmatory evidence of lithology, reef structure, faunal associations, etc.) but it cannot be assumed with the same degree of confidence as in the case of geologically younger reefs. Some rocks contain fossils only of extinct groups, for example the dark coloured graptolitic shales of the Ordovician and Silurian. In the absence of any very closely related living forms (their nearest relatives may be certain modern hemichordates), considerable speculation and even argument has taken place over the possible conditions of sedimentation these strata represent.

Of particular importance in evaluating the conditions of deposition is the faunal association, the kinds of animal which occur together. As a rule associations of different kinds of animal which share similar ecological requirements are found together in the same strata: lamellibranchs (Bivalvia) and gastropods of marine genera and brachiopods commonly occur together. They each reinforce the evidence of the others that the sediments in question were laid down in a marine neritic facies. Of course problems exist through apparently conflicting evidence. For example, the larger fossils (macrofossils) of the Chalk are largely neritic but not infrequently foraminifera of deeper water type, together with hystricospheres, occur in the same strata. In some cases of discrepancy between the kinds of fossils found together it is necessary to consider whether the organisms existed together or whether, after dying, they were collected together by currents or wave action.

Assemblages and communities

By considering the size distribution of fossil shells from a bed it is possible to deduce whether they represent a community—animals co-existing and those of any one species interbreeding—or merely an assemblage. The latter is a sorted sample of shells brought together by waves or currents. While both communities and assemblages provide some evidence of the conditions under which the deposits were formed, naturally communities give more precise data, especially where the ecological ranges of the constituent species are known.

Where shells have been distributed or collected together by waves and currents certain characters, such as the number (percentage) of shells broken up, the proportion of disarticulated bivalves and the way-up of shells, provide evidence of bottom conditions and the strength of currents (sometimes referred to, rather loosely, as turbulence). A few attempts, pioneered by Eagar, have been made to relate information so obtained to lithological or petrological evidence from the rock matrices. Work of this kind is still not extensive.

Fossils provide important evidence of climatic changes. The gradual onset of cold conditions preceding the Pleistocene Glaciation is reflected in the molluscan faunas of the Pliocene and Lower Pleistocene beds. The variations in temperature of the interglacial periods can be traced in some detail by reference to variations in percentages of several tree pollens and by fossil beetle faunas, both of which have in the last decade been proved to be extremely sensitive indicators of climate. These are discussed in Chapter 16.

II The Precambrian

3 The Precambrian

Rocks which were deposited in Cambrian and later times are generally fossiliferous and the classic work of stratigraphers from William Smith to Charles Lapworth has been the subdivision of these rocks, by means of the fossils, into a number of systems. As the rocks formed before the Cambrian Period are virtually unfossiliferous, they are not amenable to the same treatment and they are collectively referred to as Precambrian rocks. The Precambrian does not represent an era in the sense of the Palaeozoic, Mesozoic and Cainozoic Eras and even the crude division of Precambrian rocks into Archaean (Gr. 'ancient'—usually regarded as 'old' and 'crystalline') and Proterozoic (Gr. 'primitive life') eras now has little meaning since evidence of life is known in rocks more than 3,000 m.y. old. Because of the great age of the Precambrian rocks, their outcrops tend to be fragmentary or show highly complex relations to each other, and they are probably best discussed as fragments of successive orogenic cycles, since rocks completely unrelated to orogenic zones are rarely preserved from Precambrian times.

It is clear that the principles of stratigraphy discussed in the first two chapters will not always be applicable to Precambrian rocks since many of the Precambrian orogenic cycles are now only represented by the roots of mountain chains in which all the rocks are metamorphosed. The methods of correlation must therefore be based more on structural and petrological criteria than is usual in later rock formations.

Methods of Precambrian correlation

In unmetamorphosed or only slightly metamorphosed rocks, the recognition of sedimentary environments and volcanic petrographic provinces becomes important, not only for correlating rocks of the same environmental characters but also for correlations from marginal seas into geosynclinal basins of the same age. The sedimentary environments of geosynclines are therefore an important study in Precambrian rocks because of their predominance in those rocks which are normally preserved. However, it must also be remembered that in very early Precambrian time the geosynclines may not have had quite the same characters as more recent examples, and certain other assemblages such as banded ironstones—basic volcanics—chert may make up much greater thicknesses than seems to be normal in later times.

Within any one orogenic cycle, the range of sedimentary environments is large but, more important, the differences between the sediments laid down in the same episode of two different orogenic cycles may be very small. Thus it is necessary to know the relations between sedimentary, volcanic, plutonic, metamorphic and tectonic activity in orogenic belts before a clear understanding of an orogenic cycle can be obtained. The general outline of such cycles is known but they are rarely simple; in Table 3.1 only a general case is given, the most important rider being that parts of a cycle may be repeated more than once and other parts may be suppressed or eroded.

The most important methods of Precambrian correlation are those used in the deciphering of metamorphic terrain. On a broad scale the tectonic fabric (or grain) of large orogenic belts may delineate successive periods of mountain building in that older belts stop abruptly against newer belts. On such a scale the Caledonian mountain chain in Scotland cuts across the Precambrian (Lewisian) rocks which we can regard for this purpose as a mountain chain of 1,500 m.y. (Fig. 3.1). However, within both the Caledonian and Lewisian belts there is evidence of several periods of folding and metamorphism and neither represents a simple monocyclic unit. In both cases, igneous rocks older than the main period of orogenic activity show transgressive relations to a series of deformed and metamorphosed rocks which represent relics of earlier orogenic cycles. The cross-cutting nature of igneous rocks, principally basic dykes, pegmatites and granites, is one of the principal methods presently used to subdivide the Precambrian. They have the advantage, when fresh, of being directly datable by isotopic methods and where involved in later events may still, by refined methods, yield the date of intrusion. They will clearly give a minimum age for the rocks they cut but great care must always be taken in the interpretation of their cross-cutting relationships. This is particularly the case with gneissose rocks which may show the effects, in fairly small areas, of many different periods of folding and metamorphism. In these circumstances the structural geologist tries to resolve the problem of correlation by working out a sequence of structural, metamorphic and igneous events. The intersection of two fold belts is then seen to be an over simplification when the youngest fold belt contains relics of several earlier ones. It is the interference or refolding of one set of structures, with a particular orientation, style and metamorphic grade, by later structures with different characters that enables successive episodes to be recognized. Often they cannot be mapped in the normal sense since the areas involved may be too small. However, by detailed work in different areas, a complete structural sequence can be derived which, although never complete in any one area, applies in

TABLE 3.1 Elements of an Orogenic Cycle[a]

		Sedimentary Activity	*Volcanic Activity*	*Plutonic Activity*	*Tectonic Activity*	*Extra-Orogenic Volcanic Activity*
EPEIROGENIC PHASE	POST-OROGENIC EPISODE	Maximum molasse deposition merging with marginal basins of non-orogenic deposits	Andesitic and basaltic activity at maximum	Post-orogenic cross-cutting granites		Flood basalts and tholeiitic dolerites, large basic complexes
OROGENIC PHASE	LATE-OROGENIC EPISODE	Start of molasse (coarser deposits with rhythmic structureless sands and silts). Extends over foreland	Main orogenic volcanics towards end of phase. Mainly andesitic and acid	Pegmatites the latest event. Discordant late-orogenic granodiorites associated with andesite vulcanism. Discordant granites common	Late fold structures and faulting	Mixed alkaline and tholeiitic vulcanism
	MAIN SYNOROGENIC EPISODE	Maximum deposition of flysch at margins		Synorogenic pegmatitic material. Regional metamorphism leading to migmatization and synorogenic granites	Thrusting at close of episode. Continued folding usually of several phases (see Table 3.2)	
	EARLY SYNOROGENIC EPISODE	Marginal troughs with flysch (badly sorted sandstone-shale sequence). Many breaks due to tectonic activity	Andesitic vulcanism sometimes	Regional metamorphism begins. Main emplacement of ophiolite plutons	Intense folding of mobile belt. Mylonitization of basement	
GEOSYNCLINAL PHASE	MAIN DEPOSITIONAL EPISODE	Main greywacke (turbidite) deposition and thick deposits of shales (still water deposits)	Continued vulcanism, acid and andesitic particularly at margins	Ophiolitic plutonic activity (serpentinites and other ultrabasics and basics)	Downwarping of continental margins or intercontinental trough	Mixed alkaline and tholeiitic vulcanism
	VOLCANIC EPISODE	Greywackes of volcanic derivation	Island arc formation, acid-andesitic and some basic vulcanism. Ophiolite suite; spilites, keratophyres common			
	EARLY DEPOSITIONAL EPISODE	Greywackes (turbidites) in main trough. Thin shelf sedimentation outside				

[a] Adapted from J. R. HARPUM, 1960, *Int. geol. Cong.*, Rept. 21st Sess., Norden, Pt. 9, 201.

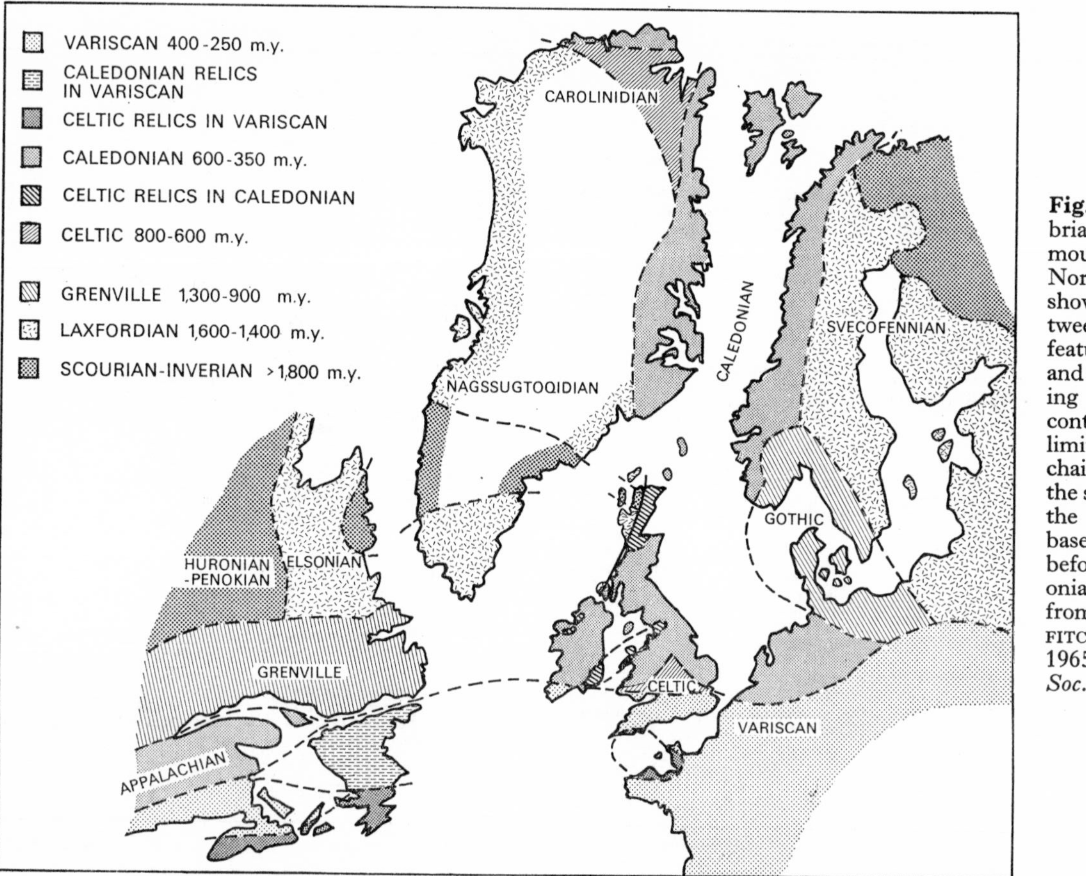

Fig. 3.1 The Precambrian and Palaeozoic mountain chains of the North Atlantic area, showing the relation between the main structural features of Great Britain and those of neighbouring continents before continental drift. The limits of the Variscan chain have been shown in the south, but in general the reconstruction is based on the position before the late Caledonian faulting. (Adapted from BULLARD, *et al.*; FITCH; and HARLAND; 1965, *Phil. Trans. R. Soc.*, **258A**, 46; 70; 191.)

TABLE 3.2 General Sequence of Deformation and Metamorphism in Orogenic Belts

Age of Folding	*Supracrustal Rocks*	*Metamorphism*	*Basement (i.e. effect on earlier cycles)*
Late	Late thrusts. Brittle, open folds, faulting	Low grade	Involved only in late thrusting
Main (often repeated)	Upright similar folds. Most intense penetrative, deformation	Migmatization may involve both basement and cover. Main metamorphic phase. High to low grade	Refolding by folds of similar type possible
Early	Often isoclinal, recumbent folds. Slides	Low grade	Mylonitization. Slices thrust into cover

general to the whole orogenic belt. The limits of an earlier orogenic belt can then be defined by the occurrence of these relics in the later belts. It may then be pointless to refer to the 'trends' of these relics as they may be expected to be of diverse orientation due to refolding in the subsequent fold episodes.

The sequence of events in a single orogenic cycle is always complex, but a typical pattern is given in Table 3.2. The importance of such sequences is apparent when an area has been subjected to several cycles. It will be obvious that care must be taken to distinguish structures or rocks of comparable type formed during different cycles in the same area.

The last important method of Precambrian correlation is isotopic dating. The principles are discussed in the opening chapter, but in metamorphosed rocks the interpretation is more difficult. The principal techniques used are those of the Rb/Sr method (on micas and whole-rock samples) and the K/A method (on hornblendes, pyroxenes, micas, whole-rock samples and feldspars). Whole-rock data are sometimes more consistent than individual mineral dates but the two may give different information. This is because a later metamorphism may homogenize the isotopes throughout the minerals in the rock without upsetting the isotopic balance of the whole rock. Figure 3.2a shows the decay paths of constituent minerals and the whole rock, which crystallized at time t_0 with an initial Sr^{87}/Sr^{86} ratio of 0·710. A rock with a simple history (point t_1) will have the whole-rock and constituent mineral isotopic growth lines converging to a point, and from a knowledge of the decay rate of Rb^{87} and the initial Sr^{87}/Sr^{86} ratio, the age can be determined. If at time t_m homogenization takes place, the Sr^{87}/Sr^{86} ratio for each mineral is changed but the Rb^{87} continues to decay at the same rate. Measurement of Rb and Sr isotope ratios at t_p will then yield an 'intersection' at point M

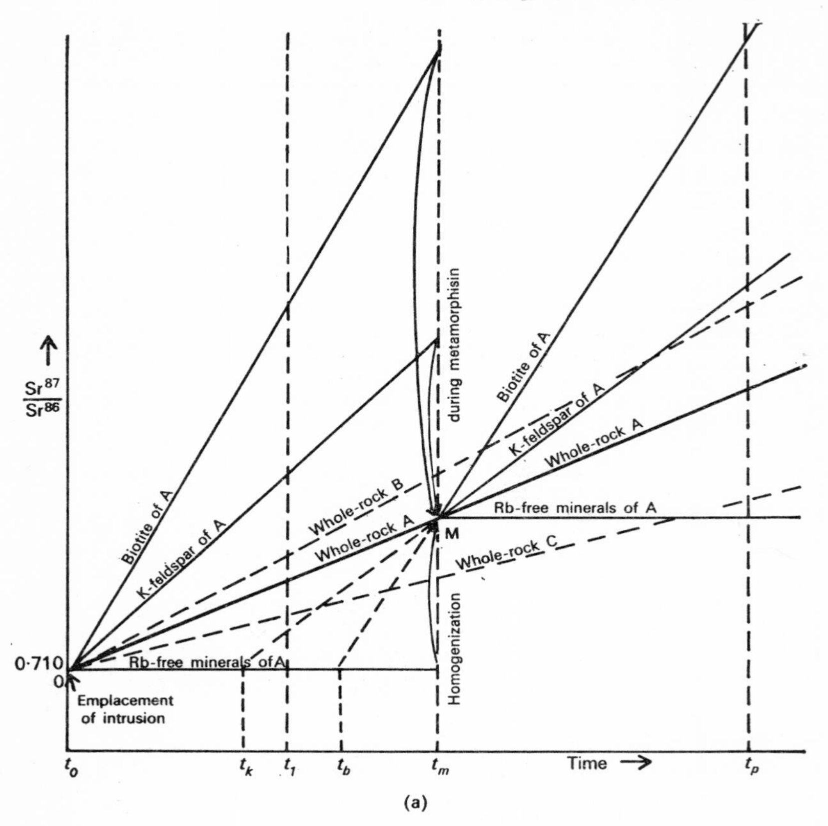

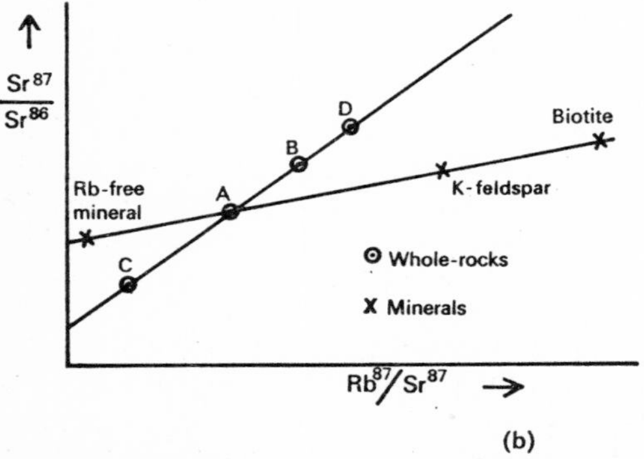

Fig. 3.2 (*a*) Schematic diagram of Sr^{87}/S^{86} growth lines for a granitic whole-rock specimen (heavy line) and its component minerals (continuous lines), which was metamorphosed at time t_m, when there was complete homogenization of the isotopes. Other whole-rock samples (B and C) from the same intrusion are shown by dashed lines. Discordant apparent ages (t_k and t_b) are indicated based on the assumption of an initial Sr^{87}/Sr^{86} ratio of 0.710. (*b*) Whole-rock and mineral isochrons for the same specimens. The slope of the whole-rock isochron is proportional to the age of the intrusion, the slope of the mineral isochron to that of the metamorphism. (After S. MOORBATH, 1965, in W. S. PITCHER and G. W. FLINN, *Controls of Metamorphism*, Oliver & Boyd, Edinburgh and London.)

indicating the age of the metamorphism (t_m) but the high Sr^{87}/Sr^{86} ratio of this intersection will suggest a complex history. Either the whole-rock values can be extrapolated to a 'normal' Sr isotope value (0·710) and the age of crystallization derived, or further whole-rock values (of rocks B and C with different Rb/Sr ratios) can be determined which will intersect at point 0, the time of initial crystallization (Fig. 3.2a). Note the projection of the metamorphosed mineral ratios to a 'normal' initial strontium ratio give quite spurious ages t_k and t_b (Fig. 3.2a). This data can also be presented as *isochrons* (Fig. 3.2b), the slope of a whole-rock isochron being proportional to the age of initial crystallization and the slope of the mineral isochron to that of the metamorphism.

Thus, in favourable circumstances, a great deal of information can be gained from analyses of several minerals and whole-rocks, but it must be realized that multiple metamorphism often results in anomalous dates being obtained because of incomplete homogenization of the isotopes. The most useful results have been obtained where a large number of rocks have been dated from a small area which has also been the subject of detailed structural and metamorphic study.

The Lewisian

The north-west Highlands of Scotland and the Outer Isles are the only extensive relic of the original continental basement on which the rest of Great Britain has been moulded. Before the break up of the continents the Lewisian rocks were probably contiguous with the Canadian-Greenland shield, one of the original continental cores (Fig. 3.1). This is composed of intersecting early Precambrian fold belts produced during successive orogenic cycles. Usually only the deepest parts of these orogens are extant and the older fold belts are generally less well defined than the later ones. The rocks in the north-west of Scotland are termed the Lewisian Complex or the Lewisian Gneiss but include relics of several orogenic belts or cycles and cannot be regarded as the equivalent of a geological system. The term is, however, useful for referring to those rocks which form the crystalline basement on which the rocks of succeeding sedimentary cycles were deposited and which provided detrital material for much of the Torridonian Series.

It was recognized at the turn of the century by the great highland geologists of the Geological Survey, C. T. Clough, B. N. Peach and J. Horne, that the Lewisian was composed of two 'complexes' separated by a period of basic dyke intrusion, but it was not until the work of Sutton and Watson in 1951 that the two complexes were recognized to be normal orogenic sequences of sedimentation, folding and metamorphism. Sutton and Watson introduced the terms Scourian for the older cycle and Laxfordian for the newer. They regarded the Lewisian as dominantly a north-west trending Laxfordian belt, with the Scourian being a relict mass of an earlier cycle at the centre of this belt. They did not recognize any Laxfordian sedimentary rocks in those areas which they mapped in detail and they regarded the Laxfordian simply as highly deformed and metamorphosed Scourian. The basic dykes which cross-cut the Scourian rocks were shown to be strongly deformed and metamorphosed in later movements and thus provided a marker enabling two cycles to be separated. More recent work has revealed the existence of a further major period of folding and metamorphism separating these two cycles. This additional event has been termed the Inverian orogeny. Further advances in the subdivision of this complex series of gneisses may be expected as isotopic dating methods become more refined and are linked more accurately to metamorphic and tectonic episodes.

There is no general agreement among those working on the Lewisian regarding the exact status of many of the rocks and there is even disagreement about the meaning of certain agreed facts, but the Lewisian still illustrates a number of important principles in the elucidation of complex metamorphic terrain.

The Scourian Cycle

The earliest rocks on the mainland appear to be banded pyroxene and hornblende granulites more than 2,600 m.y. old. Granulites are metamorphic rocks which have crystallized at very high temperatures and pressures and are often strongly banded. They represent the highest known grade of regional metamorphism. In the Lewisian they occur in isolated patches between Loch Broom and Drumbeg and are most abundant in the Scourie-Badcall district where they and some associated metasedimentary bands have been termed the Kylesku Group. Acid and basic rocks are found but basic granulites are more frequent as they have resisted the succeeding metamorphisms more readily. The banding is mainly tectonic in origin as intrafolial folds (small closed folds between the foliation planes) are common. In subsequent events, these granulites have been converted into gneisses of various types but their ultimate origin is unknown; it is possible that they were part of a geosynclinal pile. There seems no reason to call them orthogneiss (gneiss derived from igneous rocks) as their chemical variation is typical neither of igneous nor sedimentary series. Associated with the granulites is a series of ultrabasic banded rocks varying from metaperidotite to metagabbro and meta-anorthosite which are thought to be of igneous origin. They have been referred to as examples of alpine-type ultrabasic-basic intrusions and they occur as isolated pods distributed very widely throughout the Scourian granulites and their Inverian derivatives. They also have a granulite facies mineralogy and tend to be very resistant to subsequent metamorphic and structural changes. Their exact status and origin is still in dispute.

The trend of the Scourian is not known since no really large structures have been found associated with the granulites. The minor structures present sometimes trend north-easterly, cut across by the north-westerly Inverian and Laxfordian belts. North-easterly trending structures of both Inverian and Laxfordian ages are also found and diversity is to be expected in view of the great number of later fold episodes affecting these rocks.

On South Harris in the Outer Hebrides, several belts of metasediment occur: kyanite and sillimanite gneisses (pelites), forsterite marbles, calc-silicate rocks, garnet quartzites and garnet-cummingtonite-magnetite rocks (ironstones). These high grade rocks are interbanded in a few localities with granulites although generally the junction is a thrust. The most usual correlations suggest that the acid and basic granulites and gneisses which form the centre of North Uist and Benbecula and the metasediments and granulites of South Harris are equivalent to the Scourian granulites of the mainland. High grade metasediments similar to the South Harris rocks occur on Coll, Tiree and Iona together with pyroxene granulites.

The Inverian Cycle

This is not as well substantiated a cycle as the later Laxfordian, and it is possible that the Scourian and Inverian may be one long cycle. The isotopic age pattern and the structural history, however, suggest that there were two separate metamorphic events, both widespread and high grade, before the Laxfordian. The Inverian folding resulted in an almost complete re-orientation of the Scourian chain over much of the mainland outcrop. The granulites were folded into E.—W. trending monoclinal folds. Monoclinal cross-folds of various orientations are also found and the deformation ended with strong shearing on some of the vertical limbs of the E.—W. monoclines. Retrogressive metamorphism to amphibolite facies (medium grade regional metamorphism) took place, with the addition of vast amounts of water. The acid rocks were affected most and in places became almost mobile. The basic rocks reacted differently becoming agmatites, i.e. breaking up and altering to ultrabasic pods set in acid pegmatitic material. Some areas of the Inverian are almost entirely composed of agmatite. The last major effect of the metamorphism was the 'sweating out' of synorogenic pegmatites which cross-cut some of the Inverian structures, often merging into the acid gneisses. Both pegmatites and gneisses yield dates of about 2,200 m.y.

A dyke swarm—called the Scourie Dolerites or the Assynt Dyke Suite—was intruded during the waning stages of the Inverian metamorphism. The earliest and most abundant dykes are the thick quartz metadolerites ('epidiorites') which are almost entirely amphibolized. These are up to 100 yards across. They are cut by the thick bronzite-picrite and olivine-gabbro dykes, which have coarse pyroxenite margins. These margins are best explained by postulating intrusion of the magma into hot country rock. These dykes are metamorphosed only at their margins. Later dolerites are progressively narrower and progressively less altered, suggesting a marked drop in the country rock temperature over the period of intrusion. The dykes range in age from 2,190 m.y. to 1,910 m.y. The picrites occur mostly near Lochinver but there are isolated examples of similar petrographic type elsewhere. As this type is most unusual, it seems that the whole of the area from Scourie to Gruinard Bay is of Inverian age. In the Outer Isles, the metamorphism of basic dykes has been ascribed to an Early Laxfordian granulite facies metamorphism. This may correspond to the Inverian metamorphism on the mainland.

No sedimentary rocks of Inverian age have definitely been proved and much of the Inverian belt as now seen is composed of deformed and migmatized Scourian granulites re-orientated to an E.—W. trend. However, the 2,460–2,200 m.y. time interval is the time when the Huronian sedimentary sequence was laid down in eastern Canada and it is possible that either the Loch Maree Series (see below) or the South Harris metasediments (see above) are the equivalent of the Huronian rocks.

The Laxfordian Cycle

In the Gairloch area very low grade Lewisian metasedimentary rocks are found—the Loch Maree Series. This series consists of greywackes, sandstones, pelitic rocks and some ironstones and limestones. In places sedimentary structures are preserved in the greywackes. Among these metasediments are abundant and sometimes thick amphibolites which may have been lavas, sills or dykes.

This metasedimentary sequence was folded twice: firstly into large recumbent overfolds, with thrusts which brought up fragments of a granulite (probably Scourian) basement; then by NW.—SE. horizontal upright folds of very large amplitude. The whole mass was then metamorphosed at amphibolite facies with the production of acid gneisses in the north-west. The metasedimentary rocks in the south-east, enclosed by the enveloping basic bands, were not affected. To the south of Gairloch there are basic dykes cutting gneisses. The dykes are quite unfoliated at their centres, but at their margins they have a steep lineation corresponding to structures formed during a later fold episode when vertical folds were formed over a wide area of the Laxfordian. Brittle structures, including pseudotachylytes (glassy cataclastic rocks), complete the structural history.

This structural sequence cannot be matched exactly elsewhere because of the lack of supracrustal rocks, but the style of folding is certainly similar north of Loch Laxford. It was in this area that Janet Watson first described the rocks of the Laxfordian cycle. She suggested that the rocks north of Loch Laxford are the relics of a Scourian complex which has been highly deformed, metamorphosed and injected by granitic material during the Laxfordian orogeny. Recent geochemical studies suggest that the two areas are of different origin, much of the Laxfordian belt being a migmatized geosynclinal pile (though not necessarily of the same age as the Gairloch sediments). Granitic intrusions are much more abundant in the area north of Scourie than to the south of the Inverian block, but migmatites and agmatites are developed in both areas.

On the Outer Isles the Laxfordian orogeny is dominant, much of Harris and Lewis being composed of a migmatite complex of Laxfordian age. In this area, the older gneisses (and any newer supracrustal rocks that were laid down on top) have been cut by granites and pegmatites of several generations. In the Uists, the Laxfordian migmatization was not as intense, but very strongly developed Laxfordian folds can be seen to deform both dykes and the gneisses they cut—sometimes into isoclinal folds, in other cases more open in style. The Outer Isles Thrust is an important late structure which has obliterated all earlier structures in the gneisses through a thickness of hundreds of feet producing what has been termed 'mashed gneiss'. It has a great deal of epidote and pseudotachylyte associated with it, and is most probably related to the late stage brittle structures of this cycle, which are common throughout the Lewisian.

Within the Inverian block, the Laxfordian cycle is represented by a few large structures, such as the Glen Canisp Shear Belt (Fig. 3.3) which deforms

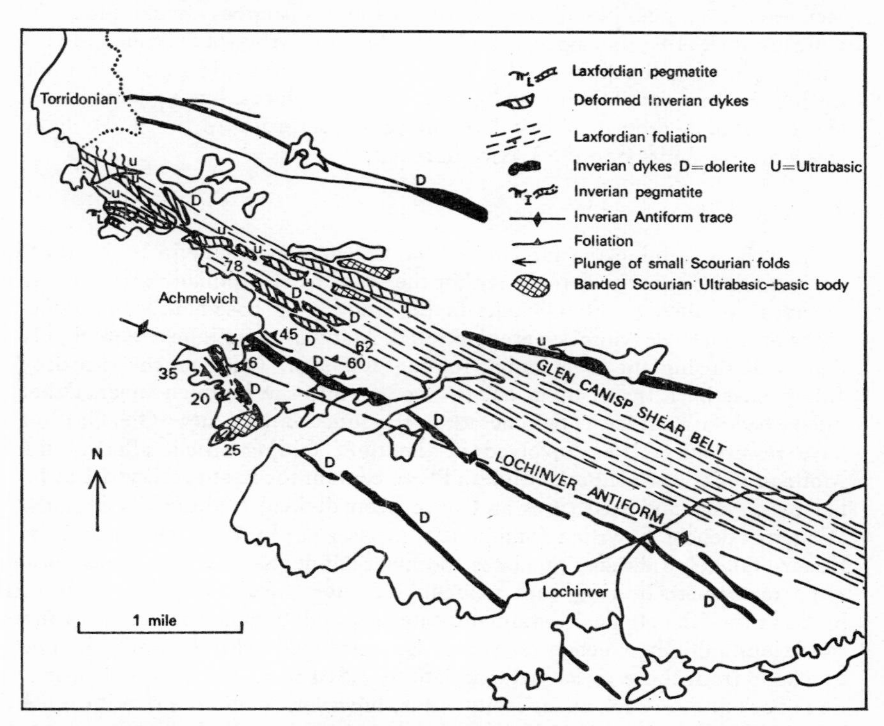

Fig. 3.3 The Lewisian of the Achmelvich area. A succinct illustration of the sequence of events which formed the Lewisian Complex. Scourian minor structures (small folds and also the foliation) are folded by the Inverian Lochinver Antiform, which is cut by the late Inverian Assynt Dyke Suite. Both Inverian structures and dykes are deformed by the Laxfordian Glen Canisp Shear Belt. (From a Ms. map by J. TARNEY.)

both the gneisses of the Inverian Lochinver Antiform and the Assynt Dyke Suite, and by abundant movements on the older Inverian monoclines, causing shearing at the margins of dykes which intrude the monoclines. In the north Laxfordian structures trending NW.—SE. are more common, and near the northern margin the Scourian and Inverian rocks have almost all been realigned by the Laxfordian deformation. A few pegmatites of Laxfordian age (1,700 m.y. to 1,600 m.y.) cut the Inverian and Scourian rocks. The ages obtained from the Laxfordian gneisses and pegmatites outside the Scourian and Inverian block range from about 1,550 m.y. to 1,300 m.y.

The Lewisian, as now exposed, thus represents the relics of the Laxfordian orogeny. In it are found fragments of earlier orogenic belts brought up as thrust slices and sometimes greatly deformed and migmatized. The central block, consisting of a fragment of the Inverian fold belt with fragments of Scourian within it, separates two areas of newer Laxfordian gneisses with some metasediments, possibly all representing a Laxfordian geosynclinal pile. The margins of the Inverian against the Laxfordian are everywhere steep and very sharp, following the lines of the long, wide belts of vertical gneiss formed during the Inverian, and it may be that the central block was a geanticline in the Laxfordian orogen or that later movements thrust up the block along steep reverse faults into the Laxfordian belt.

Lewisian inliers

As well as forming the foreland of the Caledonian mountain chain, there are outcrops of gneissose rocks within the area of the Moinian Series. These gneisses are almost certainly part of the underlying basement of Lewisian rocks on which the Moines were deposited. Many of these inliers were highly deformed during the Caledonian folding and some (such as the Scardroy Inlier) have been thrust up as narrow wedges into the Moinian cover. Other inliers have a Moinian conglomerate at the junction (e.g. near Glenelg) but have nevertheless been involved in the tight folding which affected the Moinian rocks in Caledonian times. These contain the best preserved Lewisian rock types and structures and have been divided traditionally into the Western Glenelg Lewisian (migmatitic gneisses cut by basic dykes) and the Eastern Glenelg Lewisian (gneisses and metasediments—marbles, ironstones, some manganese-bearing, and pelites). Eclogites form many of the basic bands in the Glenelg gneisses and indicate very high temperature and pressure conditions (although eclogites cannot be correlated with granulites). The pyroxene from these rocks yields a date of 1,510 m.y. The mode of formation of eclogites in such an environment is, however, so obscure that it cannot be said with certainty that all these rocks are of Laxfordian age.

The Torridonian-Moinian Series

Resting with marked unconformity upon the Lewisian of the foreland is the Torridonian Series, an enormous thickness of dominantly arenaceous rocks. Since the Lewisian also forms the basement to the Moinian Series, a dominantly arenaceous sedimentary sequence but everywhere separated from the Torridonian by the Moine thrust, the two supracrustal series have long been correlated. Work in the transitional area, between the Moine and Kishorn thrusts of Skye, has shown that the differences between the sedimentary succession in each nappe are not great and a gradual transition from typical Torridonian to typical Moinian strata can be recognized.

The age of the Torridonian is certainly Precambrian as the Cambrian beds rest on them with marked angular unconformity; the Moinian rocks are cut by intrusions dated at 740 m.y. and are affected by a metamorphism probably older than that. Fossils are very rare, only trace fossils and phosphate nodules being recognized until recently. Spores of plants or marine phytoplankton have recently been found in some of the shales and they have been tentatively correlated with similar types in the middle and upper Riphean, a sequence in Russia of late Precambrian age (1,500 m.y. to 600 m.y.). Rb/Sr age dates on the shales give ages of 885 m.y. for the Lower Torridonian and 815 m.y. for the Upper Torridonian (maximum ages).

The Torridonian sequence shows more variation from north to south than it does through the succession. In the north, a sometimes rugged landscape is filled up and overlain by a series of fan deposits of continental type and this type of environment gradually changes southwards into a geosynclinal marine trough (Table 3.3, Fig. 3.6).

Although large thicknesses of red current-bedded arkoses and conglomerates are the 'typical' Torridonian sediments of the Torridon-Assynt area, there are many other important and interesting rock types particularly in the Diabaig Group.

The Lower Torridonian

The Oronsay Greywacke Group is found only on Islay and Oronsay and is composed of turbidite greywackes, sandstones and mudstones with a basal epidotic conglomerate. On Oronsay it passes upwards into the Epidotic Grits, which form the common basal formation of the Diabaig Group over the remainder of the southern part of the Torridonian outcrop. This group has been studied in some detail in the Inner Hebrides and the Coigach-Stoer area. The marine facies of the Inner Hebrides consists of coarse grits with intervening banded silts and shales, a notable feature being the wealth of shallow-water structures. The thickness increases southwards and south-eastwards. The succession on Skye may correlate with similar sedimentary groups in the Tarskavaig Nappe, which are a median facies between the Torridonian succession and the Moinian sequence of the Loch Hourn region (Fig. 3.4).

The lowest formation, the Epidotic Grits, is always strongly current-bedded and is succeeded by rhythmic deposits of grey shales and green grits showing many load and injection structures. The grits are of shallow marine origin and are present almost everywhere. Further uplift caused more arkosic fluviatile sedimentation with microcline a prominent constituent. This group varies considerably in thickness, as would be expected, and is represented only by some arkosic intercalations in the dominantly marine succession on Colonsay. The topmost group of the Diabaig is banded silty sandstones and mudstones, the amount of fine material probably increasing south and east into the Moinian trough. Further north the Diabaig sedimentation was of completely different character and lateral variation is great (Fig. 3.5). However, the general succession of basal breccias and sandstones filling a very uneven topographic surface can be traced over a wide area. A marker horizon of

TABLE 3.3 Sedimentary Variation in the Torridonian

		Islay and Colonsay	*Skye*	*Applecross*	*Torridon*	*Coigach*	*Stoer*	*Cape Wrath*
Upper	Aultbea Group			Red sandstones, flags and shales 4,000 ft ← → 250 ft				
	Applecross Group		Sandstones and shales (shallow marine) 4,500 ft	Cross-bedded arkoses and conglomerates (continental fan deposits) 8,000 ft ← → 5,500 ft ← → 1,300 ft				
Lower	Diabaig Group	Grey, green and red sandstones and shales and calcareous lenticles (shallow marine) 6,000 ft	10,000 ft	Red sandstones, mudstones, conglomerates 9,000 ft	450 ft	Breccias and fine siltstones Unconformity Tilloid and limestone Breccias, sandstones and mudstones 1,000 ft	3,000 ft	
	Oronsay Greywacke Group	Greywackes and mudstones. Turbidites 6,000 ft						

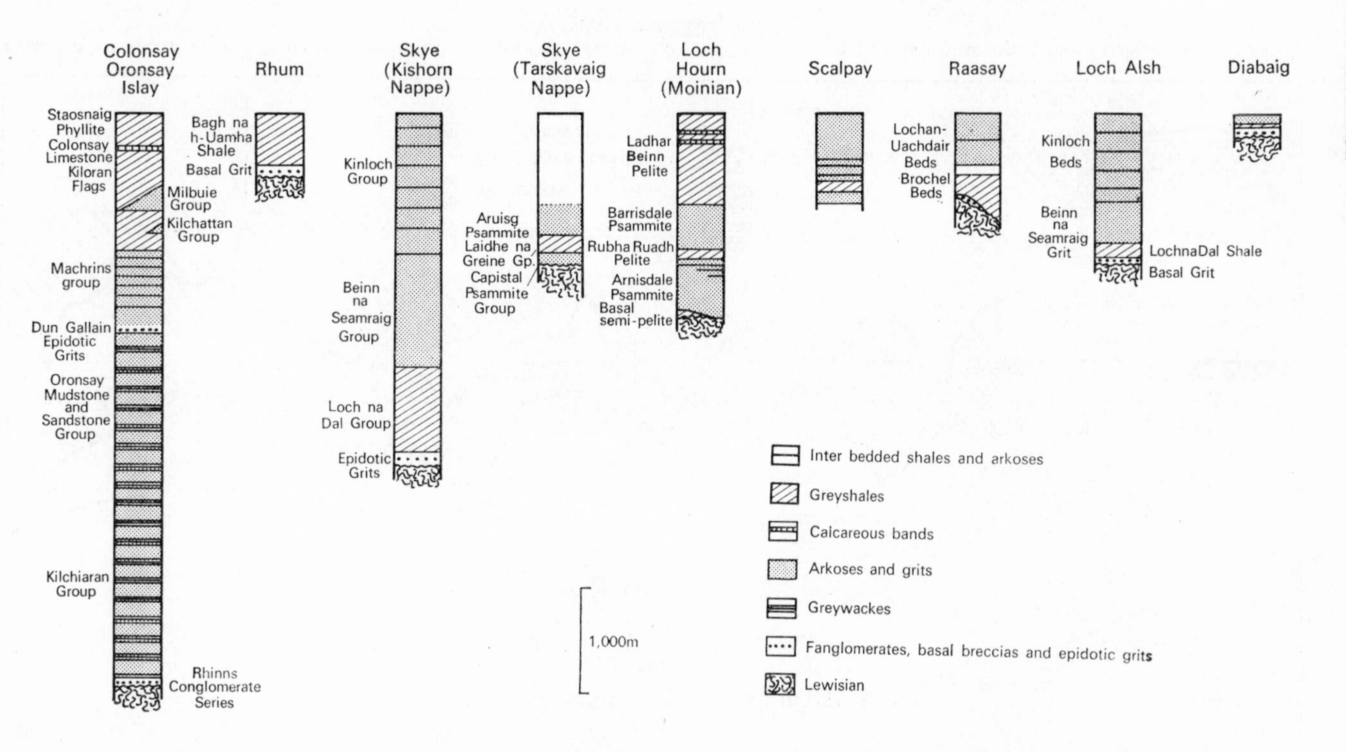

Fig. 3.4 Lateral variation in the Lower Torridonian-Moinian Series of the southern part of the outcrop.

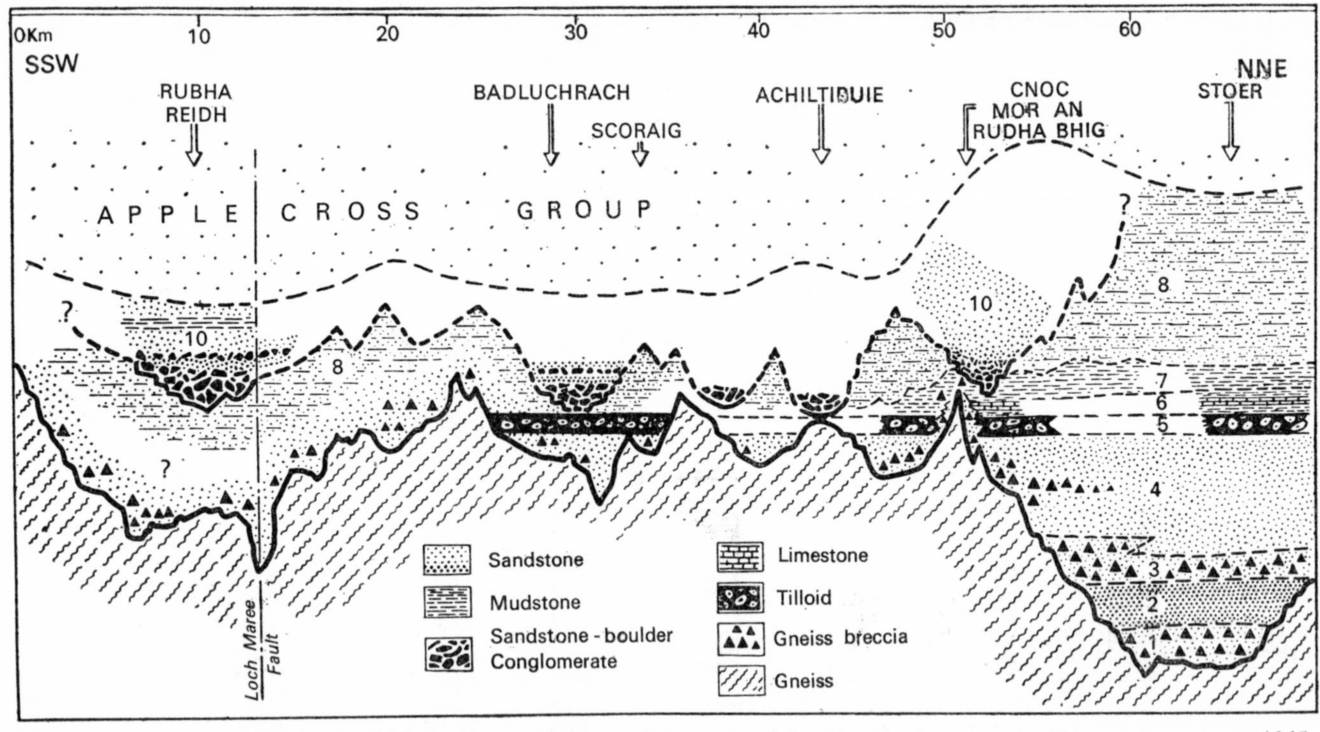

Fig. 3.5 Lateral variation in the Diabaig Group of the northern parts of the Torridonian outcrop. (From D. E. LAWSON, 1965, *Nature, Lond.*, **207**, 706.)

tilloid is found from Gruinard Bay to Stoer and consists of angular igneous fragments in a red muddy matrix. It contorts the underlying strata. It is possibly a volcanic mudflow and may pass laterally into sediments of volcanic derivation.

A period of uplift and erosion succeeded the deposition of this lower sequence, as is shown by a widespread sandstone-boulder conglomerate (Fig. 3.5). In the north the period of erosion was preceded by tilting of the lower part of the Diabaig rocks. This unconformity obviously forms a more suitable horizon for subdividing the Torridonian than the junction of the Diabaig and the Applecross Groups, but at the present time the position of this unconformity within the standard succession at Diabaig is unknown. As well as the marked angular unconformity, there is a difference in the palaeomagnetic pole position of about 50° which may indicate that this part of the Lower Torridonian is much older than the upper groups.

The Upper Torridonian

The Applecross and Aultbea arkoses are fluviatile fan deposits. The current-bedding directions suggest that these were fed by two large rivers lying to the north-west of the present mainland outcrop (Fig. 3.6). The fans filled up the valleys in a mountainous landscape of Lewisian rocks and completely covered these mountains by mid-Applecross times. This buried landscape is highly irregular, the highest Lewisian hills (A'Maighdean, north of Loch Maree) being up to 900 metres above the general level of the Lewisian surface.

Current-bedding is ubiquitous in this group and the large diapiric and cuspate structures frequently seen are thought to be due to quicksand movements taking place in rising springs in the fan deltas. The pebbles in these rocks are mostly vein quartz, quartzite, chert, grit, felsite and feldspar porphyry. These latter and a pebble of spherulitic felsite similar to the Uriconian of Shropshire indicate the existence of a late Precambrian volcanic episode further to the west or north-west. The presence of dreikanter as pebbles in these conglomerates indicates not only the rapid transport undergone by this material but also that the hinterland was a desert. No great significance should be attached to this, however, as in the absence of plants wind erosion was probably a powerful agent over the whole surface of the earth at this time. However, the high percentage of feldspar suggests a lack of chemical weathering and the red coloration of the beds indicates that oxidizing conditions were general. The climate was thus, most probably, warm, with intermittent heavy rains during a short rainy season.

The Moinian Series

The Moinian rocks are so much more highly deformed and metamorphosed than the Torridonian that not a great deal can be said about the stratigraphy or sedimentation. In various parts of the Moinian outcrop local successions have been worked out and the succession in the south-western

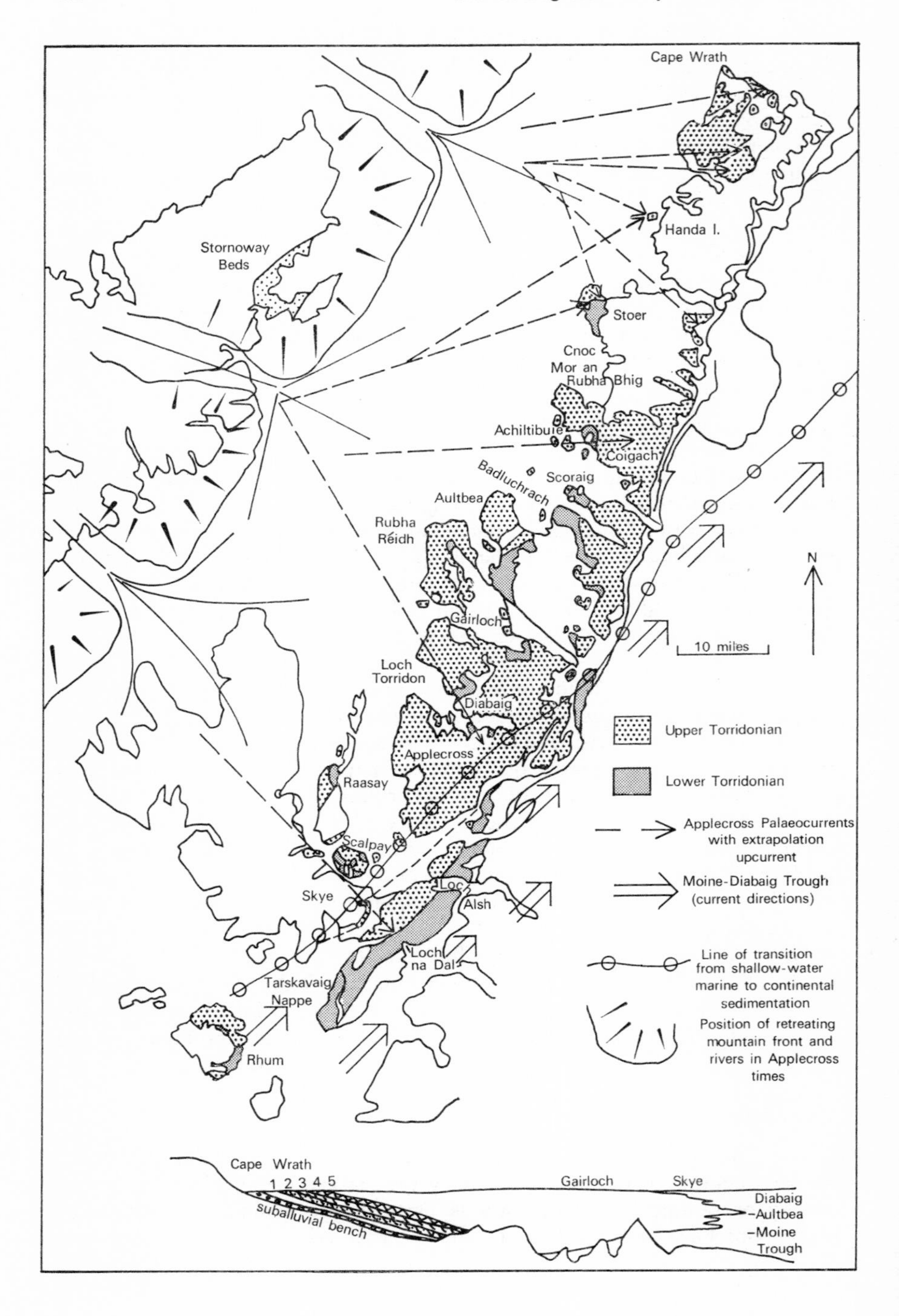
Cape Wrath
Handa I.
Stornoway Beds
Stoer
Cnoc Mor an Rubha Bhig
Achiltibuie
Coigach
Badluchrach
Scoraig
Aultbea
Rubha Réidh
Gairloch
Loch Torridon
Diabaig
Applecross
Raasay
Scalpay
Skye
Loch Alsh
Loch na Dal
Tarskavaig Nappe
Rhum
N
10 miles
Upper Torridonian
Lower Torridonian
Applecross Palaeocurrents with extrapolation upcurrent
Moine-Diabaig Trough (current directions)
Line of transition from shallow-water marine to continental sedimentation
Position of retreating mountain front and rivers in Applecross times
Cape Wrath
1 2 3 4 5
suballuvial bench
Gairloch
Skye
Diabaig -Aultbea -Moine Trough

margin of the outcrop may be correlatable with the Torridonian succession of Skye (Fig. 3.4—a higher psammite*—the Aonich Sgoilte Psammite is probably equivalent to the Applecross Group of the Torridonian). In central Ross-shire, although knowledge of the structural history is highly advanced, the rocks have suffered so much deformation that most of the local successions are not readily matched. A general succession of two semi-pelitic groups with alternating pelitic groups and an upper psammite, is of fairly wide occurrence. The inthrusting of sheets of Lewisian basement into higher parts of the succession has not made the recognition of a generally applicable stratigraphy any easier. The succession and possible correlations are given in Table 3.4. The total thickness is probably at least 19,000 feet in this central part.

The sedimentary facies of the Moinian Series is dominantly arenaceous with a few shales. It is highly feldspathic and often epidotic. Some conglomerate-like rocks have been found next to some of the Lewisian inliers, but at other boundaries a pelitic rock lies next to the Lewisian. Marbles occur but are not common. The Moinian rocks are generally metasediments showing little metamorphic segregation, although in the central migmatitic core, running from Ardgour to the north coast near Bettyhill, they are gneissose. Their metamorphic grade belongs to the amphibolite facies or a lower grade and the banding is undoubtedly sedimentary over much of the outcrop. Sedimentary structures are common, but are often strongly deformed. They are mostly turbidite structures—cross- and convolute-bedding—and, together with the immature nature of the sediments, indicate a geosynclinal origin for the Moinian Series as a whole. As has been mentioned above (see Fig. 3.6), this basin of deposition probably crosses the Moine thrust and the unmetamorphosed equivalent of the lower part of the Moine may be the Greywacke Group of the Torridonian of Islay and Oronsay.

The Moinian Series was deformed and metamorphosed in the Caledonian Orogeny: a full description of these events is given in Chapter 7. However, there is growing evidence from isotopic age data that the Moinian was also involved in a Precambrian orogeny. Whether this corresponds to the Grenville cycle of eastern Canada (900 to 1,400 m.y.) or to the later Celtic cycle

* Psammite: arenaceous rock, usually sandstone or arkose in the Moinian.
Pelite: argillaceous rock, usually highly micaceous after metamorphism.
Semi-pelite: mixed or banded shale-sandstone sequence.
These three terms are commonly used for members of the Moinian sequence.

Fig. 3.6 Palaeogeography of north-west Scotland in Applecross times. The distribution of Lower and Upper Torridonian rocks is also shown. The diagrammatic cross-section below represents the distribution of the facies of the Applecross-Aultbea Groups at the end of sedimentation. The marginal facies lying on the sub-alluvial bench are shown with greatly exaggerated thicknesses. 1. Cross-bedded cobbly facies, 0–30 ft. 2. Tabular pebbly facies, 100–150 ft. 3. Cross-bedded gravelly facies, 90 ft. 4. Cross-bedded sandy facies, 90–120 ft. 5. Contorted sandy facies, 90 ft. (After G. E. WILLIAMS, 1966. *Nature, Lond.*, **209,** 1303.)

TABLE 3.4 Sequence of the Moinian in Ross and Western Inverness

Skye Torridonian	*Tarskavaig Moinian*	*Loch Hourn*	*Morar*	*Southern Ross*	*Central Ross*
		7	Lochailort Pelitic Group with amphibolite		
		6 Aonich Sgoilte Psammite	Ardnish Psammitic Group		Tarvie Psammitic Group
		5 Ladhar Beinn Pelite	Loch Mama Pelitic Group with calc-silicate bands	Garnetiferous mica-schists of Creag na h'Iolaire	Pelites with Fannich Lewisian Meall ant-Sithe Group
				Lewisian with Moinian (Coire nan Gall)	Lewisian with Moinian (Scardroy)
Kinloch Group				Garnetiferous mica-schist series	Sgurr Mor Pelitic Group (impersistent)
Beinn na Seamraig Group	Aruisg Psammite Group	4 Barrisdale Psammite	Loch nan Uamh Psammitic Group with calc-silicate and heavy mineral bands	Siliceous granulite schist series	Inverbroom Semi-pelitic Group
Loch na Dal Group	Laidhe na greine Group	3 Rubha Ruadh Semi-pelite			Loch Droma Pelitic Group
Epidotic Grits	Capistal Psammite Group	2 Arnisdale Psammite			Achanalt Semi-pelitic Group
		1 Basal Semi-pelite local conglomerate	Beasdale Pelitic Group		
Lewisian	Lewisian	Lewisian			

[Successions from J. RAMSAY and J. SPRING, 1962, *Proc. Geol. Ass.*, **73**, 295; J. SUTTON and J. WATSON, 1962, *Geol. Mag.* **99**, 527; D. POWELL, 1964, *Proc. Geol. Ass.*, **75**, 223; 1966, *Proc. Geol. Ass.*, **77**, 79; and R. F. CHEENEY and D. H. MATTHEWS, 1965, *Scott. J. Geol.*, **1**, 256.]

represented only by relics in southern Britain and Brittany or to some other cycle is not yet clear. Pegmatites at Knoydart probably later than the earliest Moine folds and certainly earlier than the second and third fold phases, give dates of 740 m.y.; the Carn Chuinneag intrusion (at least 530 m.y. old) contains regionally metamorphosed xenoliths; and in Morar the ages obtained from Moinian metasediments show a pattern of increasing ages towards the lower grade rocks, reaching 560 m.y. in an area where only the first two sets of Moine structures are found. Evidence of the possible existence of a Moine orogenic belt before the deposition of the Dalradian Series is the occurrence of Moinian rocks to the south of the Dalradian outcrop in the Ox Mountains, around Lough Derg, Donegal (see Table 3.6), and possibly in north-east Tyrone. These outcrops may imply that the Dalradian chain cuts obliquely across a pre-existing Moinian chain.

The Dalradian Series

This series of metasediments is one of several, found in many parts of the world, where continuous deposition went on from late Precambrian times into Cambrian times. However, it is more conveniently considered with the Precambrian rocks since probably only the highest beds are of Lower Cambrian age and the whole sequence has been folded and metamorphosed as one unit. The basins of deposition, from Banffshire, through Perth and Argyll, into Northern Ireland and Connemara are probably an earlier series of basins similar to the later Caledonian geosynclinal basins of the Southern Uplands, the Lake District, Wales and Ireland.

Unlike the Moinian which must have been a relatively monotonous succession with little variation, the Dalradian represents a sequence of rocks of very varied original lithology which are now quartzites, schistose grits, slates, phyllites, mica-schists, impure limestones, amphibolites and gneisses. Since the original wide variety of rock types have been subjected to various grades of regional and contact metamorphism, the resulting metamorphic rocks are extremely diverse.

There were many controversies about the succession prior to 1930—and some persist—but Vogt showed that it was possible to make use of sedimentary structures, notably cross-bedding, to deduce whether strata are inverted or not. The succession in Perthshire was originally worked out by A. Geikie and has been generally confirmed by later investigators. There is some lateral variation in facies but the recognition of certain tillite horizons has recently enabled correlations to be made along the strike throughout Northern Ireland and SW. Scotland. However, the main variations in lithology and facies appear to have been at right angles to the strike, suggesting that original facies belts were sub-parallel to the subsequent structures.

An important structural break, the Iltay Boundary Slide, separates two contrasted sedimentary successions both in Scotland and northern Ireland. The rocks below are termed the Ballappel Foundation, forming the Ballachulish and Appin Nappes and separated in part from the Moinian by the Fort William Slide. The Iltay Nappe above the Iltay Slide is also mostly

TABLE 3.5 Successions in the Ballappel Foundation

NW. Donegal[a]	*Argyllshire*[b]	*Inverness-shire*[c]
Loughros Group (semi-pelites, quartzite and flags)		
Upper Falcarragh Pelites		
Falcarragh Limestone	Lismore Limestone	
Lower Falcarragh Pelites	Cuil Bay Slates	
Sessiagh-Clonmass Group (psammites and dolomites)	Appin Phyllite (semi-pelite) Appin Dolomite	
Ards Quartzite	Appin Quartzite (massive quartzite)	
Ards Transition Group (banded pelites and quartzite)	Striped Transition Series (banded pelites and quartzite)	
Ards Black Schists	Ballachulish Slates	
Altan Limestone	Ballachulish Limestone	Kinlochlaggan Limestone
Creeslough Schist Group	Leven Schists Glen Coe Quartzite Binnein Schist Binnein Quartzite Eilde Schist	Kinlochlaggan Quartzite Monadhliath Schist } Moine
	Eilde Quartzite	Eilde Quartzite } Moine
	Eilde Flags (Moine)	Eilde Flags } Moine

[a] From W. S. PITCHER and R. M. SHACKLETON, 1966, *Geol. J.*, **5**, 149.
[b] From M. R. W. JOHNSON, 1965, in *The Geology of Scotland*, ed. G. Y. CRAIG. Oliver & Boyd, Edinburgh and London.
[c] From J. G. C. ANDERSON, 1956, *Trans. R. Soc. Edinb.*, **63**, 15.

TABLE 3.6 Successions in the Iltay Nappe

		Lithological Groups[a]	Connemara[f]	Donegal[b,c,d,e]	Islay-Loch Awe[c,e,]	Central Highlands[e]	Deeside[e]	Banffshire[e]
DALRADIAN	UPPER	UPPER PSAMMITIC GROUP		Fahan Slates and Grits	Loch Avich Grits	Leny and Ben Ledi Grits (including Leny Limestone)		Macduff Group (greywackes and shales) Boyndie Bay Group (greywackes and shales)
		UPPER PELITIC AND CALCAREOUS GROUP		Inch Island Limestone	Tayvallich Lavas Tayvallich Limestone Group	Green Beds Aberfoyle Slates and Pitlochry Schist Loch Tay Limestone	Glen Tanner quartz and Mica-Schist Group Deeside Limestone	Whitehills Group (grits and calcareous flags) Boyne Limestone
	MIDDLE	LOWER PSAMMITIC GROUP		Upper Crana Quartzite Lower Crana Quartzite	Crinan Grits	Ben Lui Garnet Mica Schists	Queens Hill Quartzite and Mica Schist Group	
		LOWER PELITIC AND CALCAREOUS GROUP	Streamstown Group	Termon Pelites Cranford Limestone	Shira Limestone Group Ardrishaig-Craignish Phyllites	Farragon Beds Ben Lawers Calcareous Schist		Cowhythe Gneiss
		CARBONACEOUS GROUP			Easdale Slates	Ben Eagach Black Schist		Portsoy Group (impure limestones, quartzites and dark schists)
		QUARTZITIC GROUP	Bennabeola Quartzite	Slieve Tooey Quartzite	Scarba Transition Group Scarba Conglomerate Group Jura Slates Islay Quartzite	Cairn Mairg Quartzite, Killiecrankie Schists, Schiehallion Quartzite } Perthshire Quartzite		Durn Hill Quartzite
		TILLITE GROUP	Cleggan Boulder Bed	Boulder Beds	Portaskaig Boulder Bed	Schiehallion Boulder Bed		
	LOWER	BASAL CALCAREOUS GROUP	Connemara Marble	Glencolumbkille Limestone	Islay Limestone Mull of Oa Phyllites and Limestones	Blair Atholl Limestone		Sandend Group (Dark schists and impure limestone) Garron Point Actinolite Schist
MOINIAN			Barnanoraun Group	Glencolumbkille Pelites Lough Derg Psammitic Group	Mull of Oa Phyllites Maol an Fhithich Quartzite	Central Highland Granulites		Crathie Point Flags Findlater Flags West Sands Garnet-mica Schists Cullen Quartzite

Successions largely after the following authors: [a] J. G. C. ANDERSON, 1965, in *The Geologic Systems: The Precambrian*, vol. 2, ed. K. RANKAMA. Wiley, New York and Chichester. [b] J. C. G. ANDERSON, 1954, *Q. Jl geol. Soc. Lond.*, **109**, 399. [c] J. L. KNILL, 1963, in *The British Caledonides*, eds. M. R. W. JOHNSON and F. H. STEWART. Oliver & Boyd, Edinburgh and London. [d] W. S. PITCHER and R. M. SHACKLETON, 1966, *Geol. J.* **5**, 149. [e] M. R. W. JOHNSON, 1965, in *The Geology of Scotland*, ed. G. Y. CRAIG. Oliver & Boyd, Edinburgh and London. [f] P. J. LEGGO, in press, *Mem. Am. Assoc. Petrol. Geol.* (These successions are not completely established within each individual area—cf. Roberts, 1966.)

separated from the Moinian by a tectonic boundary, although conformity is suggested by some workers.

The Ballappel Foundation

The Ballappel succession is developed around Ballachulish from the island of Lismore in the south to the Grantown area in the north. The Creeslough succession of Donegal is correlated with this, although the Donegal granites have fragmented the Dalradian outcrops in that area so that many of the correlations must of necessity be tentative (Table 3.5).

The Ballappel succession is correlated with the lower part of the Iltay succession by Anderson, who equates the Ballachulish Limestone with the Islay and Blair Atholl Limestones. This correlation has not been generally accepted—the boulder beds which are so well developed in the Iltay succession have no counterpart in the Ballappel successions. The position of the Moine-Dalradian boundary is nowhere self-evident particularly as the Moine rocks south-east of the Great Glen cannot be correlated with the main part of the Moine outcrop to the north-west. The disputed succession of the Eilde, Binnein and Glencoe Quartzites with intervening schists is obviously very similar in sedimentary type to the pelite-psammite alternations of the Moinian Series and the impersistent nature of such beds is shown by the change in succession from the standard Ballachulish area into the Monadhliath area to the north-east.

The sedimentary history of the geosyncline in Ballappel times has not been investigated in detail. The lower parts of the succession are everywhere coarse-grained marine geosynclinal sediments with much of the psammite being fairly mature. The upper parts tend to form a rhythmic succession of limestone, shale and sandstone. These sandstones are very well sorted with current and scour structures in abundance and the whole succession is well laminated. No vulcanicity is known and the succession is probably indicative of deposition on an unstable shelf at the margin of a miogeosyncline.

Central Highland succession (Iltay Nappe)

Although the Perthshire succession has long been taken as the standard Dalradian sequence, there has been a wealth of new data on the rocks in the south-west Highlands and, since they are at the lowest grade of regional metamorphism, this may now be regarded as the type succession. The correlation of this sequence with Perthshire and Northern Ireland is reasonably secure, but the large amounts of granite and the high grade of regional metamorphism between Perthshire and the Deeside–Banffshire areas make these correlations highly speculative (Table 3.6). The correlation of the various members across the major structures of the Dalradian is also in dispute and the successions given must be regarded as working models rather than firmly established.

The division of the Dalradian into lithological groups suggested by Anderson is fairly consistent over the central and south-west parts of the outcrop and emphasizes the rhythmic nature of the deposition, but the division

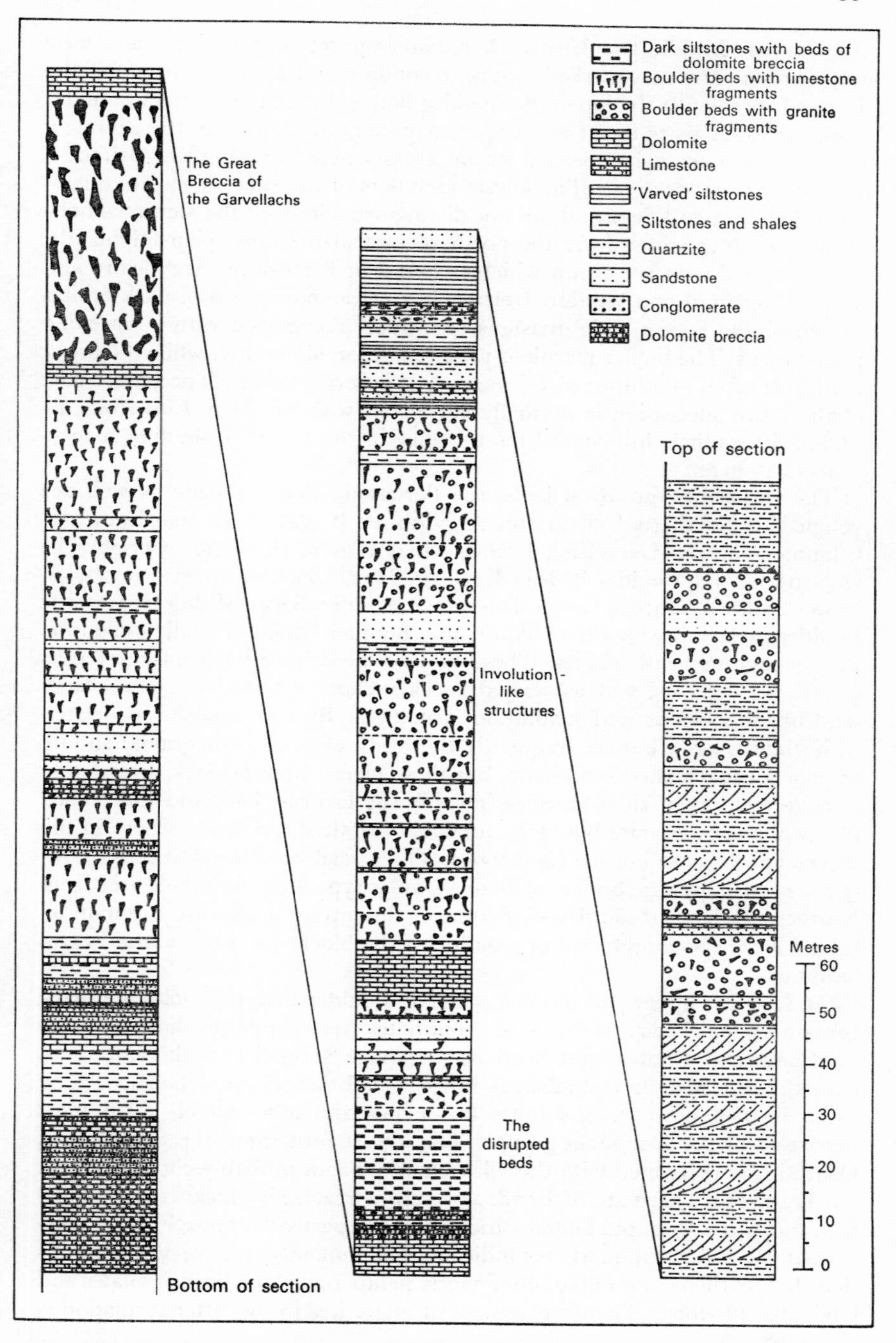

Fig. 3.7 The succession of the Portaskaig Boulder Bed in the Garvellachs. (After C. KILBURN, W. S. PITCHER and R. M. SHACKLETON, 1965, *Geol. J.*, **4**, 343.)

into Lower, Middle and Upper Dalradian is purely conventional and does not represent any very marked change in conditions (Rast puts the top of the Lower Dalradian at the top of the Boulder Beds). As in the Ballappel Foundation, the lower parts of the succession may be gradational into the Moinian, but no recent work has been done on areas where the Moinian–Dalradian Boundary is not tectonic. The lowest members of the successions are quartzites, phyllites and flags and are not distinguishable from the Central Highland Granulites which form the northern part of the Grampian Highlands. The lowest distinctive group which is found in Perthshire, Argyllshire and also in Northern and western Ireland, is a calcareous group of dark schists and marbles. Two marble divisions are widely recognized with intervening pelitic rocks. The higher marble is purer and more dolomitic, while the lower marble is often graphitic; in Connemara it is serpentinous. The lowest part of the Banff succession is normally correlated with the Islay Limestone although it has little lithological resemblance to the Lower Dalradian successions elsewhere.

The succeeding group of beds, the Portaskaig Boulder Beds, is well developed in Northern Ireland and Argyllshire. It reaches its maximum development on the Garvellach Islands where a great thickness of greywacke with intervening Boulder Beds is developed (Fig. 3.7). The normal succession in the Tillite Group is Lower Dolomitic Boulder Beds, Middle Psammitic Boulder Beds, Grey Quartzite with Upper Boulder Beds and finally Dolomitic Beds sometimes with pelites. These boulder beds have been interpreted as glacial till deposits, and associated rocks include 'varved' siltstones, large sand-filled polygons and involution structures. Breccia deposits associated with greywackes are more frequently the result of submarine gravity sliding of poorly consolidated sediments, but these glacial breccias lack several distinctive features of slide breccias (notably the loadcast base and the graded top) while large granite boulders, probably icerafted, occur elsewhere in the succession in more normal sandstones. The glacial breccias themselves contain a suite of granitic boulders, all of a similar type, over the whole area from Northern Ireland to south-west Scotland. Re-entrant angles are common on these blocks and a preferred orientation of the blocks has been found in two localities.

On Islay the upper part of the Portaskaig Boulder Bed, the Dolomite Beds, contain Fucoid Beds and Pipe-rock Quartzites, very similar to those found in the Cambrian of north-west Scotland (see page 83) and it is therefore possible that the Middle Dalradian is Lower Cambrian in age. The succeeding rocks are not, however, dolomites but a quartzite series which is the most persistent of the Dalradian groups. The lowest formation, the Schiehallion Quartzite, is correlated with the Islay Quartzite and probably correlates with the Durn Hill Quartzite of Banff. On Jura it reaches a thickness of 15,000 feet, but there are rapid lateral variations, only partly tectonic. Further conglomerates and clean quartzites indicate predominantly shallow-water conditions before the next group of dominantly pelitic rocks, the Easdale Slates and Craignish Phyllites. The development of quartzites in the latter formation is found over the same area as the maximum thickness of the Islay Quartzite. Current directions in these and later rocks and also facies variations in the Crinan Grits and Tayvallich Limestone indicate that a land-mass lay to the north-west with deeper water to the south-east (Fig. 3.8). Sedimentary structures are widespread in this part of the Dalradian and indicate a geosynclinal sequence of rapidly deposited immature psammites interbedded with finer dark phyllites and limestones. The rhythmic nature of much of the succession, as well as internal evidence in the rocks themselves, indicates that the area of deposition was quite unstable during the period of sedimentation. The detailed studies by Knill and by Sutton and Watson, of areas at opposite ends of the Scottish Dalradian, have many features in common, indicating the geosynclinal nature of the deposits, the uplift of a landmass to the north-west

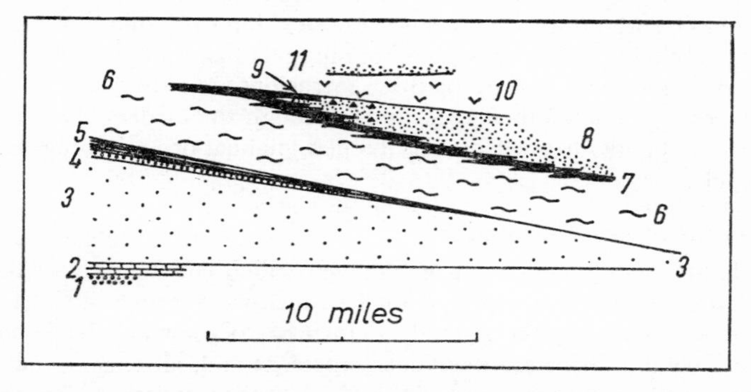

Fig. 3.8 Reconstruction of the Dalradian sedimentary pile in the Islay-Loch Awe region at the close of sedimentation. 1. Portaskaig Boulder Bed; 2. Lower fine-grained Quartzite and Dolomitic Group; 3. Islay Quartzite; 4. Scarba Conglomerate; 5. Easdale Slates; 6. Ardrishaig-Craignish Phyllites; 7. Shira Limestone; 8. Crinan Grits; 9. Tayvallich Limestone; 10. Tayvallich Lavas; 11. Loch Avich Grits. Horizontal and vertical scales equal. (From J. L. KNILL, 1963, in *The British Caledonides*, eds. M. R. W. JOHNSON and F. R. STEWART. Oliver & Boyd, Edinburgh and London.)

during the Dalradian (possibly at the beginning of the Upper Dalradian) and the rapid erosion of this material and transport into the trough by turbidity currents. In the Islay area, a succession of events has been established indicating a change from deltaic conditions (Islay Quartzite), then an epicontinental sea (Middle Dalradian) developing into a trough with greywacke deposits and a marginal shelf (Crinan Grits and Upper Dalradian). At this time, as the shelf seas encroached south-eastwards and the trough filled up, the only extensive volcanic activity of the Dalradian times took place—the Tayvallich Pillow Lavas and Green Beds of Argyll and Perthshire.

The very highest beds of the Dalradian in the south-west Highlands are the Leny and Ben Ledi Grits found along the Highland Border. They are also greywacke deposits but include a limestone, the Leny Limestone, which has yielded an agnostid trilobite, *Pagetides*, indicating a Lower Cambrian age for at least the Upper Dalradian. Roberts has suggested that the Leny Limestone may be the lateral equivalent of the Loch Tay Limestone, although the structural evidence is, as yet, ambiguous. This correlation would make the Dolomite Beds–Pipe Rock and Fucoid Beds correlation much more likely (see p. 60). The Macduff Slates of the Banffshire area may be younger than the highest Perthshire rocks, and Hugh Miller recorded a graptolite from these. Other trace fossils, stromatolites, worm tracks, tubular and globular bodies are known from the Middle and Lower Dalradian. In many areas of the world the Lower Cambrian faunas are found in rocks which conformably succeed many thousands of feet of sedimentary rocks. These are referred to an 'Eocambrian' Period and it seems probable that the Dalradian is partially equivalent to this system and is partially Cambrian in age.

In the north-eastern Dalradian area the metasediments are intruded by a series of basic and ultrabasic igneous rocks showing rhythmic layering. These lack some of the characters of ophiolitic plutons of alpine type, serpentinites, for example, are not a feature of these intrusions, but they occur as a fragmented series of related intrusions and, although the evidence is conflicting, they were probably intruded during the geosynclinal or early orogenic phase of the cycle.

The Central Donegal succession

The Kilmacrenan succession of Donegal which corresponds to the Islay succession of Scotland is replaced to the south-west, south of the Leannan fault, by a series of more strongly deformed rocks which have not been studied in detail. There are many slides present and, although some groups correspond in lithology to some of the Kilmacrenan divisions, the structural succession is quite different. Evidence of facing directions of the succession is confined to the quartzites, and the succession below must be regarded as structurally and stratigraphically tentative (Table 3.7).

TABLE 3.7 The Central Donegal Succession (Pitcher *et al.*, 1964, *Q. Jl geol. Soc. Lond.*, **120**, 241)

Lough Eske Semi-pelitic and Psammitic Group
Silver Hill Quartzite
Owengarve Calcareous Group
Gaugin Boulder Bed
Gaugin Quartzite
Craghhubbrid Dark Schists
Boultypatrick Grit
Swilly Chloritic Schist Group

South-east of the Highland Boundary Fault in Ireland the only extensive relics of Precambrian rocks are the Connemara Schists which have been mentioned along with the rest of the Northern Irish Precambrian. The Mona Complex of Anglesey and the Welsh Borderland Precambrian are part of a separate late Precambrian orogenic belt and are dealt with below. To the north, the rocks in Shetland are commonly correlated with the Dalradian.

They are made up of geosynclinal sediments, mostly highly metamorphosed and strongly deformed. Their most remarkable deposit is the strongly deformed Funzie Conglomerate, a very coarse quartzite–limestone–conglomerate with a subgreywacke matrix, usually thought to be similar in origin to the rapidly deposited Alpine Flysch conglomerates. The full thickness of sediments is very large, at least 36,000 feet, and this area represents a major extension of the Dalradian geosyncline and provides a link with the great development of Eocambrian-Lower Palaeozoic sediments laid down in the Caledonian geosyncline of Scandinavia, east Greenland and Spitzbergen.

The Celtic Cycle

A wide variety of sedimentary and volcanic rocks will be discussed under this heading. They are found as inliers in the Palaeozoic rocks of Wales, south-east Ireland and England. The sedimentary characters of the whole group indicate that a major geosynclinal sequence was laid down in the area from Rosslare, Co. Wicklow and Anglesey to Ingleton including, probably, south Wales. Acid volcanics and foredeep flysch deposits were laid down in the Welsh Borders and English Midlands and these were followed by molasse deposits and, in north Wales, by post-orogenic volcanic rocks. A sequence of fold and metamorphic episodes with subsequent granitic intrusions of Precambrian age, is recognized in Anglesey and Rosslare. In this respect this suite of rocks differs from the Dalradian rocks and represents a complete Precambrian orogenic cycle. It has been termed the Celtic orogenic cycle.

The Monian and Arvonian

The type area is that of Anglesey and north Wales where a thick sequence of typical orogenic sediments—greywackes and pelites, with spilites, cherts and limestones—is developed (Table 3.8). The South Stack Series consists of turbidite units showing graded and convolute bedding (Fig. 3.9) with one thick, mature deposit—the Holyhead Quartzite. The rocks become more pelitic upwards and the New Harbour Group is entirely chloritic siltstones.

TABLE 3.8 The Mona Sediments of Anglesey and Lleyn

Fydlyn Group	Gwyddel Beds of Lleyn
Gwna Group	Mynydd Carreg Spilitic Series
	Gwna Mélange
	Aber Geirch Phyllites and Grits of Lleyn
Skerries Group	Church Bay Tuffs
	Skerries Grits
New Harbour Group	New Harbour Beds
South Stack Series	Rhoscolyn Series
	Holyhead Quartzite
	Llwyn Group
	South Stack Moor Group

The first spilitic volcanics are found here. Pyroclastics and conglomerates form the Skerries Group, but more normal pelites and greywackes form the lower part of the Gwna Group. The Gwna Mélange is a remarkable rock,

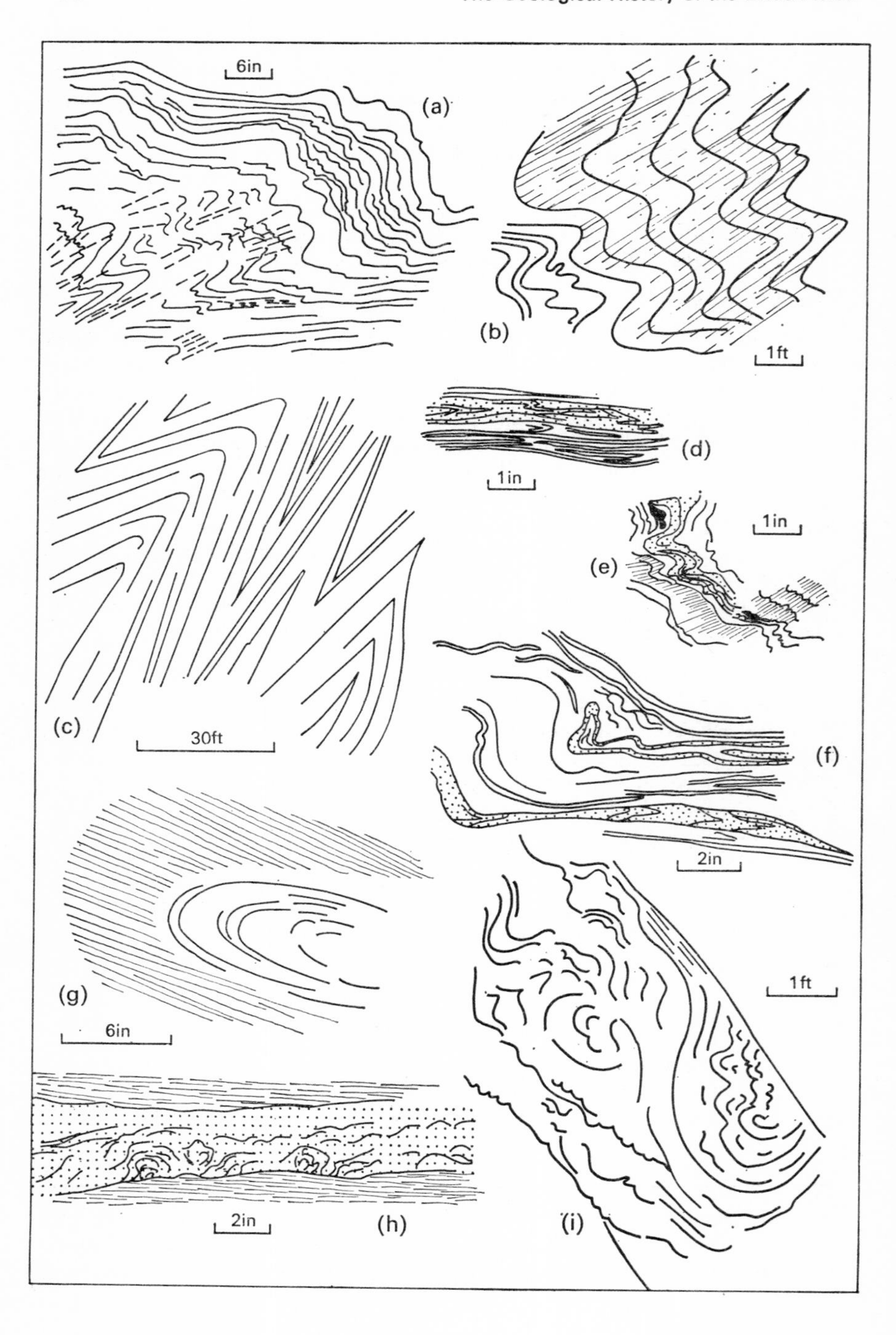
6in
(a)
(b)
1ft
(d)
1in
1in
(e)
(c)
30ft
(f)
2in
1ft
(g)
6in
2in
(h)
(i)

compositionally mature quartzite and limestone blocks set in a chlorite schist. It is interpreted as a submarine slide deposit. The spilites which follow, magnificent pillow lavas and cherts, mark a return to volcanic conditions which were maintained to the top of the succession as shown by the succeeding white sodic ignimbrites of the Fydlyn and Gwyddel Beds. The rocks have been thrust over the Ordovician, in places, during the Caledonian Orogeny but a period of deformation certainly preceded or accompanied the migmatization and low grade regional metamorphism which is dated as Precambrian. This deformation was very intense in Anglesey; intrafolial folds are widespread in the phyllitic rocks. The main phase of folding is of angular long-limb—short-limb folds with associated thrusts (Fig. 3.9). Conjugate folds are of later age. Very sharp-sided zones of migmatitic gneisses were formed in Anglesey and Lleyn and are associated with high grade sillimanite-bearing rocks. Elsewhere, the metamorphism was of greenschist facies with glaucophane-lawsonite schists developed in a very limited area. Serpentine intrusions were emplaced before the metamorphism.

The Arvonian Volcanic Suite near Caernarvon was not involved in the late Precambrian folding and is a series of sodic volcanic deposits, much of it ignimbrite, laid down towards the end of an orogenic episode. They seem to be conformable with the fossiliferous Cambrian rocks of Caernarvonshire although older than the lowest Cambrian zone. The Mona gneisses and the Arvonian Volcanics are cut by late orogenic granitic rocks, the Coedana Granite, the Sarn Adamellite and the Twt Hill Granite.

This sequence of events can be recognized in fragmentary records preserved elsewhere. In the Bray and Howth areas, Co. Wicklow a Precambrian bedded series composed of 4,000 feet of black slates (the Knockrath Series) and about 4,000 feet of greywackes and interbedded slates (the Bray Series) have been folded into major tight folds before the deposition of the Clara Series (? Cambrian). Further south, a higher grade of regional metamorphism was reached, and biotite and hornblende schists and migmatites (the Rosslare Complex) and phyllites with tight minor folds (the Cullenstown Group) are cut by a late granite (the Carnsore Granite). In south Pembrokeshire, a similar grade of metamorphism is found in some schists and gneisses. The Dutch Gin schists show retrogression from garnet amphibolite grade of a laminated quartzose metasediment and this is associated with acid and basic gneisses.

Fig. 3.9 Tectonic and sedimentary structures in the Mona Complex, Holy Island, Anglesey. (*a*) Main phase minor fold in phyllite. New Harbour Group, Trearddur Bay. (*b*) Main phase minor folds in hinge zone of major fold. New Harbour Group, Trearddur Bay. (*c*) Tight folds of main phase with strongly attenuated limbs. South Stack Series, SE. of South Stack (*d, e, f*) Intrafolial folds in psammitic bands of finely banded phyllite. New Harbour Group. They are refolded by main phase folds in (*e*) and (*f*) with a well developed fracture cleavage in the pelitic bands. Porth Isallt-bâch, Trearddur Bay. (*g*) Isoclinal fold in coarse psammite with axial plane fracture cleavage. South Stack Series, Rhoscolyn. (*h*) Convolute bedding in inverted thin psammite. South Stack Series, Graig Lŵyd. (*i*) Convolute bedding in thick psammite. South Stack Series, South Stack. (Drawn from photographs. Scales in feet and inches.)

TABLE 3.9 The Succession of the Longmyndian

Series	Group	West	East
Wentnorian Series	Bridges Group	Shale and siltstones 3,900 ft[a]	
	Oakswood—Bayston Group	Oakswood (West), 5,500 ft Sandstones Stanbatch conglomerate Sandstones Shales Darnford conglomerate Sandstones Haughmond conglomerate	Bayston (East), 6,500 ft Sandstones Lawn Hill conglomerate Sandstones Oakswood conglomerate Sandstones Shales and tuffs Radlith conglomerate Shales and sandstones with tuffs
		Unconformity	
Strettonian Series	Portway Group (Mintonian)	Shales and conglomeratic sandstones Huckster Conglomerate	3,800 ft
	Lightspout Group	Shales and siltstones Haddon Hill Grit	2,800 ft
	Synalds Group	Shales and volcanics Upper Cardingmill Grit	2,800 ft
	Burway Group	Lower Cardingmill Grit Flags and tuffs Buckstone Rock	2,400 ft
	Stretton Shales	Brockhurst Shales *Church Stretton Fault* Watling Shales Helmeth Grit	3,400 ft[a]
Uriconian Volcanic Suite	Volcanics of Hazler and Ragleth Hills		

[a] Maximum thickness seen.
(Largely after J. H. James, 1956, *Q. Jl geol. Soc. Lond.*, **112**, 315.)

Diorites and other plutonic types (Johnstone Diorites) also occur and have also probably been deformed at depth although the age of this deformation is unknown. In north Pembrokeshire a thick sequence of largely acid volcanics, tuffs and agglomerates (Pebidian) is unconformably overlain by lowest Cambrian rocks but both have suffered strong post-Cambrian deformation. They are cut by acid plutonic rocks. Similar unmetamorphosed volcanics (Bentonian) are also found in south Pembrokeshire. All these volcanics (the Pebidian Volcanic Suite) would seem to be comparable with the Arvonian Suite. The only other outcrop which falls readily into this sequence is the Ingletonian of Yorkshire. This is pre-Ordovician and has suffered a very much more intense phase of deformation than the overlying Palaeozoic rocks. It is composed of greywackes and slates with abundant detrital volcanic material, including tuffs and jaspers. Very large tight folds repeat the succession which may be 1,500 feet thick. The Ingletonian lies on a continuation of the south-east Ireland-Anglesey geanticlinal ridge (of the Caledonian geosyncline) and would seem to be comparable with some of the Mona sediments.

The Uriconian, Longmyndian, Charnian and Malvernian

In the Welsh Borderland and English Midlands (see Fig. 7.9) there are more small inliers which reveal a different sequence of rocks and events which can only be correlated in a very general way with the sequence of the Mona Complex and Arvonian. The Uriconian Volcanic Suite of Shropshire is a sequence of volcanic rocks mostly acid and pyroclastic and often submarine, which is overlain unconformably by the lowest Cambrian rocks. They are found on both sides of the Longmyndian sediments, forming the Stretton Hills and the Wrekin to the east and Pontesford Hill and other smaller outcrops to the west. The Eastern Uriconian on Hazler Hill is succeeded conformably by the Helmeth Grit at the base of the Longmyndian.

The Western Uriconian of Pontesford and Plealey has been correlated with the Eastern Uriconian on lithological grounds, but there is doubt as to the status of the Uriconian on Linley Hill which apparently oversteps the folded Upper Longmyndian and would seem, therefore, to be of later age. The Pontesford Hill rocks include rhyolites, andesites (with palagonite–breccias and laminated tuffs) and ignimbrites, and the Wrekin rocks are again of acid and intermediate pyroclastics and flows.

The Longmyndian which succeeds the Uriconian Suite has a major unconformity and a change in sedimentary type within a very thick geosynclinal pile (Table 3.9). The Strettonian Series which forms the basal group on the eastern side of the Longmynd is dominantly of flysch sandstone, starting with quiet deposition and getting coarser and with more volcanic material upwards. There is little evidence of submarine slumping. The Portway Group at the top of the Strettonian was separated off by James as a distinct Series, the Mintonian, as he postulated an unconformity below the Huckster Conglomerate. Subsequent work has not confirmed this although there is a change to molasse-type sedimentation at this horizon as the rocks are of much more conglomeratic nature from the Portway Group right through the Wentnorian. A major unconformity occurs at the base of this series which is found over a much wider area as well as in isolated outcrops from Haughmond Hill in the north to Brampton Bryan and Old Radnor in the south. It extends eastwards onto the Uriconian in the area of the Stretton Hills.

The succession of the Oakswood and Bayston Groups either side of the Bridges Group and the facing directions of these groups indicate that the Longmyndian is in the form of a very large syncline. Small-scale structures are not abundant but a series of tensional faults—followed in many cases by basic and some acid dykes—are aligned at right angles to the axial plane. These would appear to be similar in age to dolerite dykes which cut the Uriconian in many places and which have given a date of 638 m.y.

TABLE 3.10 Succession of the Charnian, Charnwood Forest

Brand Series	Swithland Slates Trachose Grits and Quartzite Hanging Rocks Conglomerate	Late orogenic molasse-type sediments
Maplewell Series	Woodhouse and Bradgate Beds Slate Agglomerate	Clastic and pyroclastic rocks
	Beacon Hill Beds Felsitic Agglomerate	Pyroclastic rocks
Blackbrook Series	Blackbrook Beds	Pyroclastic rocks

Outcrops of similar rocks occur eastwards at Lickey (? Uriconian Volcanics) and Nuneaton, where there is a series of tuffs capped by a thick welded tuff and cut by a very wide variety of acid intrusions and later basic dykes. Some of these are similar to intrusions cutting the Charnian rocks. The Charnian (Table 3.10) is a series of Precambrian sediments largely volcanic in nature, and they formed isolated hills in the Triassic desert. The lowest beds are fine followed by coarse pyroclastics which give way to conglomeratic beds. The amount of agglomerate in the Maplewell Series varies considerably and in the north-east consists almost entirely of very coarse volcanics. The Slate 'Agglomerate' is probably a slump deposit as the slate fragments were only slightly consolidated muds before the deposition of this bed. Intruded into this rhythmic succession of coarse and fine rocks are acid intrusions of varying size, the largest syenites (markfieldites) having a characteristic micropegmatite matrix. These intrusions were emplaced later than a period of folding, which folded the Charnian into a conical open anticline trending south-east and eastwards, and they give isotopic ages of 680 m.y.

There is no relation between the large Longmynd fold and the Charnwood Forest fold either in style or trend, but it is possible that the open folding of the Charnian is indicative of the earth movements preceding the wider extension of the molasse in Wentnorian times. The Charnian would then be correlated with the Uriconian-Strettonian and the similar molasse deposits of the Portway Group. There was an enormous area of subaerial volcanics laid down at this time, since volcanics of Uriconian type, often with ignimbrites, are found as far east as Norfolk, wherever boreholes have been drilled through the Mesozoic and Palaeozoic cover.

A body of Precambrian rocks with a different structural history forms the Malvern Hills to the south-east of the Welsh Borderland outcrops. They are plutonic rocks, including intrusions and gneisses, possibly overlain by the Warren House Volcanics and with Lower Cambrian rocks lying unconformably on the gneisses at the southern end of the range. The Malvernian is composed predominantly of metamorphosed igneous rocks, diorites and gabbros.

TABLE 3.11 The Celtic Orogenic Belt of England, Wales and south-east Ireland

Phase	Episode	*Geosynclinal Zone*				*Foredeep*	*Foreland*
		SE. Ireland	*N. Wales-Ingleton*	*S. Wales*	*Malvern-Hanter*	*Longmynd and Wrekin*	*Charnwood*
Epeirogenic	Post-Orogenic Volcanism		*Arvonian Volcanic Suite*	Pebidian	Warren House	Uriconian of Linley Hill	
Orogenic	Late Orogenic Plutonism	Carnsore Granite	Coedana and Sarn Granites (600 m.y.)	Retrogressive metamorphism	*Late Malvernian metamorphism* (590 m.y.)		
	Late Orogenic Folding	Retrogressive metamorphism	Retrogressive metamorphism *Main Folding* (611 m.y.)	Dolerites	Dolerites	*Longmynd Dolerites* (638 m.y.) *Longmynd Folding*	Dolerites
	Late Orogenic sedimentation				Wentnorian	*Wentnorian* (molasse)	
	Synorogenic Folding			*Johnstone Hybridization*	Hybridization	Granophyres Unconformity	*Markfieldites* (680 m.y.) *Charnian Folding*
	Synorogenic Plutonism and Sedimentation			Diorites	*Malvernian Plutons*	*Strettonian* (molasse) (flysch) Uriconian (volcanics)	(molasse) Charnian (volcanics)
	Early Synorogenic Episode	Rosslare Migmatites[a]	*Early Folding* and *Penmynnydd metamorphism*[a]	Folding High grade metamorphism[a]	High grade metamorphism[a]		
Geosynclinal	Sedimentation	Cullenstown Group	*Mona sediments* and Ingletonian	Dutch Gin sediments	Sediments (?)		

(The names which have been suggested as most appropriate for each of the episodes are in italics.)
[a]It is perhaps more likely that these include relics of an earlier basement (T. P. Crimes and N. B. Dhonau, 1967, *Geol. Mag.*, **104**, 213)

They have been hybridized by later syenitic and granitic pegmatites. Basic dykes cutting them have been altered to hornblende and biotite schists but the metamorphism did not produce a foliation in the larger plutons. Rocks of sedimentary origin occur in the southern part of the range, quartz-mica-schists and quartzites being found. The grade of metamorphism of the whole suite probably belongs to the amphibolite facies as there is a widespread development of green hornblende. Their isotopic dates have consistently given an age of 590 m.y. for the metamorphism, but dates between 700 m.y. and 600 m.y. for the age of the intrusions. The Malvernian has a weathered surface near the contact with the Warren House Volcanics leading to an interpretation of this contact as an unconformity. The volcanics have usually been correlated with the Uriconian Volcanic Suite but it seems more likely that they are a volcanic episode just preceding the fossiliferous Cambrian, similar to the Arvonian of north Wales. This block has suffered much more deformation in Hercynian times than the other Precambrian areas and most theories regarding the origin of the Malvern structure entail thrusting of this block from depth. The Malvern block, therefore, has more in common with south Pembrokeshire than any other region and the sequence bears a closer similarity to that of the Mona-Arvonian area than the Uriconian-Longmyndian. The small intrusions of Hanter Hill and Stanner Rocks, south-west of the Uriconian of the Stretton Hills, have a similar sequence of intrusive phases but lack the deep-seated metamorphism. Fragments of these intrusions are found in the neighbouring Wentnorian molasse.

An attempt at general correlations within the late Precambrian of the area south of the Dalradian is given in Table 3.11 with isotopic dates where applicable.

Crystalline rocks within the Variscan Orogenic Belt

The rocks of the Lizard and the Start Point–Bolt Head area are composed of schists, amphibolites and a serpentine-gabbro complex. Isotopic ages from the Kennack Gneiss, in the Lizard Peninsula, give ages between 350 and 400 m.y. and are thus not strictly comparable with any other plutonic metamorphic rocks in Great Britain, although this is fairly similar to the late Silurian dates obtained from the Moinian Series. It is possible that they are Caledonian plutonic rocks, as Lower Palaeozoic rocks are found in adjacent outcrops. However, they are cut by intrusions of gabbro, dolerite and granite and in this respect are similar to the Channel Islands and the North Breton coast which are mainly composed of metamorphic and plutonic igneous rocks. These are regarded as part of a late Precambrian basement to the main Variscan chain. It is therefore possible that a late Precambrian (800 to 600 m.y.) orogenic belt extended from the Moinian, through the Irish Sea area, as far south as Brittany.

III The Lower Palaeozoic Era

4 The Cambrian System

The Lower Palaeozoic Geosynclines

Although many types of geosyncline have now been defined, the term geosyncline refers generally to a region of the earth's crust which has been downwarped to permit the deposition of a vast thickness (usually some tens of thousands of feet) of chiefly marine deposits. Geosynclines are elongate regions and sedimentation was prolonged. By inference subsidence has been on a large scale. The thick geosynclinal deposits have invariably been folded by subsequent orogenesis and may contrast in structural pattern, as well as in thickness and facies, with flat-lying or relatively unfolded contemporaneous strata elsewhere.

The downwarping of a geosyncline cannot be the result of isostatic adjustment to the weight of sediment deposited—as Dana recognized almost a century ago—although the weight of sediment must be a contributory factor. That other processes must be operative can be shown by calculation and is confirmed by the fact that the initial sediments deposited are as a rule of shallow-water type, arenaceous and with a neritic fauna, followed by sediments deposited in deeper water.

The standard type of geosyncline, or Eugeosyncline, typically includes sediments which are chiefly dark shales and greywackes, the latter often deposited from turbidity currents, and these are interbedded with volcanic rocks. The volcanics are characteristically spilitic lavas extruded in submarine conditions but in many geosynclines, for example the Welsh Caledonian Geosyncline, many of the volcanic rocks are andesites found together with both basalts and more acid lavas. Tuffs are also common but it is thought that many of these rocks are ignimbrites. Geosynclines are separated by intervening upwarped areas known as geanticlines. These may be a source of supply of detrital material for the sediments of the sinking geosynclines, although minor geanticlinal features may separate geosynclinal troughs only temporarily, at times becoming emergent but at others submerged.

The Welsh Caledonian or Lower Palaeozoic Geosyncline, the concept first being applied to the Lower Palaeozoic rocks there by O. T. Jones in 1938, has something in common with the theoretical eugeosynclinal concept (although it has been said that it lacks the necessary symmetry). It was separated by a geanticline, which included Anglesey and extended south-westwards, where the southern part of the Irish Sea now exists, and north-eastwards towards Ingleton, from a similar geosyncline to the north-west. There is evidence of the latter from the Lower Palaeozoic rocks of south-east Ireland, the Isle of Man and the Lake District of northern England. The vulcanicity attains a maximum in the axial regions of geosynclines and broadly about the middle of their life or duration. One of the main objections to the Welsh Lower Palaeozoic being considered as deposited in a typical eugeosyncline is the occurrence of thick 'greywackes' of early Cambrian age, that is, deposited in the early part of the geosynclinal cycle. However, some of these Lower Cambrian rocks are typical arenites rather than greywackes and were probably deposited in shallow water, although this has been questioned; some show evidence of deposition from turbidity currents. There were considerable palaeogeographical changes in late Cambrian times; of this there is no doubt although the palaeogeography cannot be reconstructed with certainty, and it may be that the Cambrian, which lasted for approximately 80 m.y., belonged to a different, and earlier, geosynclinal cycle from the Ordovician and Silurian sedimentation.

The Lower Palaeozoic rocks of the south of Scotland and those of related areas in Ireland (the Longford-Down Massif and Murrisk) have long been considered to be also largely of geosynclinal origin. Recent work on the Ordovician and Silurian stratigraphy and the resulting reinterpretation of the structure of the Southern Uplands by Walton infers a very great accumulation of sediments, some tens of thousands of feet, and confirms this as an indisputable interpretation of their origin. The pre-Ordovician floor is unknown in the Southern Uplands and the former extent of Lower Palaeozoic rocks northwards is deducible chiefly from inliers along the south side of the Midland Valley of Scotland and from the Highland Border Series further north. There is no evidence on the north side of the Midland Valley of any sedimentary transition from Dalradian (Cambrian age) beds into the Highland Border Series (Ordovician, probably of Arenig age). In fact, only in Arran can the stratigraphical relationships be determined: there the Highland Border Series rests unconformably upon the Dalradian rocks. However, whether the northern boundary of the Southern Uplands (or Moffat) Geosyncline lay far to the north in early Ordovician times but in the region of the Highland Boundary Fault in Upper Ordovician times is still unconfirmed, although the geosynclinal axis lay within the Southern Uplands area and trended in an approximately north-east—south-west direction. The extent of sedimentation in Lower Ordovician times (Arenig) must certainly have included the north-west of Scotland where Durness Limestone was in part deposited at this time, whereas later in Ordovician and Silurian times there is evidence of sedimentation only as far north as the Highland Border Fault.

This is the basis of the palaeogeographical maps for the Arenig and Bala drawn by Wills (Plate IIB and C of Wills' *Palaeogeographical Atlas*), and work since 1952 has produced no evidence which modifies appreciably his interpretation. Williams has investigated the changes in Ordovician sediments from the Girvan shelf to the greywacke-shale environment of the geosyncline to the south of the Stinchar valley. It is possible that the Midland Valley may have been a geanticline which diminished in importance as the Lower Palaeozoic Era progressed so that in Silurian times it became the site of a basin of sedimentation. There is more positive evidence of a geanticlinal 'axial rise' in the Southern Uplands and work by several authors has elucidated the variations in the configuration of the Moffat Geosyncline through the Lower Palaeozoic.

Greywacke

Many of the rocks of Lower Palaeozoic age have been termed greywacke. Although the word was coined over a century ago, its definition is still subject to discussion and argument and there is a considerable literature on this topic. In addition to its petrological definition, authors have attempted to find an explanation for its prevalence. Its mode of transportation and the sedimentary environments in which it may be deposited have been investigated: this has resulted in additional characters being added, or at least inferred. The committee on sedimentation set up by the National Research Council in America defined greywacke on petrological and mineralogical bases as '. . . sandstone with 33% or more of minerals and rock fragments derived from easily weathered basic igneous rocks, slates and dark coloured rocks. It may or may not be intensely indurated or metamorphosed.' Despite the last sentence, metamorphosed greywackes are sometimes known as meta-greywackes. A greywacke is, therefore, essentially an impure sandstone (other than arkose) of variable grade of coarseness with grain size ranging from 2·0 mm downwards to perhaps 0·06 mm, and one of its more typical characters is that it is poorly sorted with grains occurring in a dark-coloured finer matrix. Further, graded bedding is not uncommon. The grains are usually sub-angular or angular but, since many greywackes are of marine origin and most are water transported, this does not preclude transportation of the material for some hundreds of miles. However, some of the grains may be 'first generation', i.e. derived not from older sediments but directly by the erosion of igneous or metamorphic rocks. The grains other than quartz may be felspar, pyroxene, amphibole, garnet, etc., or fragments of basic igneous origin not identifiable. The whole rock is probably of clastic origin, although Cummins questions the clastic origin of the matrix which he believes has been formed subsequent to deposition of the rock. Although greywacke may be regarded as a kind of impure sandstone or siltstone petrologically, it is unfortunate that some arenites which clearly are not greywackes have been so styled.

The depth of water in which deposition has taken place is perhaps the most difficult character of a sediment to infer. It can, for example, be stated with much less certainty than the salinity of the water. Yet depth has been regarded by some authors, incorrectly, as diagnostic in defining greywacke. Study of recent marine sediments, aided by laboratory experiments, has shown that coarse sands are not restricted to the shallow shelf areas of the sea; they may be carried down the continental slope—and beyond—by fast-flowing suspension currents of high density known as turbidity currents. Such currents impart characteristic sedimentary structures such as groove casts, flute casts and channelling of existing unconsolidated sediment. Highly contorted bedding may occur due to rolling and sliding of sediments laid down on the continental slope, which has an average gradient of about 1 in 10 compared with the low normal depositional dip of sediments deposited on the continental shelf (about 1 in 100) or on ocean floors. Naturally, sediments deposited on the continental slope or carried down it and deposited in deeper troughs would contain little neritic benthonic fauna. Many greywackes characteristically contain scarcely any fossils. Some Lower Palaeozoic greywackes appear to be unfossiliferous; some include, rarely, graptolites or they may be interbedded with graptolitic shales. The presence of only a pelagic fauna of this kind does suggest that these greywackes were deposited far from shore and, by inference, they may have been deposited in deep water. On the other hand, some greywackes are of shallow-water origin since they include a neritic shelly fauna and exhibit cross-bedding.

The former idea that grain size (grade) is a guide to the depth of a sedimentary environment or even to the distance from shore does not hold for arenaceous rocks, including greywackes, any more than it does for fine-grained clastic argillites. The nature of the material of the grains of a greywacke associated with a fine matrix (and the paucity of material of intermediate grade) are essentially diagnostic. Many Lower Palaeozoic greywackes may have been transported by turbidity currents, that is they are turbidites, but the terms greywacke and turbidite cannot be invariably linked for, as Basset has neatly summed up the position in *The British Caledonides*, with reference to sedimentation of greywacke facies '. . . all Welsh "greywackes" are not turbidites: and furthermore, all Welsh turbidites may not be greywackes; it has been suggested, for instance, that the calcareous siltstones of Ludlow age are the products of turbidity flow.'

Although greywackes are certainly not exclusively of Lower Palaeozoic age —they occur from Precambrian times up to at least the Miocene, they are prevalent in the Palaeozoic generally and in that group have been subject to intense study.

The Cambrian System

The Cambrian System, the earliest of the three which make up the Lower Palaeozoic, was first described by Sedgwick from north Wales, from which the name is derived. It succeeds the Precambrian rocks in many instances in Britain with a major unconformity. This means, in terms of geographical changes, that across the ancient continent of Precambrian rocks, by that time greatly eroded, the sea transgressed over a wide area from Scotland to south-eastern Europe and from Scandinavia to eastern North America. This was the beginning of a very long period of marine deposition over much of Britain. Here, the earliest deposits are coarsely arenaceous: although subsidence was not uniform, and hence they show considerable variation, the Lower Cambrian strata are everywhere chiefly of arenaceous facies.

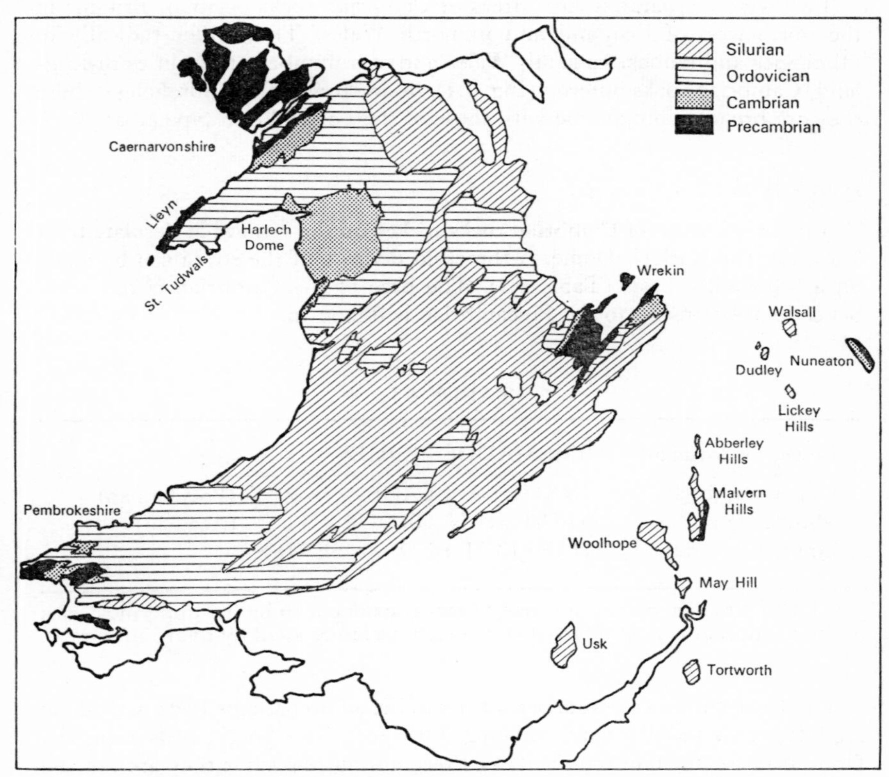

Fig. 4.1 Lower Palaeozoic outcrops of Wales and the Welsh Borders.

While the Precambrian–Cambrian boundary in some areas is conveniently situated at this major break in the geological record, this is not a world-wide picture. In many parts of the world Cambrian age sediments follow conformably on a great thickness of unmetamorphosed Precambrian strata known as the Eocambrian. The base of the Cambrian can only be defined by reference to the appearance of typical Lower Cambrian fossils (and in this it is, of course, analogous to the way in which the bases of later systems are defined); in practice the base is often placed at the unconformity below the beds containing Lower Cambrian fossils. Difficulties arise, primarily, in almost unfossiliferous successions like the Dalradian, but even in north Wales the Lower Cambrian is sufficiently lacking in fossils to make the position of its base uncertain. The currently held view is that locally there may be a transition from underlying Precambrian bedded volcanics and sediments while elsewhere there is unconformity and part of the Lower Cambrian may be absent.

Two widely separated large areas of Cambrian rocks occur in Britain: in the north-west of Scotland and in north Wales. They differ radically in lithologies and faunas. In south Wales and at a number of localities in England, Cambrian rocks outcrop (Fig. 4.1), or are known from boreholes, where they are broadly comparable with those of the north Welsh type area.

North Wales

A great thickness of Cambrian rocks makes up the somewhat desolate tract known as the Harlech Dome. A threefold division of the strata can be made on a lithological basis (Table 4.1). The base of the Cambrian is not seen, hence its relationship to the Precambrian is unknown.

TABLE 4.1

Lower Ordovician	TREMADOC SERIES[a]	
Upper Cambrian	LINGULA FLAGS	(Potsdamian)
Middle Cambrian	MENEVIAN GROUP	(Acadian)
Lower Cambrian }	HARLECH BEDS = SERIES	

[a] The Tremadoc Series, for many years considered to be the uppermost part of the Cambrian is now referred to the Ordovician System by many authors.

The Harlech Series comprises a thick series of arenaceous beds with some argillites, now usually slaty, totalling 7,000 feet. The lowest beds seen, the Dolwen Grits, include fragments of Precambrian rocks but they are not the basal member of the Cambrian, being correlated to the north of Snowdon in Caernarvonshire—where Cambrian rocks again outcrop—with the Glog Grit. Beneath the latter is a conglomerate which rests on the volcanic rocks of the Padarn Ridge. The Rhinog Grits, which form the high hills of the Harlech Dome, are up to 2,500 feet in thickness but they thin to the north-west where the succession is much slatier. Shales containing primary manganese follow the Rhinog Grit and are themselves succeeded by a further grit and shales. The presence of manganese ore suggests contemporaneous lateritic weathering of the area which was the source of supply of material. Eastwards the Harlech Beds become more coarsely arenaceous and this, together with the slatier facies in the Llanberis area to the north-west, has suggested that the material was largely transported from the east. This, however, conflicted with the evidence from the occurrence of Monian-type pebbles in some of the conglomeratic beds. Very recent work on sedimentary structures has helped to elucidate palaeocurrent directions. Since the Cambrian rocks represent a period of about 100 million years' duration and the arenaceous beds some appreciable part of this, no doubt the direction of transport of material varied considerably from time to time: the possibility of different sources of material for different members of the succession has long been realized. There is, as yet, not complete agreement on palaeocurrent directions and different criteria, cross-bedding, groove casts, flute casts, pebble orientation, etc., have inferred different directions. Crimes and other workers have recently collated various criteria and, while different horizons show different palaeocurrent directions (and here problems of correlation intrude), several grit horizons show that the palaeocurrents came predominantly from the south or south-west. Isopachytes for the Cambrian as a whole (Fig. 4.2b) must be interpreted with caution, since there has been post-Cambrian deformation and erosion,

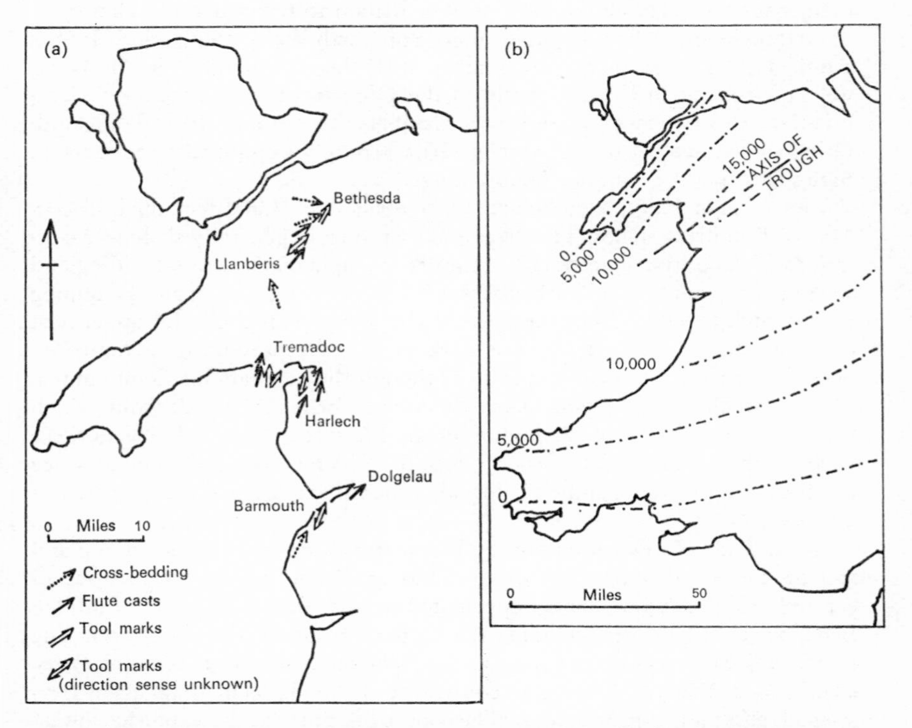

Fig. 4.2 (*a*) Evidence of palaeocurrent directions in the Maentwrog and Ffestiniog Beds (*Lingula* Flags) of north Wales. (*b*) Isopachytes for the Cambrian rocks of Wales. (*a* from A.M. EVANS, P. GARRETT and J.H.MCD. WHITTAKER, 1966, *Nature, Lond.*, **209**, 1230.)

but they provide some indication of the form of the Cambrian geosyncline in Wales. Although it is clearly extended far to the east, its axial region was north-east—south-west and transport directions may represent axial flow. The Harlech Beds show a rhythmic pattern of sedimentation of alternations of coarse and fine material which might reflect depth of water at the time of deposition. It is no longer agreed that the Harlech Beds were entirely deposited in a shallow-water environment and some of the uniformly bedded greywackes which exhibit sole-marks are thought to have been transported by turbidity currents.

The uppermost beds of the Harlech Series are finer grained and pass upwards into the Gamlan Shales, shales with ribs of flagstones and grit. Possibly in part of Middle Cambrian age they pass up into the Menevian Group = the Clogau Shales, pyritous mudstones with a fauna of inarticulate brachiopods and trilobites including *Paradoxides.* The latter confirms the age as Middle Cambrian. The high percentage of manganese at some horizons in the Cambrian shales infers derivation by lateritic weathering from acid gneiss. A hot climate during the Lower Palaeozoic is not improbable since palaeomagnetic evidence suggests the proximity of Britain to the equator at that time. The strata below this in the main outcrop of North Wales, the Harlech Dome, are devoid of trilobites but by analogy with the succession in south Wales were considered to belong chiefly to the *Olenellus* division = Lower Cambrian. The discovery of a late Lower Cambrian trilobite in the Hell's Mouth Grits, the equivalent of the Rhinog Grits, strongly implies that the Harlech Series is in part of Middle Cambrian age.

The *Lingula* Flags, arenaceous flaggy rocks over 4,000 feet thick, follow Middle Cambrian shales. They represent a return to generally shallower conditions of deposition since many examples of ripple marks, cross-bedding and worm casts occur in the Penrhos Slates (= Maentwrog Slates) and Ffestiniog Beds which together form the bulk of the series. However, the uppermost division, the Dolgelly Beds, are dark argillites representing a return to deeper-water conditions of deposition, though they contain a trilobite fauna. The inarticulate brachiopod *Lingulella davisii* (formerly *Lingula*) from which the group takes its name is common in the Ffestiniog Group. Volcanic rocks make an appearance in the *Lingula* Flags, the first renewal of vulcanicity since the beginning of the Cambrian Period, which is characterized by its lack of igneous activity.

In St. Tudwal's Peninsula (of the Lleyn area) Cambrian rocks outcrop and can be correlated with those of the Harlech Dome with a high degree of certainty. The *Lingula* Flags, represented by the Maentwrog and Ffestiniog Beds, occur, the former including the Upper Cambrian trilobite *Olenus.* The underlying Menevian, here known as the Nant-pig Mudstones, contains *Paradoxides* and other trilobites. The equivalent of the Harlech Series again consists of alternating grits and argillaceous beds and the base of the lowest member, the Hell's Mouth Grits, is not seen so that the analogy with the Harlech Dome is very close.

North of the Ordovician rocks of Snowdonia, where Cambrian rocks out-

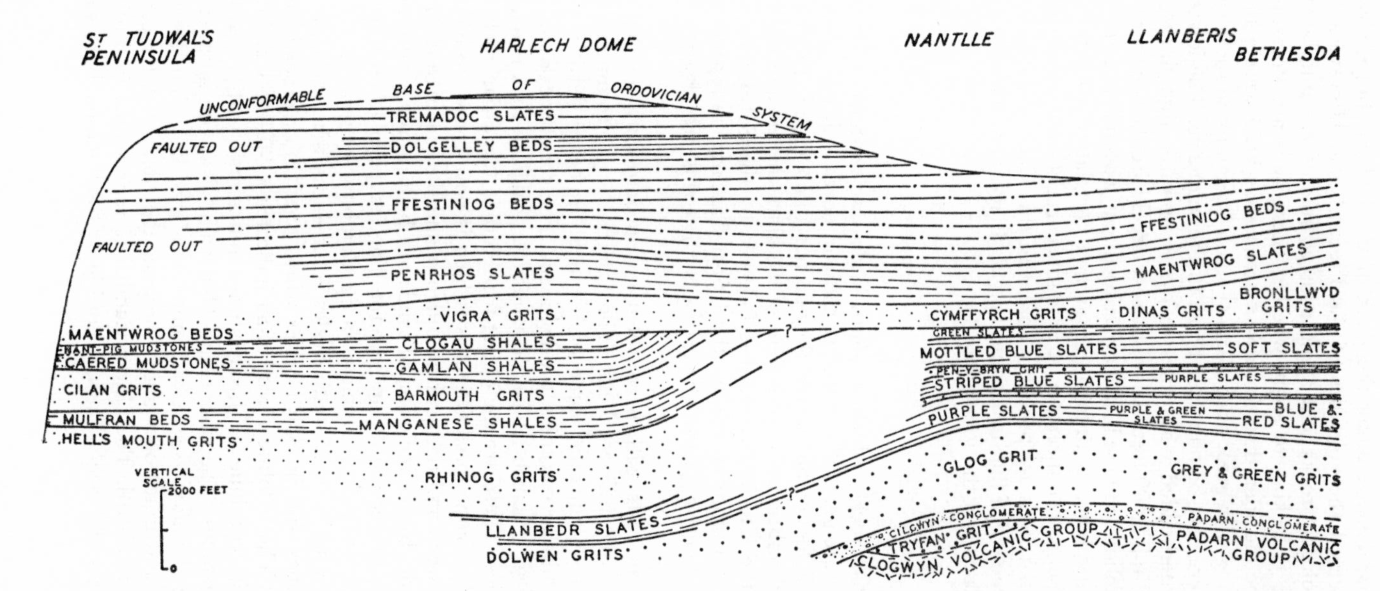

Fig. 4.3 Diagram illustrating an interpretation of lateral variations in the Cambrian rocks of north Wales. (From B. SMITH and T. N. GEORGE, 1961, *British Regional Geology, North Wales*, 3rd edn. by T. N. GEORGE. Crown Copyright Geological Survey Diagram. Reproduced by permission of the Controller of H.M. Stationery Office.)

crop in Caernarvonshire along the flank of the Padarn Ridge, coarse conglomerates succeed the volcanics of the Precambrian. However, it is now known that volcanic rocks and arenaceous sediments are interbedded, but fossil evidence is lacking to show whether these beds are Precambrian or Cambrian. The old idea that the lowest arenaceous beds were a basal conglomerate resting with great unconformity upon Precambrian rocks cannot be upheld and there may be no major break in the sequence. The lowest arenaceous beds (interbedded with volcanics) pass upwards into finer grits and quartzites totalling 2,000 feet. Above these a largely argillaceous series, somewhat thicker, follows. These beds have subsequently been converted to slates. Much of the sequence is unfossiliferous and correlation with the Harlech Dome succession is still uncertain. A probable correlation is given in Fig. 4.3.

The Menevian Beds are probably absent in this Caernarvonshire outcrop through the overstep of the succeeding *Lingula* Flags, here represented by the Maentwrog Slates and the Ffestiniog Beds.

South Wales

The Cambrian rocks, which in south Wales are confined to outcrops in Pembrokeshire, are generally similar to those of north Wales in character but the succession is less complete and they do not attain such a great thickness.

The Lower Cambrian, the equivalent of the Harlech Series of north Wales, is here called the Caerfai Series. It is followed by the Solva and Menevian Series of Middle Cambrian age but the Upper Cambrian is represented only by the Lower *Lingula* Flags. Higher beds than this are absent, as is the Tremadoc Series.

The Cambrian rests with a conspicuous unconformity on Precambrian rocks, overstepping different beds. The Caerfai Series commences with a reddish basal conglomerate, some of the pebbles being derived from the adjacent Precambrian volcanics, the Pebidian. The conglomerate is followed by unfossiliferous greenish felspathic sandstone which in turn is overlain by bright red shales. These are the oldest fossiliferous Cambrian rocks and include the trilobite zone fossil *Olenellus* (which is now known to be not confined to the Lower Cambrian), worm burrows and the brachiopod *Lingulella*. The shales grade, by the inclusion of sandstone beds, into the overlying grey sandstone. The Lower Cambrian here totalling less than 1,000 feet is much thinner than in the Harlech Dome as is the Cambrian generally.

The Middle Cambrian of south Wales commences with the basal beds—pebbly sandstones—of the Solva Series which rests on an eroded surface of Caerfai Series. The trilobite *Paradoxides* indicates a Middle Cambrian age. The Solva Series continues with sandstones, mudstones with a rich trilobite fauna, and flagstones. The succeeding Menevian, also with abundant trilobites including *Paradoxides* species, commences with flags, but the presence of dark shales and mudstones above suggests that deeper water conditions soon became established. The Upper Cambrian, represented in south Wales by only 2,000 feet of *Lingula* Flags, is not highly fossiliferous, but its equivalence to the Maentwrog Beds of north Wales has been established and it represents a return to shallower water conditions of deposition before the widespread contraction of the area of deposition, which accounts for the absence of higher beds here and elsewhere.

Welsh Borders and English Midlands

As in both north and south Wales, the Lower Cambrian rocks are arenaceous and everywhere the Cambrian commences with a shallow-water basal deposit. Called the Wrekin Quartzite in Shropshire, the Malvern Quartzite and Lickey Quartzite at those localities, and the Hartshill Quartzite at Nuneaton it represents, despite its many local names, a widespread marine incursion. In Shropshire it rests on either Uriconian Volcanic rocks or Rushton Schists and at Nuneaton on the Caldecote Volcanic rocks. The latter contributed boulders to the Lower Cambrian inferring the proximity of at least a temporary shoreline.

In Shropshire the Lower Comley Beds, sandstones with thin fossiliferous limestones, follow the basal quartzite and in the Malverns a similar sequence occurs: the Hollybush Sandstone with calcareous bands follows the basal quartzite. In the Nuneaton outcrop, the Hartshill Quartzite has been divided into four and occupies much of the Lower Cambrian. Near the top of the quartzite succession the *Hyolithus* Limestone occurs, and shaly conditions commenced before the close of the Lower Cambrian, since the overlying Purley Shales (the lowest beds of the Stockingford Shales) are partly referable to that division.

LOWER CAMBRIAN	Lower Purley Shales	Lr. Stockingford Shale
	Camp Hill Grit	Hartshill Quartzite
	Tuttle Hill Quartzite	
	Park Hill Quartzite	
	Basal conglomerate	

While in north Wales Harlech Grit sedimentation probably continued into Middle Cambrian times, in Shropshire local uplift took place so that the Upper Comley Series rests unconformably on eroded Lower Comley beds and includes pebbles derived from them. This break in sedimentation was followed by the deposition of marine glauconitic sands and shales. Although the break may be approximately contemporaneous with that between the Solva and Caerfai Beds in south Wales there is no evidence of interrruption in sedimentation at Malvern: the Hollybush Sandstone is partly of Middle Cambrian age.

At Nuneaton, the furthest east outcrop of the Cambrian (though it is known from boreholes further east), muddy sedimentation which deposited the Stockingford Shales commenced in late Lower Cambrian times and continued through Middle and Upper Cambrian times. The presence of the Middle Cambrian zones, recognized in Sweden, has been confirmed. During the Upper Cambrian the deposits were flaggy and generally resemble the

Lingula Flags of equivalent age. The sequence here differs considerably from that of north Wales, not least in thickness for here it is only 4,000 feet, perhaps only one-third the maximum attained in the type area. Phosphates and glauconite both suggest that sedimentation may have been slow and the sea bed current swept.

In Upper Cambrian times there is evidence of contraction of the area of deposition: in south Wales only the lower part of the *Lingula* Flags is present, and no beds are definitely referable to this age in Shropshire, where the Middle Cambrian Upper Comley Series is followed by the Shineton Shales which are believed to be entirely of Tremadoc age. However, in the Malverns the White Leaved Oak Shales (black in colour) are of Upper Cambrian age succeeding the Hollybush Sandstone.

The approximate equivalence of the main stratigraphical divisions is given in Table 4.2.

TABLE 4.2

	NORTH WALES	SOUTH WALES	SHROPSHIRE	MALVERNS	NUNEATON
TREMADOC	Tremadoc Beds		Shineton Shale	Bronsil Shale	Stockingford Shales: Merevale Shales
UPPER CAMBRIAN	*Lingula Flags*: Dolgelley, Ffestiniog, Maentwrog	*Lingula* Flags		Whiteleaved Oak Shales	Stockingford Shales: Oldbury Shales
MIDDLE CAMBRIAN	Menevian; Harlech Series (Middle and Lower Cambrian)	Menevian Solva	Upper Comley	Hollybush Sandstone	Stockingford Shales: Purley Shales
LOWER CAMBRIAN	Harlech Series	Caerfai	Lower Comley; Wrekin Quartzite	Malvern Quartzite	Hartshill Quartzite

The Cambro-Ordovician rocks of Scotland

Although Lower Palaeozoic rocks occur in the English Lake District, Southern Uplands and Highland Border Series, the presence of Cambrian rocks has not been established. However, in north-west Scotland there is an extensive outcrop of rocks of Cambrian and Early Ordovician age.

To the west of the Moine Thrust in the Caledonian foreland and in the sub-Moine nappes the Lower Palaeozoic rocks are found resting unconformably upon Precambrian rocks, of both Lewisian and Torridonian age, which had been planed by marine erosion. The outcrop is long and narrow, extending from Durness on the north Scottish coast to the Sleat area of Skye.

Apart from a lower series of arenaceous beds, quartzites, there is little resemblance to strata of equivalent age in England and Wales in either lithology or fauna. About 3,500 feet, the greater part of the succession, consists of dolomitic limestones while elsewhere in Britain limestones are rare in the Cambrian and Ordovician sequences. The succession can be divided into three:

1. ARENACEOUS SERIES, 500–600 FEET

The Lower Quartzites consist of a basal conglomerate only a few feet thick which passes upwards into 300 feet of cross-bedded quartzites and grits which are generally unfossiliferous. The Upper Quartzites or 'Pipe Rock' of about the same thickness follows. The latter again consists of quartzites but has been called Pipe Rock because of the profusion of vertical worm burrows chiefly of the genus *Skolithus*. The former recognition of five 'zones' dependent upon the type, size and profusion of worm burrows can no longer be sustained. They are not traceable laterally nor can the same sequence everywhere be found. It has recently been suggested that the two main kinds of worm burrows may reflect differences in the rate of sedimentation.

2. PASSAGE BEDS, 50 FEET

The Planolites or 'Fucoid' Beds are dolomitic shales containing flattened worm casts, once erroneously thought to be fucoids, and in thin beds trilobites of the genus *Olenellus* as well as *Olenelloides* and chitinous brachiopods. The *Salterella* Grit (formerly known as the Serpulite Grit) follows and, although thin, is a persistent horizon. This bed has also yielded *Olenellus*, *Salterella*, *Skolithus* and brachiopods.

3. THE DURNESS LIMESTONE SERIES

This is essentially a great thickness of dolomitic limestone, but variations in lithology occur and some horizons are cherty. Divided on a lithological basis in the northern part of the outcrop at Durness into seven divisions (Fig. 4.4), the lowest four divisions totalling 850 feet have been recognized in the Isle of Skye, but local names have been given.

The lowest part of the Durness Limestone, the Ghrudaidh division, contains *Salterella* and is, with the arenaceous rocks beneath, referable to the Lower Cambrian. The second division, the Eilean Dubh division, has yielded only algae and worms, the Sailmhor division a sparse molluscan fauna, chiefly gastropods, while above this the Sangamore division is unfossiliferous. Thus there are several hundred feet of strata with no fauna which give precise

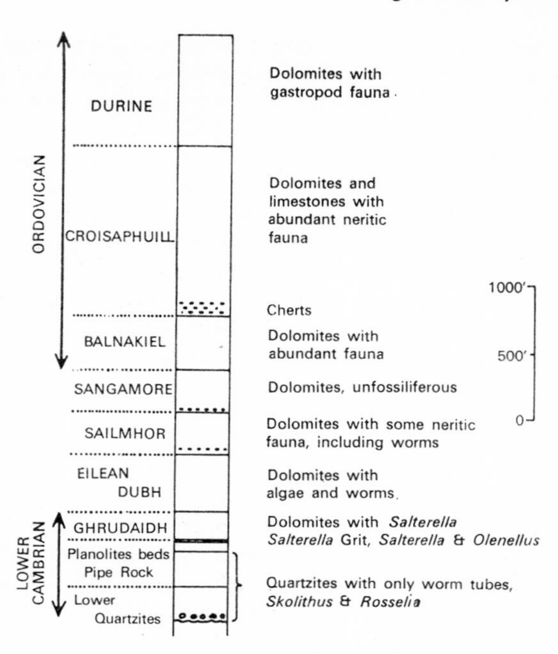

Fig. 4.4 The Cambro-Ordovician succession in the Scottish North-west Highlands. (Simplified from E. K. WALTON, 1965, in *The Geology of Scotland*, ed. G. Y. CRAIG. Oliver & Boyd, Edinburgh and London.)

information as to their age. However, the highest three divisions—the Balnakiel, the Croisaphuill and Durine—comprising some 2,500 feet, contain a richer, more varied, fauna including gastropods, trilobites, brachiopods and sponges which, by comparison with North American and East Greenland faunas, establish the age as Lower Ordovician. The greater part of the Durness succession is, therefore, of Ordovician age (though the lower part is Lower Cambrian) so that the Upper and Middle Cambrian cannot be represented by a great thickness of strata although there do not appear to be any major breaks in sedimentation.

The north-west of Scotland, in Lower Palaeozoic times, appears to have been part of a wide shelf sea which extended westwards into North America (which may have been very much closer to Europe at that time). Initially, arenaceous beds were laid down, on a marine eroded surface, in shallow water. Ripple marks and worm burrows both infer this. The supply of terrigenous material waned and a very lengthy period of limestone deposition took place on a very slowly sinking sea floor. Varying lithologies and faunas reflect differing depths of water so that the possibility of emergence causing non-deposition in Middle and Upper Cambrian times cannot be ruled out despite the lack of confirmatory evidence.

One of the most puzzling problems of Palaeozoic geology is the relationship between the Scottish Durness sequence and contemporaneous deposits in Wales and their very different fossils, although they are now separated by a distance of no more than 300 miles. The evidence of Cambrian rocks in the intervening regions is sparse, yet complex. As has been suggested in Chapter 3, the Dalradian rocks must, in part, be of Cambrian age though they contrast strongly with the Cambrian of the Durness area. After deposition including limestones, a succession of greywackes and generally arenaceous rocks, though with some argillites, were laid down probably in a subsiding trough.

The Highland Border Series (discussed on p. 125) which occurs in wedges immediately to the south of the Highland Border Fault appears to be entirely Ordovician in age. Possibly of Arenig age, its black shales and cherts with spilitic lavas are followed unconformably by more arenaceous beds, and the type of sedimentation is very different from the approximately contemporaneous upper part of the Durness Limestone. Further, it bears little relationship to the Upper Dalradian, though the metamorphism of the latter makes comparison difficult. Possibly comparison is irrelevant since Dalradian sedimentation may have been terminated before the deposition of the Highland Border Series. How far south Cambrian age sedimentation spread is conjectural since no Cambrian rocks are known from the south of Scotland. In fact Lower Ordovician rocks, chiefly lavas and cherts, are found in only two areas there and in neither is the base of the Ordovician seen. It is possible that in Cambrian times a geanticlinal landmass may have lain between the north Welsh area of deposition and the Dalradian geosyncline. However, by Ordovician times sedimentation was more widespread in Britain, and the Durness Limestone probably represents a shelf sea facies which bordered on a complex area of geosynclines and geanticlines.

Ireland and the Isle of Man

The Manx Slates, a great thickness—perhaps 25,000 feet—of greywackes, siltstones and pelitic schists, are almost devoid of fossils except for worm burrows and trails, but the discovery of *Dictyonema* half a century ago in beds recently confirmed as high in the succession proves that the greater part must be of Cambrian age. Lithologies and sedimentary structures suggest a generally deep-water geosynclinal environment. The Manx Slates must be rather older than much of the Skiddaw Slates of the northern Lake District with which they were once thought to be correlatable. On the other hand, they may be of approximately the same age as the Clara Group of south-eastern Ireland. The Manx Slates have been affected by polyphase folding and low-grade metamorphism of Caledonian age.

Typical Cambrian rocks are absent from north-western Ireland where the Lower Ordovician rests on deformed and metamorphosed Dalradian rocks, the Connemara Schists. In south-eastern Ireland, in East Leinster (the counties of Wicklow and Wexford), pre-Ordovician strata several thousands of feet thick, called the Bray Series, outcrop. Consisting largely of greywackes, commonly with sole markings, they are believed to be of Precambrian age, although there is no conclusive palaeontological evidence. The overlying Clara Group, a great thickness of mudstones, is also pre-Ordovician in age but is unconformable upon the Bray Series and so is assumed to belong to the Cambrian Period.

The absence of the Cambrian System from Anglesey (Fig. 4.1) and its apparent absence from the Pennines—where inliers of Precambrian and Ordovician and Silurian rocks occur (but no Cambrian rocks), while further south in the Pennines a deep borehole found a basement of Uriconian type volcanic rocks—both point to the possible presence of an Irish Sea Landmass as early as Cambrian times. This must have lain between the Welsh outcrops and those of south-eastern Ireland and the Isle of Man. However, any attempt

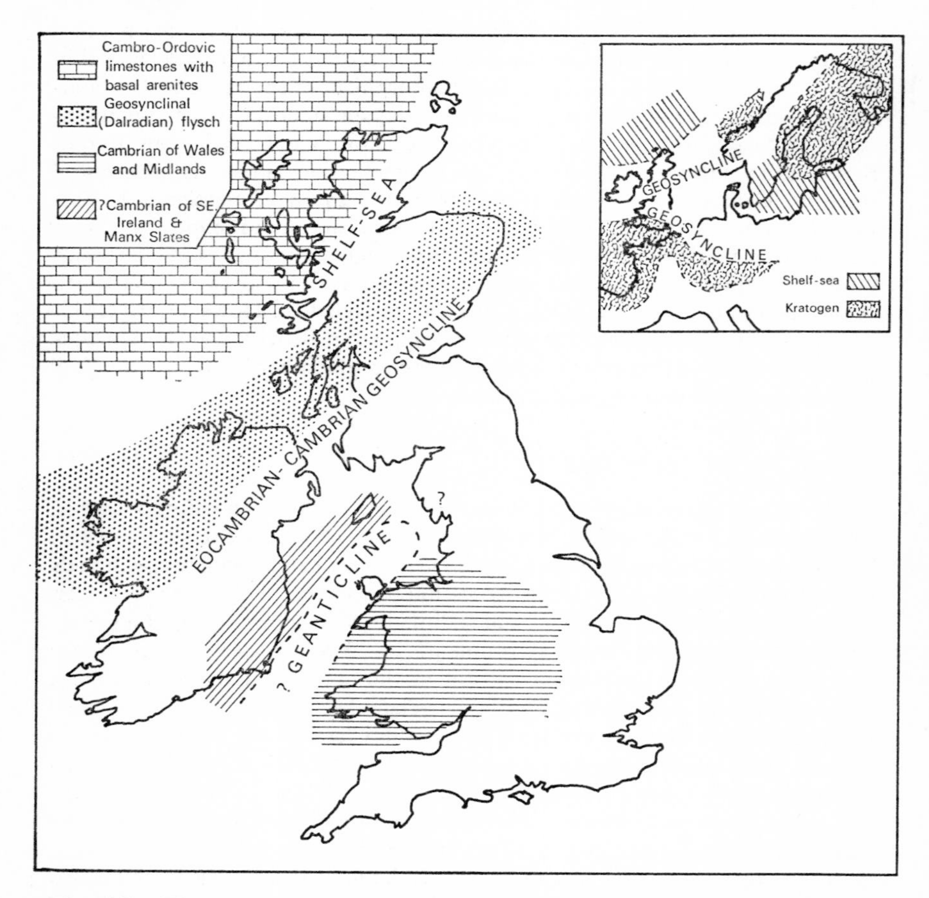

Fig. 4.5 The palaeogeography of Britain in Cambrian times. *Inset:* the chief geographical features of Europe in Cambrian times.

to reconstruct the palaeogeography of the British area in Cambrian times (Fig. 4.5) must necessarily be tentative.

Brief reference to selected outcrops of Cambrian rocks in northern Europe helps to elucidate the more general palaeogeography. In the Oslo district Lower Cambrian arenaceous beds are succeeded by a dominantly argillaceous succession ranging from the Middle Cambrian, with *Paradoxides*, up into the Tremadocian. The Cambrian succession is, therefore, broadly comparable with that of the English Midlands, and the wide extent and generally south-west—north-east elongation of the geosyncline is confirmed (Fig. 4.5 inset). In Brittany and in inliers in the Ardennes Cambrian rocks are known, but paucity of the fossil record combines with complex Variscan tectonics to make detailed stratigraphy difficult. Comprising schists, phyllites and slates, sedimentation may have continued from Precambrian times into the Cambrian (as in the Dalradian area) and no closely analogous succession is known from southern Britain.

Cambrian fossils

Despite the paucity of the Precambrian fossil record, the succeeding Cambrian faunas are abundant and varied. Even in earliest Cambrian strata there is a considerable variety of fossil remains, although in Britain much of the Lower Cambrian is arenaceous and fossils have not been preserved in some areas. All Cambrian sedimentary rocks appear to be of marine origin, certainly all Cambrian fossils are marine, chiefly shallow-water (neritic) invertebrates and some marine plants of low grade.

As formerly defined, to include the Tremadoc Series, most invertebrate animal phyla were present in the Cambrian. However, several made their appearance in the Tremadocian. Trilobites predominate in Cambrian faunas; 60% of all Cambrian fossils are trilobites, and all the zone fossils of the Cambrian (Welsh Type) are trilobites, with the exception of two brachiopods and one annelid worm. The latter achieved their acme in late Cambrian–early Ordovician times. The fauna of the Lower Cambrian chiefly comprises Olenellid trilobites, but all orders of trilobites except the Proparia occur from early Cambrian times onwards, together with chitinous brachiopods and Hyolithids. The Middle Cambrian is typified by *Paradoxides*, *Microdiscus* and some dozen species of *Agnostus*. The Olenid trilobites of the Upper Cambrian (Olenidan = *Lingula* Flags) have no direct affinity with the *Paradoxides* fauna below but they occur with new species of *Agnostus*.

Brachiopods occur from early Cambrian times; for example, *Paterina* and *Obolella* are found in the Lower Cambrian. The majority of Cambrian forms are inarticulate but primitive articulata (the Palaeotremata) and Orthida occur. While coelenterates make their appearance, represented by Hydrozoans and stromatoporoids, corals are unknown from the Cambrian. Annelids are common, sometimes profuse. Molluscs are chiefly of the gastropod class and the little-known Monoplacophora, but Nautiloidea make their appearance before the close of the period. If the Tremadoc Series is excluded from the Cambrian, graptolites are not represented.

In the Durness sequence, in the calcareous facies, trilobites are not common and the commonest fossils are gastropods such as *Maclurea* (a Euomphalid), *Hormotoma* (a Murchisonid), *Euconia* and *Raphistoma* (Pleurotomarids) together with orthocone cephalopods, lamellibranchs, echinoderms and sponges. The affinity of this fauna with North American faunas was recognized over a century ago and a number of species occur both in the Durness Group and the Beekmantown Group of the Appalachians.

5 The Ordovician System

The Tremadoc Series

The Ordovician System, proposed by Charles Lapworth in 1879 to include strata previously included both in the Cambrian by Sedgwick and in the Silurian by Murchison, takes its name from an ancient British tribe, the Ordovices. Commencing with the Tremadoc Beds, it almost everywhere rests unconformably upon older strata which had been eroded and in some cases folded.

The Tremadoc Beds are a thick series of cleaved mudstones which occur on the periphery of the Harlech Dome in north Wales. In the area around Tremadoc they may be transitional between the Upper Cambrian and the Arenig: sedimentation may have continued without interruption. Elsewhere the Arenig grits (Lower Ordovician) rest with marked unconformity on Tremadoc or older rocks, overstepping on to successively older strata towards the north-west. As a result, the Tremadoc Series is absent in St. Tudwal's Peninsula, north-west of Snowdon in Caernarvonshire and in Anglesey.

Although the Tremadoc Series is missing too in south Wales, it is represented in the Welsh Borders at Malvern by the Bronsil Shales and in Shropshire by the Shineton Shales some 3,000 feet thick. The latter have been divided into zones:

Zone of *Shumardia pusilla*
(Brachiopod beds)
Zone of *Clonograptus tenellus*
Zone of *Dictyonema flabelliforme*

The Tremadoc is represented further to the east in the Nuneaton area by the Merevale Shales, the upper part of the Stockingford Shales. The Tremadoc distribution is broadly the same as that of the Cambrian (despite recession of the seas in Upper Cambrian times) and not that of the overlying Arenig which apparently did not extend eastwards beyond west Shropshire.

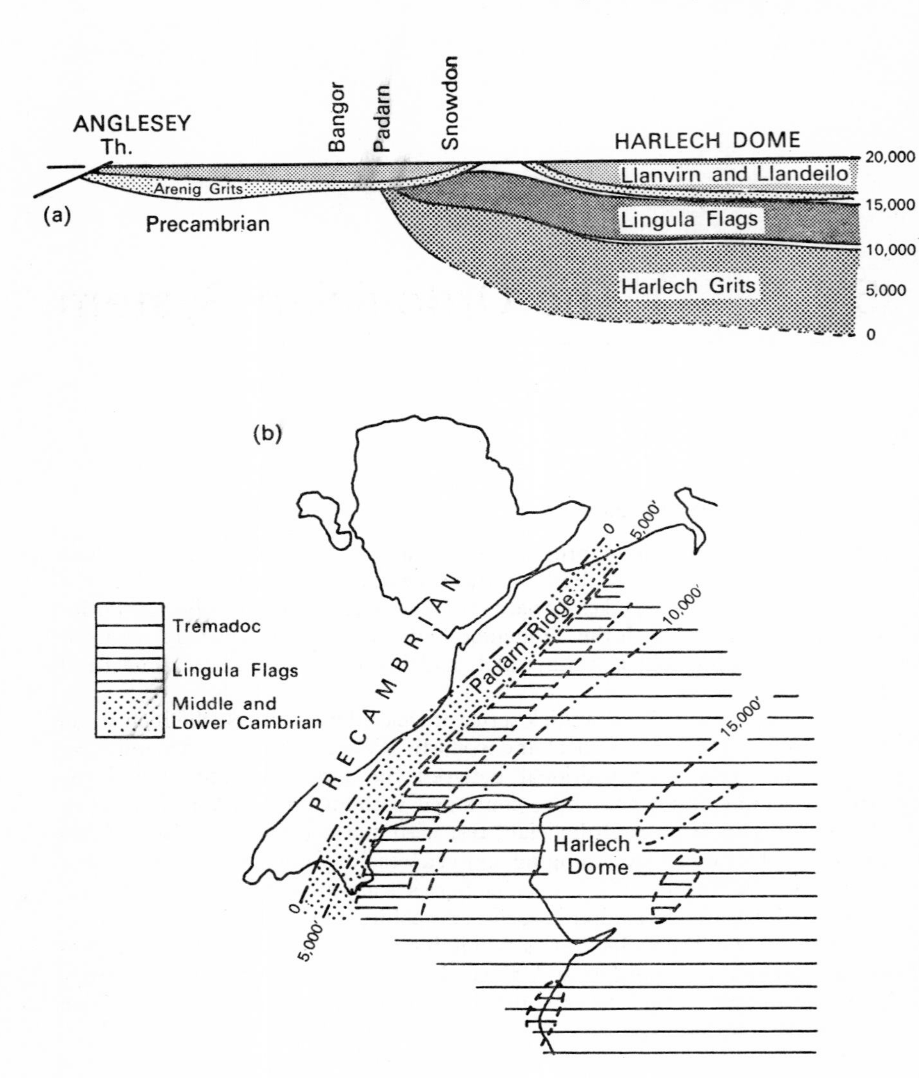

Fig. 5.1 The Arenig overstep in north Wales. (*a*) Reconstructed section to illustrate the later Ordovician geology of north Wales. Recurrent overstep (at the base of the Cambrian, of the Upper Cambrian, of the Arenig and the Bala) is cumulative northwards. (*b*) Outline map illustrating the Ordovician overstep in north Wales. (*a.* from T. N. GEORGE, 1961, in *The British Caledonides*, eds. M. R. W. JOHNSON and F. R. STEWART. Oliver & Boyd, Edinburgh and London. *b.* from B. SMITH and T. N. GEORGE, 1961, *British Regional Geology, North Wales*, 3rd edn. by T. N. GEORGE. Crown Copyright Geological Survey Diagram. Reproduced by permission of the Controller of H.M. Stationery Office.)

The Tremadoc beds include a considerable trilobite fauna with Olenids such as *Angelina* and forerunners of typical Ordovician families, such as *Asaphellus*. Perhaps most significant, graptolites make their appearance, *Dictyonema* in the lowest beds but, also, many branched, anisograptids.

The late Tremadoc *Shumardia pusilla* zone is marked by the rapid spread of shallow-water sedimentation depositing arenaceous beds, the Garth Grit and its equivalents, over a wide area. They follow the pelagic muddy sediments of the Tremadoc Shales at Tremadoc, and elsewhere overstep on to older beds. This is particularly well demonstrated in north Wales (Fig. 5.1) and the overstep is especially apparent in St. Tudwal's Peninsula where, in a distance of only a mile, the Ordovician oversteps from the *Lingula* Flags to the Hell's Mouth Grits, while in Lleyn and Anglesey the Arenig is faulted against and may rest on the Precambrian rocks. Southwards the unconformity is also conspicuous in contrast to the Tremadoc area.

Ordovician sedimentation

The general geography can be summarized as a broad geosynclinal trough with a north-east—south-west trend, perhaps bounded by fault structures so that downwarping occurred in a graben. Certainly, the Midlands of England formed a kratogen which persisted throughout the Ordovician and Silurian, although at times it became submerged beneath a shallow epeiric sea. The Precambrian and Cambrian of Shropshire formed, in post-Tremadoc times, a landmass bounded by the Church Stretton-Pontesford fault system and it is only to the west of this that a full Ordovician sequence can be found. The region immediately to the east of this was not inundated again by the sea until Upper Ordovician (Caradocian) times so that in south Shropshire, east of the Longmynd, a much attenuated Ordovician succession occurs. Still further east, in the Midlands, the Ordovician is absent in outcrops, probably chiefly through non-deposition although Llanvirn age beds have been located in a borehole.

After early widespread arenaceous sediments were deposited there was general deepening of the Ordovician geosyncline which resulted in two contrasting facies of deposition. In deeper water areas shales with graptolites and greywackes with few graptolite fossils (and little else) were deposited, while in shallow marginal areas more arenaceous sediments, often with rich faunas of benthonic animals, were laid down. The shallow-water sediments attain a thickness of thousands of feet, demonstrating a gradual subsidence of the area of deposition. Naturally, crustal warping of this kind is not at a uniform rate in any one locality and may be differential from one area to another. As a result in a particular locality the facies may at one time be graptolitic (perhaps bathyl, a point discussed on p. 97) and at another time neritic. In other words, the succession may be of mixed facies. The distribution of graptolitic and neritic facies changed considerably during the 60 or so million years' duration of the Ordovician Period.

In areas of deeper water sedimentation is more continuous, since minor uplifts, producing changes in the position of shorelines, do not cause such areas to be elevated to a position where already formed sediments become subjected to subaerial erosion or even to the effects of wave action. From such a sequence of sediments should come the most reliably complete sequence of faunas. It is from rocks of the mid-geosyncline that graptolites are obtained and they were, during the Ordovician (and Silurian), a rapidly evolving group. They therefore show great changes in form with the progression of time and make excellent zone fossils. The graptolite zones are combined to form larger divisions, series or stages, which take the names of earlier lithological groups (Table 5.1). All but the Ashgill stage are named from locations in Wales where strata of that age typically occur.

TABLE 5.1

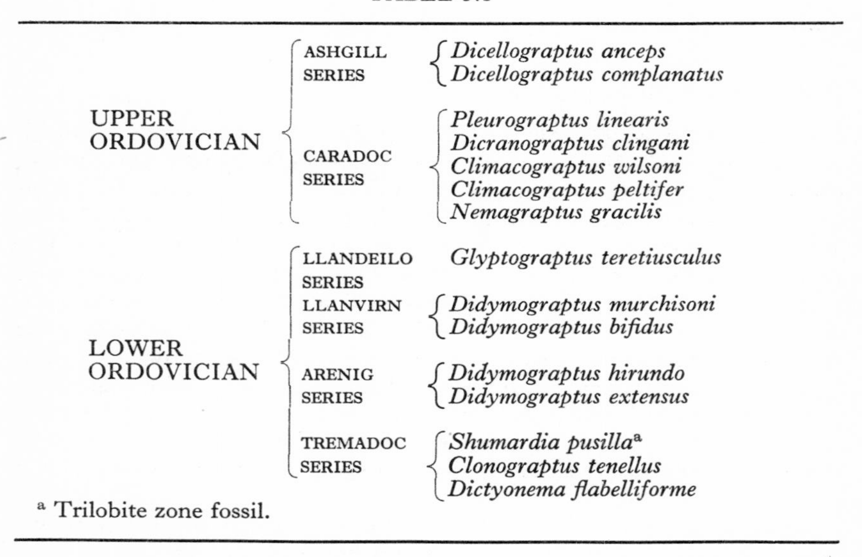

UPPER ORDOVICIAN	ASHGILL SERIES	*Dicellograptus anceps*
		Dicellograptus complanatus
	CARADOC SERIES	*Pleurograptus linearis*
		Dicranograptus clingani
		Climacograptus wilsoni
		Climacograptus peltifer
		Nemagraptus gracilis
LOWER ORDOVICIAN	LLANDEILO SERIES	*Glyptograptus teretiusculus*
	LLANVIRN SERIES	*Didymograptus murchisoni*
		Didymograptus bifidus
	ARENIG SERIES	*Didymograptus hirundo*
		Didymograptus extensus
	TREMADOC SERIES	*Shumardia pusilla*[a]
		Clonograptus tenellus
		Dictyonema flabelliforme

[a] Trilobite zone fossil.

Shallowing of the geosyncline caused contraction of the area of deep-water sedimentation and a spread of neritic facies. The converse is naturally true and the occasional extension of deep-water conditions gave rise to occasional beds of graptolitic shales in dominantly neritic sequences, thus facilitating correlation between the contrasting facies.

The Ordovician volcanic rocks

One of the chief contrasts between the Ordovician System and the preceding Cambrian is the widespread vulcanicity. On the south side of the Harlech Dome, in the Dolgelley district—and in South Wales—volcanic activity began in pre-Arenig times, and the earliest volcanic rocks rest upon the *Lingula* Flags and are themselves succeeded by the basal Arenig Garth Grit. Subse-

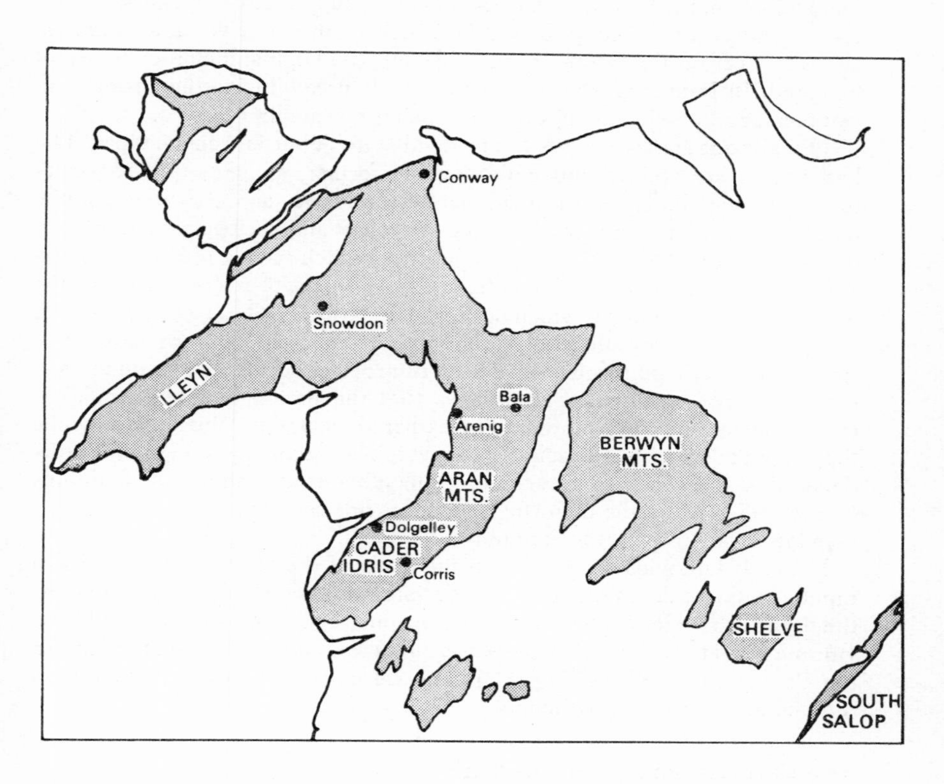

	S. WALES		CORRIS		DOLGELLEY		BALA		SNOWDON		CONWAY		BERWYNS
ASHGILL													
CARADOC									VVV VVV VVV VVV VVV	Snowdon Volcanic Series	VVV VVV VVV VVV	Conway Volcanic Series	VVV VVV
LLANDEILO			VVV VVV	PERIOD	VVV VVV	OF		RELATIVE		QUIESENCE			
LLANVIRN	VVV		VVV VVV		VVV VVV								
ARENIG	VVV VVV VVV		VVV VVV		VVV VVV VVV		VVV VVV						

Fig. 5.2 Centres of Ordovician vulcanicity in north Wales. (Outcrops of Ordovician rocks, based on Crown Copyright Geological Survey Map. By permission of the Director, Institute of Geological Sciences.)

quently vulcanicity became widespread and many centres of activity are known from north Wales. In the Lake District an enormous thickness of volcanic rocks accumulated in Lower Ordovician times and volcanic rocks are also known from the Southern Uplands and the Highland Border Series of Scotland. In Ireland igneous activity was widespread during the Ordovician, commencing in early Arenig times and being repeatedly renewed.

The volcanic rocks of Wales can be divided into basic and acidic types. The basic rocks are spilites and andesites, the former, showing pillow structures, were clearly deposited in submarine conditions, as were the palagonite lavas and tuffs. The acidic rocks include rhyolites and dacites and ignimbrites together with bedded tuffs. The ignimbrites, which have a readily recognizable texture due to flattened shards, are the product of nuées ardentes—the fiery cloud of lava and gas which occurs when lavas are silica rich and viscous. They are formed in subaerial conditions, in contrast with the submarine spilites (although ignimbrites have been recorded as occurring in sequences of entirely marine sediments). It follows that the nature of the volcanics is closely related to sedimentation. The earlier volcanics are chiefly basic while the later ones are generally acidic. However, in a specific instance from Snowdonia, Rast has observed a transition from submarine to subaerial conditions clearly indicated by the following sequence: fine sediment—coarse sediment—pillow lavas—ropy lava—ignimbrite.

The early Ordovician volcanics on the flanks of the Harlech Dome thicken rapidly southwards, as do Llanvirn age volcanics, and the Bala Fault may be the deep fault which tapped subcrustal magma. Following an interval of little vulcanicity in the Llandeilan there were renewed outbursts of vulcanicity in the Caradocian in Wales (Fig. 5.2). The volcanics are described along with their contemporaneous sediments.

The Ordovician System in Wales

The Arenig Series

Following the outburst of volcanic activity in the Dolgelly area and in south Wales, the sea spread across the eroded Cambrian and Precambrian rocks. The earliest deposits are arenaceous, known in north Wales as the Garth Grit but in Shropshire as the Stiperstones Quartzite. These early beds are particularly coarsely conglomeratic in Pembrokeshire and in north Wales, there becoming coarser and much thicker in a north-westerly direction: while they are only about 300 feet thick to the east of the Harlech Dome they attain a thickness of 3,000 feet in Anglesey.

The *Tetragraptus* Shales succeed the Garth Grit, typically reaching a thickness of about 1,200 feet but they show considerable variation. In the Lleyn the lowest few hundred feet are arenaceous and flaggy, and south-west of Cader Idris, where they attain their maximum thickness, sandy beds are well developed. Interbedded volcanic rocks, both tuffs and rhyolitic lava flows, total about 1,000 feet in the upper part of the Arenig of Cader Idris indicating great volcanic activity contemporaneous with sedimentation. This pile of sediments reinforced by the interbedded volcanic rocks, which are responsible for increasing their resistance to erosion, have strikingly rugged relief. The shales include numerous graptolites, but interbedded thin limestones such as the *Ogygia* Limestone have a trilobite-brachiopod fauna which includes, as well as *Ogygiaocaris*, *Neseuretus* and *Ampyx*. Some of the volcanic tuffs are fossiliferous, containing trilobites, indicating that they also accumulated in the sea.

Arenig vulcanicity was profuse in south Wales also. Lavas are interbedded with sediments on the mainland in Pembrokeshire, and on Skomer Island, where they reach a thickness of 3,000 feet, they have been called the Skomer Series. The flows show a remarkable diversity of rock type; some are acidic (rhyolite and felsite) and they range to basic (olivine dolerite).

In Shropshire the equivalent of the *Tetragraptus* Shales is a lithologically similar group called the Mytton Shales which contain *Ogygiocaris* and brachiopods at some horizons. The higher beds, typified by extensiform graptolites, infer deposition in deeper water (or at least further from the shore) since neritic fossils are absent.

The Llanvirn Series

Similar tranquil conditions of deposition continued from the Arenig into the Llanvirn times, widely depositing dark graptolitic shales in which 'tuning fork' *Didymograptus* occurs. In south Wales the base of the series is marked, at least in some areas such as north Pembroke, by a tuff called the *Bifidus* Ash. Variations in thickness are considerable and the series thickens from 400 to 500 feet on the south side of the Harlech Dome to three times this thickness on the north-west side in Snowdonia. However, it is in south Wales that the Llanvirn Series is most typical (and from which the name is derived) and most fully developed, attaining its maximum thickness of 2,000 feet. Volcanic activity took place on a vast scale, especially in the Cader Idris area but also in the Arenig area to the east of the Harlech Dome and in south Wales. In the Arenig area two series of volcanic rocks have been described: a lower series which can be traced into the Cader Idris area, and a higher series lying above *D. murchisoni* shales (so that it is possibly of Llandeilo age) which cannot, and despite a thickness of 4,000 feet seems to be more locally developed. However, there were volcanics in the Cader Idris area contemporaneous with this higher series though they were largely spilitic and frequently exhibit pillow structures.

A relationship between thickness of sediments and vulcanicity has been remarked by T. N. George: sediments tend to be thickest where vulcanicity was slight, for example in Snowdonia. It has been suggested that at times vulcanicity was so prolific that outpourings of lava and deposition of tuff maintained the volcanoes above sea level despite a steadily downwarping crust. In conflict with this is the evidence from the nature of the sediments which are associated with the volcanic rocks; typically they are of graptolitic facies. In general, in eugeosynclines maximum vulcanicity occurs near the axial region.

It is clear that the considerable lateral variations in thickness of sediment and the intensity of vulcanicity are inter-related, at least in so far as they are both consequences of instability of the downwarping crust.

Broadly similar conditions existed in west Shropshire in Llanvirn times, for graptolitic shales with *D. bifidus*, known as the Hope Shales, are followed by volcanic tuffs and lavas which, with interbedded shales, are called the Stapeley Volcanic Series. (Activity here was approximately contemporaneous with that in the Cader Idris and Arenig areas.) Succeeding sediments are of mixed facies: the Stapeley Shales contain a trilobite-brachiopod fauna but there are some graptolitic horizons which confirm the age as *Bifidus* Zone. The entirely neritic group which follows (the Weston 'Stage' comprising grits, flags and shales) is rich in trilobites, brachiopods and molluscs with graptolites too. The overlying Betton 'Stage' is again of mixed facies with trilobites and brachiopods and, in some beds, graptolities of *Murchisoni* Zone. It is, perhaps, surprising that the Arenig and Llanvirn Series of this Shelve area of west Shropshire are not more completely neritic in facies since the margin of the area of deposition may have lain not far to the east. Strata of this age are unknown east of the Longmynd and are absent from the Caradoc area of south Shropshire, the Midlands Lower Palaeozoic inliers, the Malverns and the Gloucestershire inliers (although, as has been noted, beds of Tremadoc age are found at some of these localities). However, further east Llanvirn age strata are recorded in a borehole in Huntington, and an area of deposition occurred east of the Midlands Kratogen.

The Llandeilo Series

Uplift at the beginning of Llandeilo times resulted in change from the widely uniform conditions which had existed so far in the Ordovician Period. At Llandeilo in south Wales the Llanvirn age graptolitic shales are succeeded by coarse grit followed by calcareous flags and limestones, the Llandeilo Limestone, attaining a thickness of 2,500 feet. Westwards deeper water conditions persisted and the basal bed, the *Asaphus* Ash, is succeeded by the Hendre Shales. The latter contain a graptolitic fauna, *Glyptograptus teretiusculus* is the zone fossil and scandent biserial forms abound with some trilobites in the lower beds (*Ogygiocarella*). This genus and many others, including *Calymene* and *Basilicus*, together with Orthids and other brachiopods are common in the Llandeilo Limestone and Flags. At some localities the post-Llanvirn uplift caused the Llandeilo Beds to be deposited unconformably on earlier strata.

As defined in recent years, the Llandeilo Series comprises only the beds of the *Teretiusculus* Zone and it is probably not present throughout much of north Wales, although it occurs in Anglesey and in the Berwyn Hills. In the latter, 2,000 feet of neritic strata are of closely similar facies to the Llandeilo Flags and Limestone of the type area and contain similar fossils. Further east, in Shropshire, similar strata are called the Meadowtown Beds. Beds of tuff, and rarely lava flows, occur in the Llandeilo succession but it was a period of relative volcanic quiescence in Wales by Ordovician standards.

The Caradoc Series

Although the Llandeilo Series is, on graptolite zonal evidence, believed to be absent from much of north Wales, the earliest Caradoc Beds (of the *Gracilis* Zone) are of similar facies to the underlying Llanvirn strata. There is neither change in the type of sedimentation nor a readily observable break in the sedimentary record except in Anglesey. There the discordance is readily demonstrable for the Caradoc beds overstep on to Precambrian rocks.

The Caradoc beds are of graptolitic facies comprising mudstones and shales in the Snowdon area and generally around the Harlech Dome. Variations both in thickness and lithology have been described and there is considerable variation in nomenclature. In the Bala country the Derfel Limestone, with a rich trilobite fauna, includes forms new to Britain in the Ordovician and confirms the changes in geography of which there is widespread stratigraphical evidence. It is followed by a largely argillaceous succession but which has been divided into several named members. In some areas much of the Caradoc succession is sandier and known as the Ceiswyn Beds, for example in the Berwyn Hills.

Vulcanicity was renewed on a huge scale during Caradoc times commencing at Conway, where the succession is largely made up of volcanics (2,500 feet), and north-east of Snowdon in the Capel Curig area. Snowdon became a major centre of volcanic activity, rhyolite flows and tuffs totalling 2,000 feet have been divided into a lower and an upper series. Volcanic rocks of this age are also known in the Lleyn and in the Berwyn Hills.

In south Wales the Caradoc strata are again chiefly of graptolitic geosynclinal type, the Mydrim Shales, similar to the Hendre Shales of the preceding Llandeilo Series. They are well exposed from Builth south-westwards towards Haverford West and again in Pembrokeshire. Impure limestones are developed at the base in some areas while locally, in the core of the Towy Anticline, volcanics occur. Vulcanicity was, however, neither widespread nor prolonged in south Wales during Caradoc times.

Meanwhile in Shropshire, while somewhat similar strata were being deposited to the west of the Longmynd, and are now exposed in the Shelve area and in the Breidden Hills (graptolitic shales with some interbedded volcanic rocks), shallow seas spread eastwards over the Caradoc area (Fig. 5.3). Here a highly fossiliferous series of conglomerates, sandstones and shales was deposited over an eroded surface of Cambrian and Precambrian rocks. Probably the Church Stretton Fault had, until this time, formed a faulted coastline but now in Caradoc times submergence allowed sedimentation to encroach upon the margin of the Midland landmass (although seas did not apparently spread further east into the Midlands until much later). Naturally, subsidence was far from uniform and, although much of the Caradoc succession is of neritic facies, graptolites are found in some beds. *Nemagraptus gracilis*, the lowest Caradoc zone fossil, has been found in the Hoar Edge Grits. The overlying Harnage Group, formerly placed in the *Clingani* Zone, is now referred to the *Peltifer* Zone and there is no evidence for a major non-sequence or that seas receded temporarily from the Caradoc area. It seems probable that sedimentation was not interrupted.

The succession in west Shropshire is as follows:

MARRINGTON STAGE	Whittery Shales
HAGLEY STAGE	Whittery Ash Hagley Shales and lower volcanic rocks
ALDRESS STAGE	Aldress Shales Spy Wood Grit
RORRINGTON STAGE	Rorrington Flags and Shales

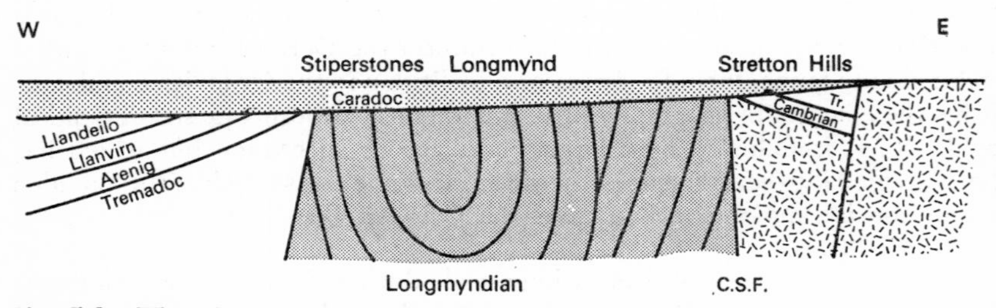

Fig. 5.3 The Caradoc transgression in Shropshire. (Before post-Silurian earth movements the dips in the Longmyndian may have been as shown, although they now dip generally westwards.)

In south Shropshire, in the Caradoc Series of neritic facies, a zonal system of 'stages' based on trilobites was worked out by Bancroft and emended by Dean (Table 5.2). In view of the scarcity of graptolites it is of considerable value, especially in relating localities in the type area. Naturally, and fortunately, this zonal system bears a relationship to the old-established lithological divisions but eliminates some of the former ambiguities.

Table 5.2 does not indicate thicknesses of the main lithological divisions, the Hoar Edge Grit and its basal conglomerate may attain a thickness of up to 300 feet and the overlying Harnage Shales over 1,000 feet, whereas the remainder of the succession does not amount to more than 500 feet.

The Ashgill Series

The Ashgill Series bears some resemblance to the Caradoc in lithologies and lateral variation. They are often referred to jointly as the Bala Group, following Sedgwick's early nomenclature; the Caradoc Series = Lower Bala and the Ashgill Series = Upper Bala, both series are regarded as Upper Ordovician but there is frequently a stratigraphical break at the base of the Ashgill. Certain geographical changes took place between the two: the spread of greywacke facies which, in Caradoc times, Jones has shown was largely restricted to north Wales but extending into Carmarthenshire now became confined to west mid-Wales (Fig. 5.4), if indeed these latter greywackes are of Ashgill age. They exhibit typical turbidite structures, which much of the Ashgillian sediments do not, and comprise shales, sandstones, siltstones and silty mudstones. Some of this sedimentation took place in relatively shallow water and shows cross-bedding and ripple-drift bedding. Detailed stratigraphical work has shown that in Ashgillian times and during the succeeding Silurian Period sedimentation was controlled by uplift along axes which were to become, in

TABLE 5.2

Graptolite Zones	*Lithological Divisions*	*Stages*	
Dicranograptus clingani	*Onnia* Beds	Onnian	14 13 12
	Acton Scott Beds	Actonian	11
	Upper Cheyney Longville Flags	Marshbrookian	10
	Lower Cheyney Longville Flags and *Alternata* Limestone	Upper Longvillian	9
	Upper Chatwall Sandstone	Lower Longvillian	8
Climacograptus peltifer	Soudley Sandstone, Lower Chatwall Sandstone, Horderley Sandstone	Soudleyian	7 6
	Harnage Shale	Harnagian	5 4 3
Nemagraptus gracilis	Coston Beds Hoar Edge Grits Basal Conglomerate	Costonian	2 1

the Caledonian Orogeny, major structures. The Towy Anticline formed a barrier to sedimentation, confining it to a basin to the west.

Caradocian vulcanicity may have elevated some volcanic areas such as Snowdon above sea level since the succeeding Ashgill Series is absent. It is also missing in Anglesey, but in the extreme north of Wales it consists of black shales such as the Deganwy and Bodeidda Mudstones which are, of course, much thinner than the greywacke succession.

Lateral changes cannot be traced in detail since the Ordovician is covered by Silurian rocks in the Central Wales Syncline—the Silurian outcrops over about half the area of Wales—but it comes to the surface in inliers at Plynlimmon and Llanidloes (Fig. 4.1). In south Wales the black shale facies passes southwards into more calcareous beds, the Redhill and Slade Beds, and some of the limestones have an abundant and varied trilobite and brachiopod fauna. In the Caradoc area of south Shropshire varying amounts of the

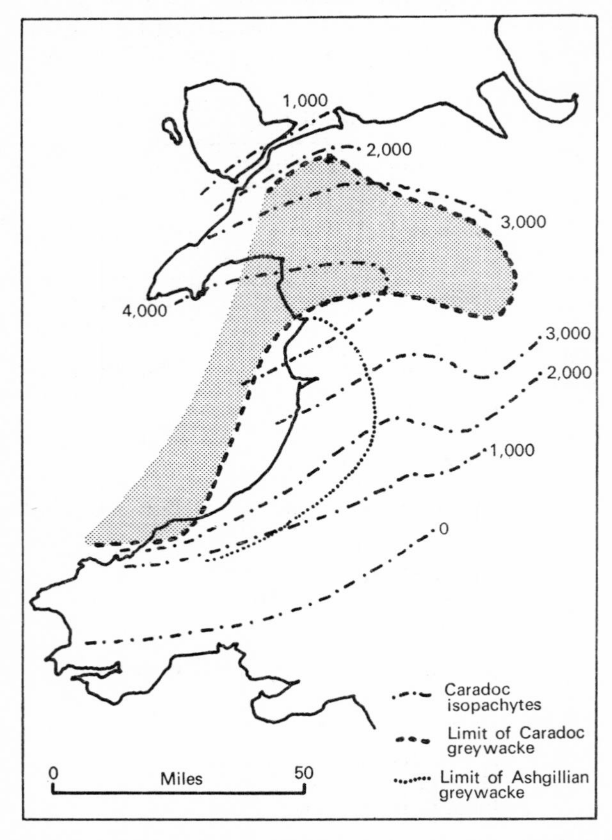

Fig. 5.4 Caradoc isopachytes and the extent of Caradoc (and Ashgill) greywackes. (Based partly on O. T. JONES, 1955, *Q. Jl geol. Soc. Lond.*, **111,** 323, Pl. XVII.)

Onnian Stage of the Caradoc are present due to pre-Silurian erosion. The Silurian follows unconformably and beds of Ashgill age are not found.

Lower Palaeozoic rocks of the Lake District

The Lake District forms an area of mountainous terrain carved out of Lower Palaeozoic rocks which are surrounded by younger rocks, chiefly

Carboniferous but, in the west, by Triassic. The region is naturally divisible into three main areas, the northern third being the outcrop of the Skiddaw Slates, the central and highest part of the Lake District formed of volcanic rocks known as the Borrowdale Volcanic Series, while the southern third is formed of Bala and Silurian sedimentary rocks. The Ordovician succession is:

Ashgill Series	Applethwaite Beds	100 ft
Caradoc Series	Coniston Limestone Series	*c.* 1,000 ft
Llandeilo Series and Llanvirn Series	Borrowdale Volcanic Series	10,000 ft
Arenig and Tremadoc Series	Skiddaw Group	*c.* 6,000 ft

Relatively little recent work has been done on the Palaeozoic sedimentary rocks of the Lake District but there has been in the last decade continuous addition to our knowledge of the Borrowdale Volcanic rocks, the structure, the Caledonian intrusions and mineralization.

The oldest strata present are the Skiddaw Group (Slates), the base of which is nowhere seen. The presence or absence of the Cambrian cannot be confirmed. The Skiddaw Group has been shown to be as old as early Arenig (*Extensus* Zone) although the lowest beds are unfossiliferous and could belong to the Tremadoc. Fossils are scarce but graptolites are commoner than the very rare trilobites and brachiopods. Over a period of many years a considerable faunal list has been assembled.

The structure of the Skiddaw Slate area is not known with certainty but clearly the strata are highly folded and, in places, cleaved. The total thickness has been estimated at several thousand feet, and sedimentation which commenced in early Ordovician times (or earlier) persisted through to Llanvirn times. The rocks show some variation in lithology and not all are slates. Grits, flags and mudstones are present, the lower earlier strata perhaps being coarser. Sedimentary structures, including flute casts, load casts, graded bedding, etc., have been described from the arenites. Flute casts in the Loweswater Flags suggest a dominantly southerly source of material. Although the stratigraphy is imperfectly known, certain lithologically distinctive horizons have been mapped and a succession of four lithological groups suggested:

Latterbarrow Sandstone
Mosser–Kirkstile Slates
Loweswater Flags
Hope Beck Slates

Similar rocks occur in the Cross Fell Inlier which lies between the Inner and Outer Pennine Faults and they are undoubtedly Skiddaw Slates, found together with representatives of the Borrowdale Volcanic Series and Caradoc and Ashgill sedimentary rocks.

The close proximity of fossils of different zones in outcrops on the west side of Bassenthwaite Lake in the Lake District was interpreted by Dixon as evidence of co-mingling. However, in the absence of structural evidence there seems no need to suppose that the fossils occur in anything but the usual sequence and that intense folding may account for their geographical proximity.

The Skiddaw Slates have been slightly metamorphosed (hornfelsed) by later intrusions, chiefly by the Skiddaw Granite. However, the early Ordovician seems to have been devoid of igneous activity—unlike Wales—and only late in Skiddaw times did it commence: tuffs and lavas are locally interbedded with what are believed to be the youngest Skiddaw Slates. Thereafter followed the most impressive vulcanicity of Lower Palaeozoic times, a vast pile of lavas, tuffs and agglomerates accumulating to a thickness of perhaps 10,000 feet or more. The duration of this volcanic episode can be only approximately fixed since the junction of the volcanic series with the older Skiddaw Slates is seldom a normal junction, being faulted or thrust. However, it may have commenced in Llanvirn times and almost certainly continued throughout the Llandeilan. No sedimentary rocks of this age are present. The Borrowdale Volcanic Series outcrops in three areas in the Lake District, the central area which includes such well known peaks as Helvellyn and Scafell, Eycott Hill (a small outlying outcrop of the lower volcanic rocks) and Binsey Hill between Bothel and Caldbeck on the northern side of the Skiddaw Slate area. In the central Lake District structures are complex and the succession difficult to elucidate, although this has been done in several areas. In the Binsey Hill outcrop the dips are steep, often close to the vertical, but a straightforward sequence of volcanic beds is seen outcropping over a width of a mile. The total thickness is believed to be of the order of 10,000 feet.

The lavas are chiefly andesites, weathering to a grey-green colour. Some are coarsely porphyritic with felspar phenocrysts but, more commonly, they are fine grained. In well exposed areas vesicular tops of individual lava flows can be seen and, even where exposures are poor, successive flows can be discerned from the step-like topography. This is especially conspicuous where the flows are flat lying or have a low dip as on Gowbarrow Hill north of Ullswater. Rhyolite lavas are not uncommon, especially in the higher part of the series, and they often show flow-banding and many show autobrecciation. The latter may be mistaken for tuffs in the field. In general, the lavas and tuffs, which are interbedded, can be distinguished as a result of subsequent deformation, for the tuffs readily acquired slaty cleavage (and have been used locally as roofing slate) whereas the lavas do not. The tuffs vary very greatly in grade from very fine grained material to agglomerate. Undoubtedly much was laid down in water for the tuffs show graded bedding, channelling and other sedimentary structures.

Rocks of volcanic origin would naturally be expected to vary greatly laterally since there must have been many centres of activity, although their locations are not known. Individual flows of lava may be extensive and bedded tuffs are likely to be widespread, although they thin rapidly away from centres of activity unless the material is spread by currents. The sequence of rocks varies greatly and interpretation is complicated by structures imparted chiefly during the Caledonian Orogeny. Early workers on the Lake District suggested that a threefold division was recognizable:

Rhyolites and andesites
Tuffs with some lavas
Chiefly andesites, with some tuffs.

The volcanic rocks of the Borrowdale Series are followed in the Cross Fell Inlier by the *Corona* Beds, shales with calcareous ribs, which include the brachiopod *Trematis corona*. These beds cannot be equated with any in the Lake District, where sedimentation apparently commenced slightly later, with the Stile End Beds, which are the same age as the lower part of the Dufton Shales of the Cross Fell Inlier. The whole succession in the latter area was shaly while in the Lake District beds of more variable lithology were deposited and a rhyolitic lava flow was extruded, vulcanicity lingering on in that area.

TABLE 5.3

	Lake District	*Cross Fell Inlier*
ASHGILL SERIES	Ashgill Shales *Staurocephalus* Limestone	
CONISTON LIMESTONE SERIES ≡ CARADOC SERIES	Applethwaite Beds, calcareous shales and limestones	Dufton Shales
	Conglomerate	
	Stockdale Rhyolite	
	Stile End Beds, fossiliferous bedded ash	
		Corona Beds

Unlike Wales, the Ashgill Series follows the Caradoc Series conformably (in the Lake District, however, there was a slightly earlier break in the sequence after the extrusion of the Stockdale Rhyolite, when folding and erosion occurred). A basal bed of limestone, the *Staurocephalus* Limestone, is succeeded by about a hundred feet of shales, ashy in the lower part. A fauna of trilobites,brachiopods and gastropods distinguishes it from the overlying Silurian shales which include a graptolitic fauna, reflecting a marked environmental change at the end of the Ordovician Period, although there is no break in the succession.

In the Cautley District, east of Kendal, six small areas of Ordovician rocks occur which are of great importance, since they provide a more complete sequence of the beds of Upper Ordovician age than can be seen elsewhere and form a better type sequence for the Ashgillian than does Ash Gill, Coniston. These outcrops have been described as recently as late 1966. The succession is given in Table 5.4.

The succession is much more complete than that of the Coniston District

TABLE 5.4

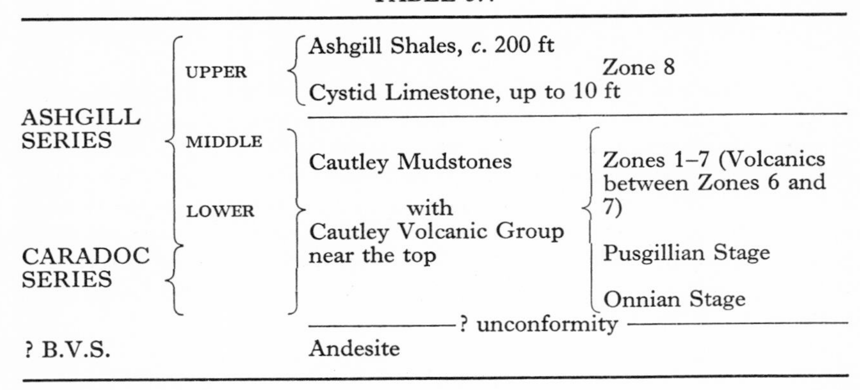

Series	Division	Beds	Zones / Stages
ASHGILL SERIES	UPPER	Ashgill Shales, *c.* 200 ft Cystid Limestone, up to 10 ft	Zone 8
ASHGILL SERIES	MIDDLE LOWER	Cautley Mudstones with Cautley Volcanic Group near the top	Zones 1–7 (Volcanics between Zones 6 and 7)
CARADOC SERIES			Pusgillian Stage Onnian Stage
		? unconformity	
? B.V.S.		Andesite	

of the Lake District, though the Ashgill Shales of the latter may be equivalent to Zone 8 of Cautley: the underlying ash may have been formed at the time of the Cautley Volcanic episode of activity. The succession in the Cross Fell Inlier is similarly incomplete, for the *Staurocephalus* or Swindale Limestone (equivalent to Zone 6 of Cautley) rests on beds of Pusgillian age unconformably. The Keisley Limestone of the Cross Fell Inlier, occurring in fault-bounded outcrops, is probably equivalent to a lower horizon in the Cautley sequence than is the *Staurocephalus* Limestone.

The Southern Uplands Ordovician

Peach and Horne recognized three geographical subdivisions of this region of Lower Palaeozoic rocks. Termed the Northern, Central and Southern Belts, these are still useful concepts. Lower Palaeozoic rocks also occur to the north of the Southern Upland Boundary Fault (which defines the northern margin of the Uplands geologically) in a number of inliers, including a triangular area with Girvan near the apex. This latter area is invariably discussed along with the Southern Upland rocks although, strictly defined, it lies outside. To the south, the Southern Uplands are bordered by Upper Devonian and Carboniferous strata.

Cambrian rocks are unknown in this region and it is debatable whether rocks of Cambrian age (possibly of Dalradian type) may lie at depth in the Midland Valley of Scotland or even beneath the Ordovician rocks of the Southern Uplands. While there is no evidence from the south of Scotland, the presence of the Connemara Schists of Dalradian type in Ireland, lying to the south of the continuation of the line of the Highland Border Fault, infers that sedimentation of this type and age did extend southwards. However, the Southern Uplands comprise only Ordovician and Silurian rocks: Ordovician rocks form the Northern Belt and inliers in the Lower Silurian Central Belt, while the Southern Belt is composed of Wenlock and Ludlow (Upper Silurian) age strata.

Arenig Beds

The oldest rocks present are of Arenig age and they occur within the Southern Uplands at Leadhills and Polshill, but more extensively south of Girvan in the Ballantrae District. They comprise spilitic lavas showing excellent pillow structures which are followed by agglomerates and tuffs with thin interbedded black shales containing graptolites. These have been recognized as being from the *Extensus* Zone and confirm the age as Arenig. As in certain areas of north and south Wales, vulcanicity commenced early in Ordovician times, but in Scotland it was not to be extensive or prolonged. The Arenig succession was completed by the deposition of red, green and grey cherts, with Radiolarians and thin interbedded tuffs. The Arenig rocks of the Girvan area have been intruded by plutons of serpentine, gabbro and granite.

Volcanic rocks, lavas sometimes exhibiting pillow structures, are seen in the cores of isoclinal or periclinal folds in the Southern Uplands as far south as Abington in the Central Belt, but may not all be of Arenig age as was formerly supposed.

The Caradoc and Ashgill Series

Rocks of Llanvirn and Llandeilo age have not been recognized in the south of Scotland, but the Caradoc and Ashgill Series are extensively represented and occupy the Northern Belt. There they are largely greywackes together with graptolitic shales. In the Central Belt, in which Moffat lies, they outcrop in inliers (of which the one at Dobb's Linn was made the classical outcrop by Lapworth's work) and are represented by very thin groups of black shales. In the Girvan area there is, on the other hand, a thick succession of strata which accumulated in shallow marine conditions. The great contrast between the thin beds of graptolitic facies at Moffat and the thick neritic succession of Girvan is one of the most notable features of the Lower Palaeozoic geosyncline in the south of Scotland, for this contrast persisted throughout the Lower Silurian as well.

In the Moffat area the Caradoc and Ashgill Series are represented by only 120 feet of dark graptolitic shales; the Glenkiln Shales, 20 feet; succeeded by the Hartfell Shales, 100 feet, representing slow accumulation. At the present day such a slow rate of accumulation is known from oceanic depths and this, together with the absence of neritic fossils such as brachiopods and trilobites (only graptolites are found), suggests that these sediments may have been deposited in deep water. The absence of a benthonic fauna (the graptolites were planktonic or pseudoplanktonic) infers sea-bottom conditions which could not support life. It cannot be ruled out that sediments of graptolitic facies may have been laid down in water of only moderate depth, but at least deposition must have taken place far from any shore. The present distance between rocks of graptolitic facies and the contemporaneous neritic rocks of Girvan is no more than 50 miles, but post-Lower Palaeozoic folding may account for considerable crustal shortening. On the other hand, present-day foredeeps (trenches) are quite close to the shallow seas of island arcs.

The Glenkiln Shales contain the fossils of the lowest two zones of the Caradoc Series, *Nemagraptus gracilis* and *Climacograptus peltifer*, while the Hartfell Shales contain the zone fossils of the remainder of the Caradoc and and those of the Ashgill Series, i.e. from the *Climacograptus wilsoni* Zone to the *Dicellograptus anceps* Zone. Essentially both shale groups are of similar lithology and represent similar conditions of deposition. No break in sedimentation is evident.

Along the strike direction the Glenkiln Shales show little variation in character, although cherts and thin volcanics are present at some localities. However, north-westwards—within the Northern Belt—the dark shales become interbedded with grey and green mudstones and pass into a succession of grey shales and greywackes with only thin graptolitic dark shale beds. The lower part of the Hartfell Shales follows the Glenkiln conformably and is lithologically similar, containing many graptolites. The upper part of the Hartfell Shales (of Ashgill age), though conformable, includes more coarse material and fossils are not so abundant. Towards the east there is limited evidence from the inlier in the Central Belt at Ettrickbridgend, near Selkirk, that conditions of deposition were somewhat different due to the proximity of the edge of the geosyncline. Grits attain a thickness of 180 feet in the Upper Hartfell Shales. However, it is to the north-west that the changes in lithology can better be seen as one moves away from the axis of the geosyncline. As described above, the dark shales of the Glenkiln become interbedded in a thicker sequence of coarser material. Similar transition is observable in the Hartfell but the change in conditions of deposition 'set in' further to the south than in Glenkiln times.

In the Girvan area the Caradoc and Ashgill Series (or Bala) are represented by a thick succession of conglomerates, sandstones, mudstones and occasional limestones. These have been divided into a lower series, the Barr Series, of lower Caradoc age and chronologically equivalent to the Glenkiln Group of Moffat, and a higher series, the Ardmillan, equivalent in age to the Hartfell Group. The Barr Series, originally thought to be about 800 feet thick, has been shown by Williams to be considerably thicker. The Ardmillan Series, too, is known to be vastly thicker than earlier figures which have been given and the Bala totals about 8,000 feet. This is in the Girvan area, the shelf area of the geosyncline. Passing into the geosynclinal trough the sediments thicken to twice this amount (Fig. 5.5).

Williams' work has also shown that the succession is more complex than Lapworth believed. Limestones thought to be correlatable were shown not to be, and there are probably three quite distinct horizons at which thick limestone occurs. Furthermore, a number of conglomerates of very similar lithology were formerly all mapped as Benan Conglomerate. The Bala succession in the Girvan area is given in Table 5.5.

The Kirkland Conglomerate rests unconformably upon Arenig beds and its thickness is very variable. The lower part is of purplish colour and it contains

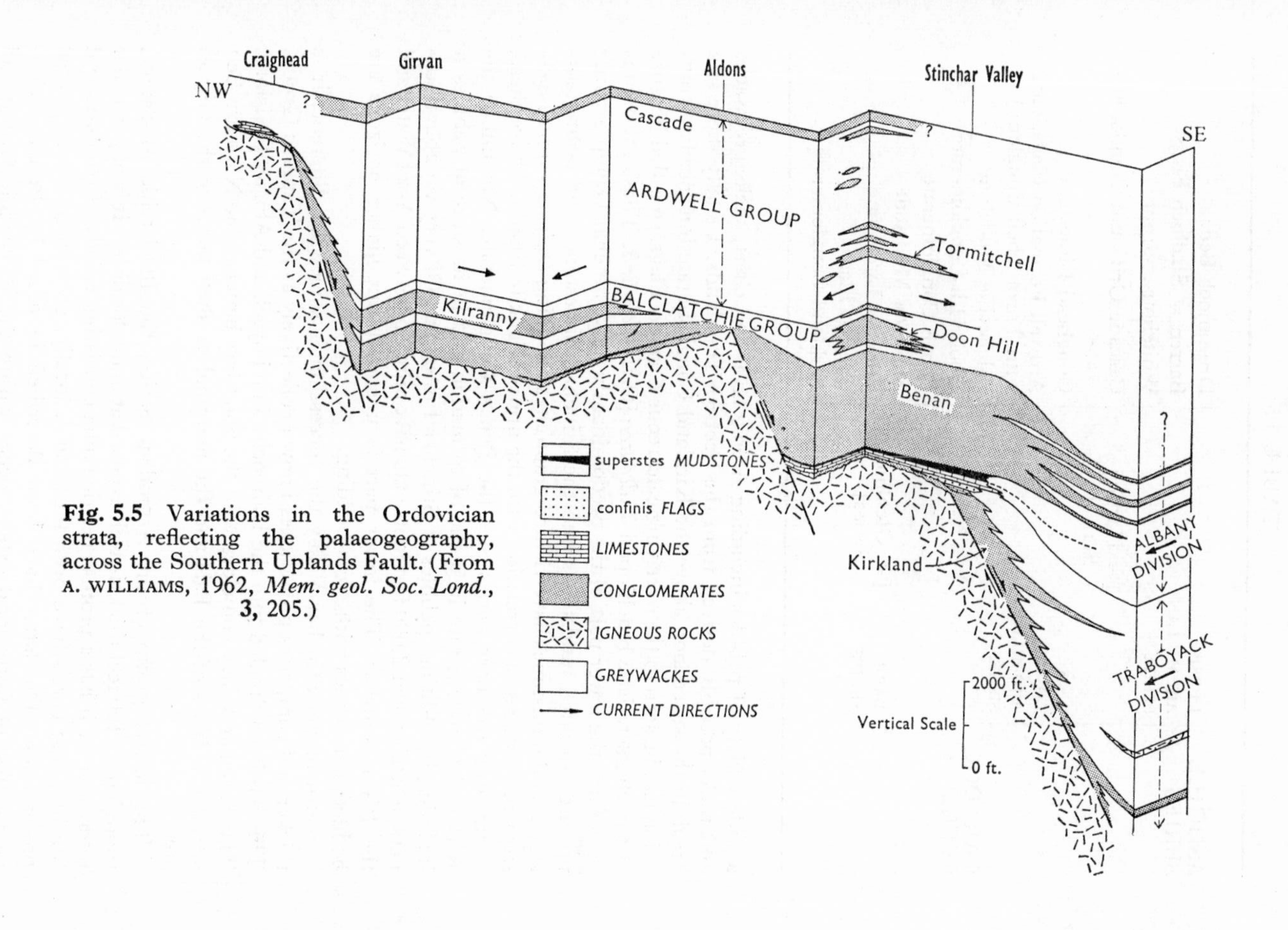

Fig. 5.5 Variations in the Ordovician strata, reflecting the palaeogeography, across the Southern Uplands Fault. (From A. WILLIAMS, 1962, *Mem. geol. Soc. Lond.*, 3, 205.)

TABLE 5.5

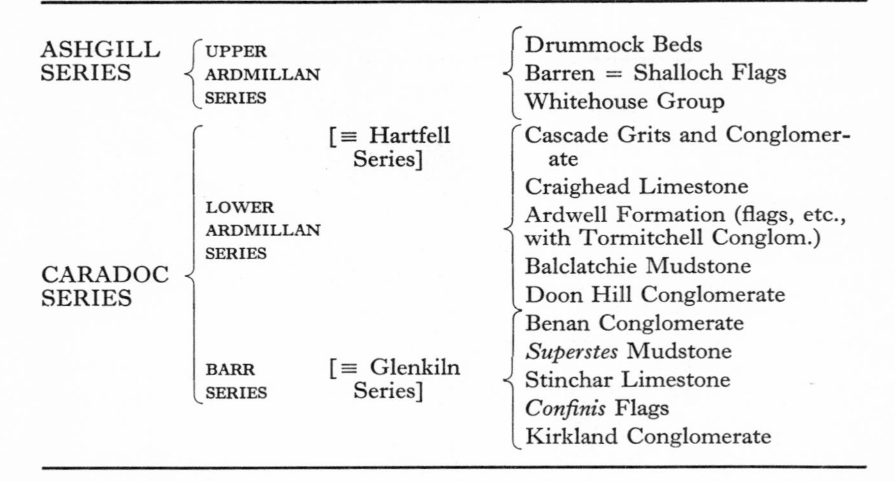

Series	Subdivision	Equivalent	Groups
ASHGILL SERIES	UPPER ARDMILLAN SERIES		Drummock Beds Barren = Shalloch Flags Whitehouse Group
CARADOC SERIES	LOWER ARDMILLAN SERIES	[≡ Hartfell Series]	Cascade Grits and Conglomerate Craighead Limestone Ardwell Formation (flags, etc., with Tormitchell Conglom.) Balclatchie Mudstone Doon Hill Conglomerate
	BARR SERIES	[≡ Glenkiln Series]	Benan Conglomerate *Superstes* Mudstone Stinchar Limestone *Confinis* Flags Kirkland Conglomerate

a wide variety of pebbles including Arenig lavas and chert, shale, greywacke and rocks possibly derived from the Scottish Highlands. Usually there is a break in the succession above the Kirkland Conglomerate but locally a transition can be seen to the overlying calcareous *Confinis* Flags, so called because of the characteristic brachiopod *Valcourea* [*Orthis*] *confinis*. These calcareous flags form the lower part of the predominantly limy Stinchar Group grading upwards through mudstones into the Stinchar Limestone. Succeeding siltstones and mudstones have yielded graptolites including *Didymograptus superstes* which suggest a correlation with the upper Glenkiln Shales. Above these, completing the Barr Series, lies the Benan Conglomerate. Not unlike the Kirkland Conglomerate in mode of formation and in its suite of pebbles it includes, in addition, pebbles of Stinchar Limestone. Of very variable thickness, it attains over 2,000 feet in its extensive outcrop between Assel Water and the River Stinchar. The lower part is unsorted but higher horizons are bedded, sometimes with cross-bedding.

South of the Stinchar Valley the succession is somewhat different; it is thicker and comprises greywackes known collectively as the Tappins Group. This has been divided into the Dalreoch, Trayboyack and Albany Divisions. The Stinchar Valley coincides with the boundary between the Northern Belt greywacke facies of the Bala and the marginal or shelf facies of the Girvan area.

The lowest group of the Ardmillan Series, the Balclatchie Mudstone, though only 100 feet thick, is well exposed at many localities. It is of particular interest since it has a profuse neritic fauna consisting chiefly of trilobites and brachiopods; corals have been found and, locally, graptolites of exceptional preservation. The latter have been described in detail by Bulman. Conglomerates, of fairly local extent, occur within the Balclatchie Group and above, within the Ardwell Group, and are very similar to the Benan Conglomerate. The Ardwell Group, especially well seen in the shore at Ardmillan, consists of striped siltstones and flags. The highest beds of the Ardmillan Series, the Cascade Grits, well exposed in the waterfalls of Penwhapple Burn (hence the name) are not seen in the shore section.

To the north of the main outcrop of Lower Palaeozoic rocks, north-east of Girvan, lies the Craighead Inlier of Ordovician and Silurian strata. The Craighead Limestone was thought to be correlatable with the Stinchar Limestone which it broadly resembles, but later workers have assigned it to higher horizons on fossil evidence (see Table 5.5).

The Whitehouse Group is divisible on a lithological basis, as recognized by Lapworth, into the lower grey and green shales and flaggy beds (with occasional graptolitic shales only inches thick) and the upper shales with calcareous beds and conglomeratic bands. Washouts and sole marks are common. The Caradoc–Ashgill boundary lies within the Whitehouse Beds; it may possibly be marked by a conglomeratic horizon but, essentially, sedimentation was continuous and there is no significant break. Positive evidence of beds of the *Pleurograptus linearis* Zone (top of the Caradoc) and the *Dicellograptus complanatus* Zone (lower of the Ashgill zones) has long been known. As well as graptolites, a considerable trilobite fauna has been described.

The Barren Flagstones, also known as the Shalloch Flagstone Group, as the name implies, are scantily fossiliferous. Following the Whitehouse Group on the Girvan shore section, they comprise flagstones with shaly partings and beds of greywacke. They are the highest part of the Ordovician succession seen there, being faulted against Silurian beds. The group also outcrops inland in Penwhapple Burn and in the Craighead Inlier, and in the latter even higher Ordovician beds, the Drummock Group, outcrop. Formed chiefly of mudstones, this group is very highly fossiliferous at some localities such as South Threave, where a profuse neritic fauna occurs including as well as trilobites and brachiopods, molluscs and echinoderms. In addition, graptolites have been found, of which *Dicellograptus anceps* indicates the age as upper Ashgillian.

The Ordovician rocks of Ireland, showing many similarities to those of Scotland, but important differences, are described together with the Silurian in Chapter 6.

Ordovician fossils

Fossils are very much commoner, in general, in Ordovician strata than in beds of Cambrian age. Graptolites made their appearance in Tremadoc times and rapidly became diverse in type and often very numerous in the dark shaly facies. In neritic facies beds, trilobites and brachiopods were still by far the commonest forms of life but there was, during the Ordovician, a great development of echinoderms and bryozoans, while ostracods and foraminifera appeared.

The Ordovician System has been zoned by means of graptolites (p. 92) but neritic shelf sediments have been zoned by means of brachiopods and trilobites, for example in south Shropshire and in the Girvan area. The occasional incursion of graptolitic facies into the generally neritic areas permits correlation of the diverse successions.

In the Tremadoc Series trilobites are the commonest fossils. Families persisting from the Cambrian, Olenidae, Agnostidae and Ptychoparidae, existed together with early members of typical Ordovician families such as Asaphidae and Trinucleidae. Trilobites attained their acme in the Ordovician Period (though they are seen to have declined little in the succeeding Silurian), and in general Ordovician forms are more advanced than those of the Cambrian. They possess short genal spines or have rounded genal angles (the Trinucleids are exceptions) and in many there had been loss of pleural spines. Large-tailed forms, like the Asaphids, are typical with the development of a flat border and loss of evidence of segmentation. In the family *Calymene* there is an increase in the number of glabellar lobes in forms from successively younger beds.

Ordovician shelly faunas are rich in brachiopods. The inarticulate forms are less important than in the Cambrian, and Orthids and Strophomenids predominate. Some Orthids were of long time-range (*Heterorthis* and *Dinorthis*) while some species of '*Orthis*', '*Strophomena*' and *Macrocoelia* [*Rafinesquina*] were of short time range and vie with the trilobites in stratigraphical importance. In some beds brachiopods are sufficiently numerous to be considered rock-forming organisms, for example in the *Alternata* Limestone of the Caradoc Series. In the Ashgill the larger Caradoc brachiopods disappear to be replaced by species such as *Dicoelosia* (which persist into the Silurian) and '*Strophomena*'. Rhynchonellids and Spirifers appear, as do Pentamerids.

The brachiopod faunas of the Ordovician rocks of the Girvan area, while comprising chiefly Orthids and Strophomenids, are quite distinct from those of Wales. No species appears to be common to both areas. However, the ninety or more species described by Williams from the Girvan area have close affinities with the brachiopods of the Appalachian Ordovician rocks, now 4,000 miles distant. Not only are there morphological similarities and many species are common to both areas, but Williams cites biometric proof of similar variability. On the other hand, while the trilobites of Girvan include a number of species not found elsewhere in Britain (but found, for example, in North America and Bohemia), they do not indicate a complete isolation of the Girvan area from the Welsh Geosyncline.

Of the molluscs, gastropods are not uncommon fossils in Ordovician rocks but, apart from *Bellerophon*, they are not important. Bivalvia (lamellibranchs) such as *Modiolopsis* are common and cephalopods, for example *Orthoceras vagans* (Ashgill age), are locally abundant. Primitive echinoderms appeared in Ordovician times and by late in the period Cystids had become common. Crinoids also had made their appearance. On the other hand, Coelenterates remained a relatively unimportant phylum, although some genera of tabulate corals date back to the Ordovician.

Branched graptolites are typical of the early Ordovician. A rapid reduction in the number of stipes (branches) from the many branched forms of the Tremadoc took place among the Dichograptidae (*Bryograptus*, *Clonograptus* and *Dichograptus* are succeeded in the Arenig by *Tetragraptus* and then *Didymograptus*). Note that horizontal forms evolved most rapidly in this reduction of stipes, hence the extensiform *Didymograptus* precedes the pendent *Didymograptus* in the stratigraphical succession. The Dichograptid faunas were succeeded by Leptograptids and these in turn were followed by Diplograptids, which are found in strata of Bala age. *Nemagraptus* attains a maximum in the *N. gracilis* zone, of world-wide occurrence. The unusual *Dicranograptus* is also found in Bala age beds.

6 The Silurian System
(and Lower Palaeozoic rocks beyond Great Britain)

The Silurian System was first established by Murchison and is named from a Welsh Border tribe. Silurian rocks outcrop over half the area of Wales, concentric with the Cambrian and Ordovician of the Harlech Dome, and, extending southwards from the north Wales coast, almost reach Carmarthen. Within the Silurian area, Ordovician rocks already described outcrop as inliers in the cores of anticlinal folds of Caledonian age. The earliest Silurian rocks are overlapped by younger beds which rest unconformably on Precambrian, Cambrian and Ordovician rocks in the marginal regions of the Welsh Geosyncline. Elsewhere there is a major palaeontological break so that the base of the system is often easily recognizable and is readily definable. Upwards the Silurian beds pass into overlying Devonian beds without any such abrupt change, as a rule, and considerable discussion has taken place about the choice of the Siluro-Devonian boundary. This is considered on p. 167.

Despite a widespread break in sedimentation at the end of the Ordovician Period, similar conditions existed during the Silurian; a broad geosyncline lay across Wales, and deep-water graptolitic pelagic shales, greywackes and marginal facies existed. Initially, sedimentation did not spread eastwards beyond the Longmynd and Old Radnor and, at that time, fine-grained (deep water) sediments were deposited in north Wales—the Gyffin Shales—while deposition of marginal facies occurred in mid-Wales and south Wales. Later, sedimentation spread eastwards and this transgression was of greater extent than that of Caradoc times. Evidence of this is provided by the occurrence of Silurian rocks of neritic facies in south Shropshire and many inliers in the Welsh Borders and Midlands (Tortworth, Usk, May Hill, the Malverns, Woolhope and the several inliers of the South Staffordshire Coalfield, of which the ones at Dudley are best known). The strata of these inliers are of particular importance in attempting to deduce the palaeogeography of the period.

The Silurian has been divided into the Llandovery (of which the Valentian is a synonym), the Wenlock and the Ludlow Series. These are divisible, in Britain, into 20 graptolite zones, most of them named after species of *Monograptus* (Table 6.1). These zone fossils are characteristic of the deposits laid down further from the shores of the geosyncline, and the shelf deposits of the Welsh Borders have been zoned by means of brachiopods and trilobites.

TABLE 6.1

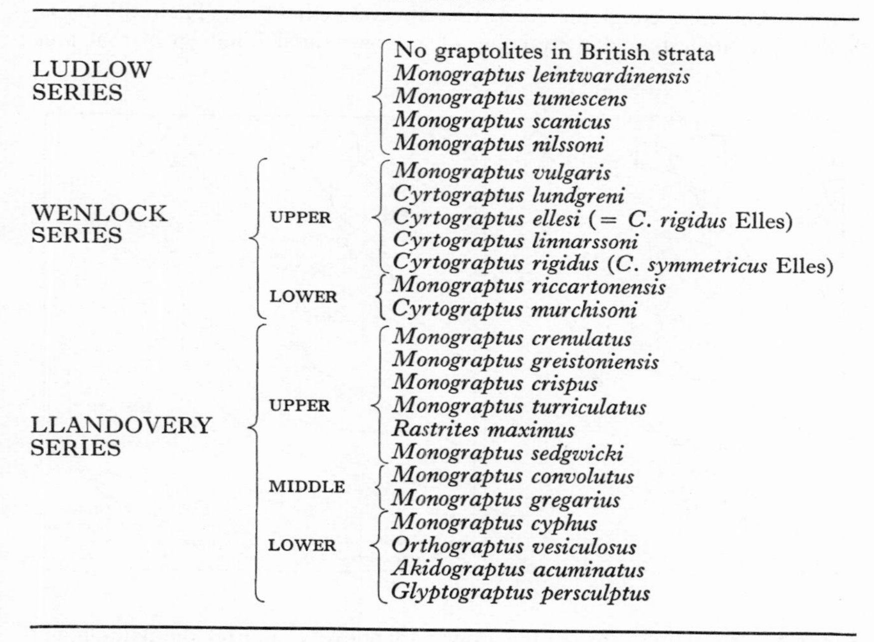

Series		Zones
LUDLOW SERIES		No graptolites in British strata
		Monograptus leintwardinensis
		Monograptus tumescens
		Monograptus scanicus
		Monograptus nilssoni
WENLOCK SERIES	UPPER	*Monograptus vulgaris*
		Cyrtograptus lundgreni
		Cyrtograptus ellesi (= *C. rigidus* Elles)
		Cyrtograptus linnarssoni
		Cyrtograptus rigidus (*C. symmetricus* Elles)
	LOWER	*Monograptus riccartonensis*
		Cyrtograptus murchisoni
LLANDOVERY SERIES	UPPER	*Monograptus crenulatus*
		Monograptus greistoniensis
		Monograptus crispus
		Monograptus turriculatus
		Rastrites maximus
		Monograptus sedgwicki
	MIDDLE	*Monograptus convolutus*
		Monograptus gregarius
	LOWER	*Monograptus cyphus*
		Orthograptus vesiculosus
		Akidograptus acuminatus
		Glyptograptus persculptus

In late Ordovician times there had been a great decline in vulcanicity, a trend which continued so that volcanic rocks are almost absent from the Silurian—one of the main contrasts between the two periods.

The Silurian rocks of Wales and the English Midlands

The Llandovery

In north Wales (and elsewhere) the beds of Llandovery age have been called Valentian; consequently, there were different names for the neritic and graptolitic facies respectively. Since it is now possible to correlate the contrasting successions with some certainty, one of these terms is superfluous. The term Llandovery is retained but it is emphasized that it is not used in any restricted sense referring to shelf facies alone. In fact it is not an adequate term to define a particular facies, since facies boundaries transgress time-planes.

The sediments which lay nearest to the axial trough of the geosyncline, the Gyffin Shales of Conway and the Bala country, are only 300 feet thick and represent a very slow rate of accumulation. Similar beds occur in Anglesey. Southwards, towards Machynlleth, while the Lower and Middle Llandovery rocks are of similar character, the Upper Llandovery thickens up to as much as 2,000 feet and there is evidence of deposition from turbidity currents. Still further south towards Plynlimmon these beds become principally greywackes (but with conglomerates), the Aberystwyth Grits. Their thickness, 8,000 feet, suggests a downwarping of a geosynclinal foredeep in that area

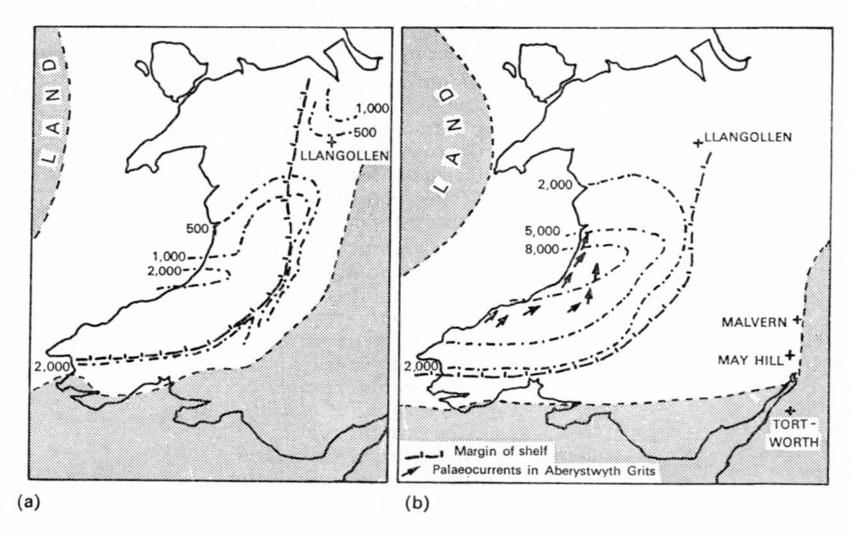

Fig. 6.1 Isopachytes for (*a*) the Lower Llandovery, and (*b*) the Middle and Upper Llandovery Beds in Wales. (Based on O. T. JONES, 1955, *Q. Jl geol. Soc. Lond.*, **111**, 323, and A. M. ZIEGLER, 1965, *Nature, Lond.*, **207**, 270; palaeocurrent directions in the Aberystwyth Grits after A. WOOD and A. J. SMITH, 1958, *Q. Jl Soc. Lond.*, **114**, 163.)

(Fig. 6.1). The Llandovery strata are much thinner in the east around Corwen and Llangollen and are coarse to fine grained with thin graptolitic shales, although they chiefly possess a shelly fauna of trilobites, brachiopods and corals. Variations in thickness have been recorded south of the Berwyn Hills in the Llanfyllin and Welshpool area from 400 feet to about 1,000 feet in a distance of only a few miles. Post-Ashgillian erosion is evident here and successive beds of the Llandovery overlap to the south-east: in early Llandovery times this was the margin of the geosyncline. However, although the lowest beds are invariably arenaceous, the succession is not wholly of shallow water type of deposition and there is evidence from detailed work in many localities (by different authors) of differential subsidence of the geosyncline.

The typical area of shallow-water sediments of Llandovery age is in the Llandovery area which lay to the south of the geosyncline in north and mid-Wales, already described. This Llandovery area is now separated by erosion along the Towy Anticline from contemporaneous graptolitic strata. The type area succession comprises: a basal series of conglomerates, sandstones and shales followed by 1,200 feet of mudstones (with a dominantly brachiopod fauna), referred to the Lower Llandovery; a further 800 feet of mudstones (with brachiopods and trilobites) of Middle Llandovery age; and sandstones, shales and mudstones of Upper Llandovery age. Some bedding planes are crowded with brachiopods and *Pentamerus oblongus* is the most abundant fossil. This shelly succession totalling over 3,000 feet contrasts with the contemporary graptolitic shales, only 300 feet thick, at Conway.

A similar succession is seen in the Haverford West area but the Gasworks Mudstone is overlain by the Gasworks Sandstone, which indicated shallowing of the area of deposition towards the close of the Lower Llandovery. This was followed by emergence and erosion as the Middle Llandovery beds are absent and the shelly Upper Llandovery follows unconformably.

In the Welsh Borders Silurian sedimentation did not commence until Upper Llandovery times when there was an expansion of the area of sedimentation. Palaeogeographical changes at this time were widespread—even in north Wales they are reflected by the thicker development of Upper Llandovery beds—and in the region of the Towy Anticline by the introduction of conglomerates and sandstones. In the Welsh Borders direct evidence of the shoreline is found, and the presence of deposits of littoral facies resting on older rocks (from Ashgill to Precambrian in age) has been known for over a century. At the south end of the Longmynd details of the shoreline, including sea-stacks and headlands, with Upper Llandovery conglomerates representing old beach deposits, have been traced. The succeeding strata, formed as subsidence continued, are calcareous argillaceous beds with *Pentamerus oblongus* the commonest brachiopod (as it is also at Llandovery in contemporaneous beds). The Hughley Shales follow.

Generally similar beds of this age have been found in the Woolhope Inlier where the Llandovery (Upper only) is flaggy and shaly, in the May Hill Inlier where after a conglomeratic base it is sandy, and in the Malverns. In this last area the upper limy beds pass into the overlying Woolhope Limestone.

The Llandovery beds of the Tortworth Inlier, north of Bristol, are noteworthy since interbedded with fossiliferous sandstone is a lava flow and some of the sediments are ashy. Tuffs have also been noted in the basal beds of littoral facies in Pembroke and these two occurrences, together with pre-Devonian volcanics in the Mendips, seem to be the main indications of vulcanicity in Britain during Silurian times, with the exception of some later activity in south-west Ireland. However, bentonitic clays are known from widely separated areas of N. Wales, the Lake District and Scotland.

The geographical picture in Llandovery times, deduced from the outcrops described above and their faunas, is of an early shoreline to the west of the

Longmynd and Old Radnor running south-westwards towards Carmarthen (Fig. 6.1a). Adjacent to this, thick shallow-water deposits were laid down in the Llandovery area while northwards into mid-Wales greater thicknesses of greywacke and shales were deposited in the foredeep. Still further north, in north Wales, and remote from the shoreline only thin pelagic shales occurred. The widespread changes which marked the beginning of the Upper Llandovery caused the spread of shallow-water deposits over the Welsh Border area and Midlands (as far as May Hill and the Malverns) which had previously been land (Fig. 6.1b).

Still later in Upper Llandovery times the *Lingula* and *Eocelia* littoral faunas of these areas gave way to *Pentameroides*, etc., as the shoreline moved still further east and the Tortworth area was included in the area of deposition. Elsewhere the Upper Llandovery rests unconformably on Middle Llandovery or older beds, for uplift, folding and erosion preceded this widespread inundation. However, in the geosynclinal trough, although much greater thicknesses of greywacke were laid down than earlier in the period, in the north little change in sedimentation is apparent and deposition of graptolitic shales continued.

In the Midlands the Upper Llandovery is found at Rubery, Birmingham (Lickey Hills), where it rests on Cambrian rocks as it does at Malvern and at the Wrekin. At Rubery there is evidence of the proximity of a shoreline since the basal Upper Llandovery is conglomeratic where it fills hollows in the old land surface and neptunean dykes occur where fissures in the underlying Lickey Quartzite (Cambrian age) have been infilled with Llandovery material. The beds are generally of shallow water type: the Rubery Sandstone is succeeded by the Rubery Shales. The base of the Silurian is not exposed in the other Midland inliers and the Llandovery rocks are not exposed, although the Midlands area must be widely floored by Silurian rocks. Evidence of derived material (in Upper Coal Measures conglomerates of Warwickshire) indicates the even further eastwards extent of the Upper Llandovery transgression.

The Wenlockian and Ludlovian (*Salopian*)

The Welsh Geosyncline persisted in Wenlock times and both geosynclinal and shelf-sea facies are known. In fact their distribution is similar to that of the Upper Llandovery and, while shallow-water deposits were laid down in south Shropshire and the Midlands (also in Pembrokeshire), greywackes and pelagic shales were deposited in mid- and north Wales. A great deal of relatively recent work has been done on the limestone and shale shelf-sea facies of Wenlock and Ludlow age in the Welsh Borders. Here, naturally, graptolites are not common and the strata, originally divided up on a lithological basis, have been divided into faunally based divisions (or zones) recognized by their characteristic assemblages of fossils, chiefly brachiopods. Zone fossils have not been designated and the divisions take their names from localities in the Welsh Borders.

In north Wales the Wenlock Series largely consists of arenaceous rocks, the Denbighshire Grits, which attain a thickness of 4,000 feet. Similar beds occur

	Ludlow Zones	Lithological Divisions
LUDLOW SERIES	WHITCLIFFIAN	Whitcliffe Flags
	LEINTWARDINIAN	*Dayia* Shales
	BRINGEWOODIAN	Aymestry Limestone
	ELTONIAN	Lower Ludlow Shales
WENLOCK SERIES		Wenlock Limestone
		Wenlock Shale
		Woolhope Limestone
LLANDOVERY SERIES		Purple Shales or Woolhope Shale = Rubery Shales, Yarleton Beds, etc.
		Llandovery Sandstones

in the Builth area and it seems that this greywacke facies was widespread—although evidence is lacking from the region of the Central Wales Syncline where post-Llandovery sediments have been eroded away. Great variations in thickness accompany lateral changes in facies, and in the extreme north near Conway and eastwards between Corwen and Llangollen chiefly shaly beds occur and they are much thinner, of the order of 300 feet. Palaeocurrent directions in the greywackes indicate direction of transportation of material (Fig. 6.2). South-eastwards in the Welshpool area a transition to shelf-sea depositional conditions took place and calcareous shales with an abundant neritic fauna are found there. By late Wenlock times shaly sedimentation was widespread.

It is in Shropshire that the shelf-sea facies is most characteristically developed, the Silurian strata with an easterly dip make a cuesta topography and the west-facing escarpment of the Wenlock Limestone forms the well known Wenlock Edge. In the Welsh Borders the Woolhope Limestone follows the Llandovery beds both in the Woolhope Inlier and in the Malverns and also at May Hill, but further north calcareous shales, the Buildwas Beds, are contemporaneous. Reef limestones in the Radnor District, where an archipelago existed, may be of this age. A representative of the Woolhope Limestone occurs as far east as Great Barr north of Birmingham, where it occurs in an inlier at the eastern edge of the South Staffordshire Coalfield. There it is known by the local name, the Barr Limestone.

The Wenlock Shales are grey-green, calcareous and highly fossiliferous with brachiopods, trilobites, etc., and they exhibit little variation. They are the oldest strata seen in the inliers in the Dudley area of the South Staffordshire Coalfield where they are followed by the Wenlock Limestone (once known as the Dudley Limestone). Both here and in Wenlock Edge, as well as at Woolhope and May Hill, the Wenlock Limestone has a prolific fauna, and as well as a great variety of trilobites and brachiopods, crinoids, bryozoans, corals, stromatoporoids, gastropods and cephalopods abound. For unknown palaeoecological reasons trilobites are less common in the Wenlock Edge outcrop than elsewhere, for example at Dudley. Reef structures occur widely and interrupt the bedding, suggesting that they were formed by upgrowths of coral and stromatoporoids from the sea bed. They have been called 'ball-stones'. Some lateral variation in the limestone is known, a threefold division into a Lower Bedded Limestone, a middle Nodular Limestone and an Upper Bedded Limestone is especially well seen at Dudley and May Hill (Longhope). Pisolitic and oolitic developments have been recorded at May Hill and in the Malverns outcrop.

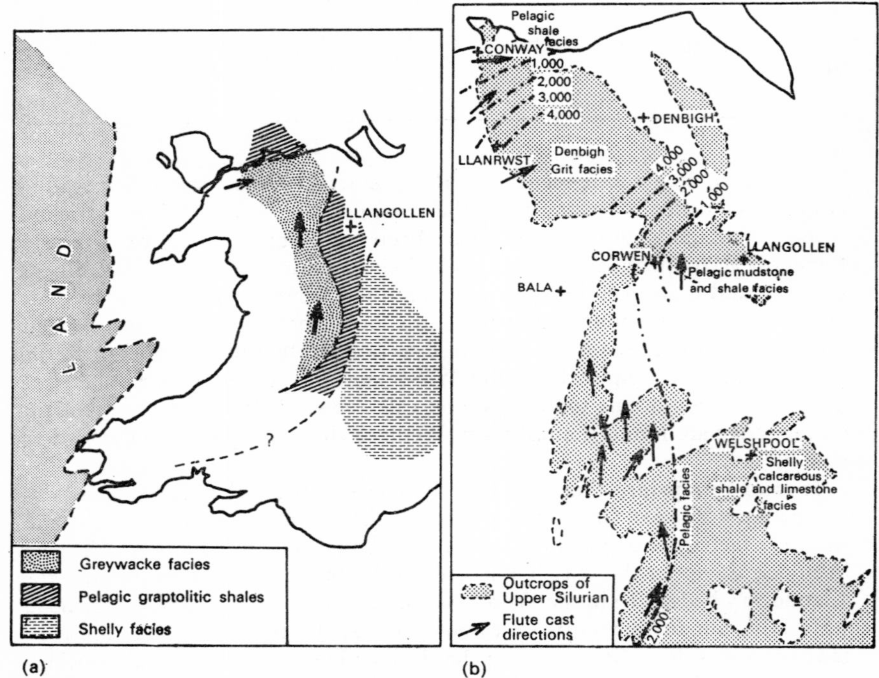

Fig. 6.2 Wenlock palaeogeography of Wales with isopachytes for the Wenlockian and palaeocurrent directions in the Denbigh Grits. (Based on W. A. CUMMINS, 1957, *Geol. Mag.*, **94**, 423.)

The original classification of the Ludlow Series into the Lower Ludlow Shales, Aymestry Limestone and Upper Ludlow Shales, is not sufficiently widely applicable to merit its continuance, and the Aymestry Limestone facies as well as being impersistent may be diachronous (though it corresponds approximately to the Upper Bringewoodian). The well known large pentamerid brachiopod, *Conchidium knightii*, is very restricted in area and is best seen at View Edge where it occurs in shell banks. The shelf-sea facies without typical development of the Aymestry Limestone facies was first described by Lawson from the May Hill Inlier, and the stages or zones of the Ludlovian recognized there, based chiefly on brachiopod faunas, are widely agreed in the Welsh Borders.

In its typical development the Ludlow Series is about 1,000 feet thick.

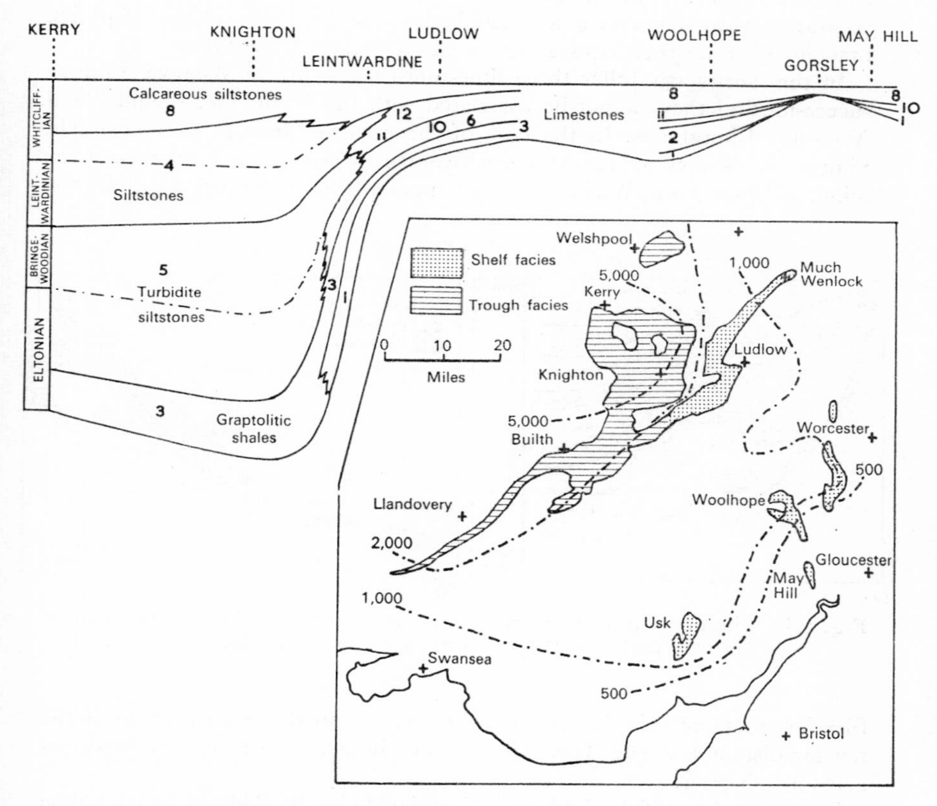

Fig. 6.3 Facies variations in the Ludlovian of the Welsh Borders. Figures for facies as in the original: 1. shelly olive mudstone; 2. muddy siltstone; 3 graptolitic shale; 4. laminated siltstone; 5. turbidite siltstone; 6. *tumescens* siltstone; 8. calcareous shelly siltstone; 10. strophomenid siltstone; 11. massive limestone; 12. flaggy limestone. (Simplified from C. H. HOLLAND and J. D. LAWSON, 1963, *Geol. J.*, **3**, 269, Fig. 16.)

The Lower Ludlow, succeeding the Wenlock Limestone, consists of shales with a fauna of graptolites and cephalopods, though further south graptolites are rare and a stunted shelly fauna is characteristic. Calcareous siltstones, with a brachiopod fauna, grade upwards into Aymestry Limestone in which corals and stromatoporoids are present. Above this are flaggy calcareous siltstones.

Southwards and westwards the strata of the basin facies are thicker, up to 5,000 feet or more, and show great variation (Figs. 6.3, 6.4). Partly comprising dark graptolitic shales, grits similar to those of the Denbigh Moors occur, while in south-central Wales (Llandovery area) red sandstones and quartzites, the Trichrug Beds (a local deltaic development) occur (Fig. 6.4): a landmass to the west in the Llandeilo area has been postulated. While *Monograptus scanicus* suggests a low Ludlovian age for these deltaic deposits, brachiopods infer their equivalence to the Aymestry Limestone.

In the Tortworth Inlier three limestones occur in the Wenlock-Ludlow succession but they cannot be correlated with the Woolhope, Wenlock and Aymestry Limestones. In the core of the eastern Mendips periclinal fold Silurian mudstones and sandstones with andesitic lava and tuff outcrop. Other minor inliers occur at Wickwar and Sharpness. It has been suggested that the

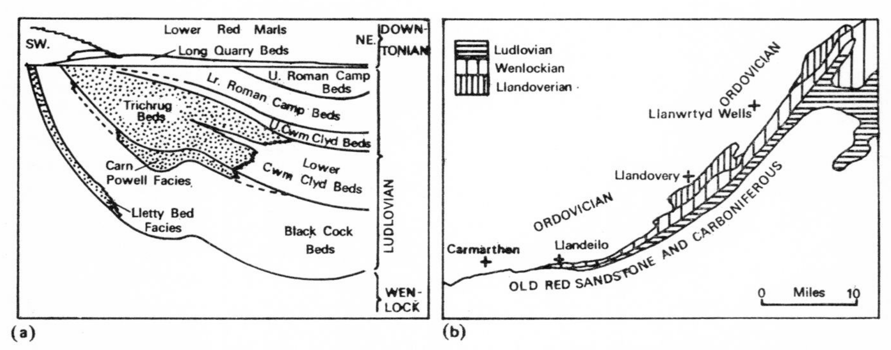

Fig. 6.4 The development of arenaceous facies in the Ludlovian. (From J. F. POTTER and J. H. PRICE, 1965, *Proc. geol. Ass.*, **76,** 379.)

Silurian strata of this Somerset-Gloucester region differ from those of the not far distant Usk and May Hill outcrops because the Lower Severn Axis acted as a barrier.

The upper beds of the Ludlow Series, the Whitcliffe Beds, are fossiliferous green calcareous flags in the Welsh Borders. The uppermost beds include thin organic conglomeratic bands, one of which is called the Ludlow Bone Bed. This has yielded many fish fragments of the species *Onchus murchisonii* and *Thelodus parvidens* as well as ostracods and *Pachytheca*, possibly an alga. A readily recognizable horizon, the Ludlow Bone Bed, has often been taken as the Siluro-Devonian boundary. It is, however, of only local significance and cannot be traced even along the outcrops north of the South Wales Coalfield. The now widely recognized top of the Silurian succession, the base of the *Monograptus uniformis* Zone of Europe, is of course not discernible in the strata of continental facies which follow the Ludlow Bone Bed in Shropshire. It may be near in horizon to the Dittonian–Downtown boundary here marked by only slight lithological change, so that it is logical to include the Downtonian Beds in the Silurian, although in facies the Downtonian closely resembles the overlying Old Red Sandstone of the Devonian System. The base of the Devonian is here defined, therefore, as the beginning of the Dittonian, which was marked by the appearance of the fossil *Pteraspis.*

The Downtonian

Following the Ludlow Bone Bed are the beds of the Grey Downtonian or Temeside Group which in Shropshire consists of the Downton Castle Sandstone, 40 feet thick, overlain by the Temeside Shales, over 100 feet. These are overlain by up to 1,000 feet of purple marls and red sandstones known as the Red Downtonian or Ledbury Beds. In the Shropshire–Hereford area there is, therefore, a transition from the Silurian to the Devonian. North-westward the former extent of the Downtonian is unknown, but in Anglesey Old Red Sandstone conglomerates rest unconformably on older rocks and some 4,000 feet of strata have been removed by erosion, so that a long period of pre-Devonian erosion is postulated to explain this.

Southwards from Shropshire lateral changes are seen in the Downtonian, and the Downtonian Castle Sandstone becomes flaggy and is known as the Tilestones. It has a scanty molluscan fauna and the bone beds found in Shropshire are missing. Accompanying these changes in fauna and lithology, the Downtonian is found to overstep on to successively lower beds and its base in south Wales is an unconformity. This has been another reason for including it in the Devonian. The Caledonian structures of Wales (discussed in the next chapter) were probably initiated as early as Llandovery times as we have noted that one, the Towy Anticline, separates different facies. Further uplift of this area is inferred from the Downtonian unconformity. The Tilestones are overlapped by the Red Downtonian, known here as the Raglan Marls. These have a sparse fauna but fish fossils occur at some horizons such as the *Psammosteus* Limestone with the genus *Phialaspis.* The red marls are widespread but probably thin south-eastward towards the Lower Severn axis, for in the E. Mendips pericline Upper Old Red Sandstone is found resting directly upon Wenlock age mudstones. Eastwards into the Midlands of England, while there are no outcrops of Devonian (but it is known from boreholes), the Downtonian occurs in the Netherton Anticline of the South Staffordshire coalfield, and at Gornal on the Western margin, where the succession closely resembles that of Shropshire.

The red marls of the Downtonian and those of the overlying Dittonian of Devonian age are similar, often indistinguishable. There is, then, an almost perfect transition in the Welsh Borders from the Ludlovian into the Dittonian reflecting a gradually changing environment from marine to non-marine. Firstly, in the uppermost Ludlovian we note the absence of graptolites, then the impoverishment of the fauna so that, of the neritic marine forms, only molluscs persist. Then non-marine conditions set in with the deposition of marls (in sheets of shallow water) with calcareous bands and sandstones. Fish beds occur at many horizons, the Trimpley as much as 1,400 feet above the Ludlow Bone Bed.

The Silurian rocks of Northern England

In the Lake District and in the Cross Fell Inlier the Silurian rocks follow the Ashgill Series of the Upper Ordovician with no break in sedimentation. Nevertheless, there must have been a marked change in environmental conditions since the Llandovery age strata (Silurian)—here comprising the Stockdale Shales—have a graptolitic fauna which contrasts with the brachiopod-trilobite fauna of the Ashgill Beds. The Stockdale Shales are divisible into the Skelgill Beds, black graptolitic shales representing slow accumulation since their 50 feet represents 7 graptolite zones, followed by the Browgill Shales, grey-green and black shales 200 feet thick corresponding to only 3 graptolite zones. Rarely, trilobites are found in the Browgill Beds.

Following the generally slow accumulation of the Llandovery strata a great change in environment took place. The Wenlock and Ludlow Series are here grits, flags and greywackes, sparsely fossiliferous and attaining a combined thickness of perhaps 13,000 feet. The succession given by Marr and adapted by the Geological Survey is:

LUDLOW SERIES	Kirkby Moor Flags	*c.* 1,500 ft
	Bannisdale Slates	over 5,000 ft
	Coniston Grits	4,000 ft
	Upper Coldwell Beds Middle Coldwell Beds Lower Coldwell Beds	+ 1,500 ft
WENLOCK SERIES	Brathay Flags	1,000 ft
LLANDOVERY SERIES	Stockdale Shales	250 ft

The Brathay Flags are shales and flags with graptolites, deposited, it is presumed, in shallower water than the preceding Stockdale Shales. The Coldwell Beds are grits, siltstones and flags and, except for the lower division, contain a shelly fauna with few graptolites. The Coniston Grits are greywacke grits and fossils are sparse. Upwards alternations of mudstone and siltstone, often showing graded bedding and other interesting sedimentary structures, form a passage to the Bannisdale Slates. These upper beds of the Coniston Grits may have been deposited relatively rapidly: Lapworth compared them with flysch. It has been suggested that during the later Ludlovian the area of deposition was becoming generally shallower but minor oscillations caused temporary periods of deepening. The Downtonian is absent.

As well as the main outcrop of Silurian rocks in the southern Lake District there are, in northern England, several inliers of Lower Palaeozoic rocks, in the Crummackdale, Horton and Malham areas north of the Craven Faults. These inliers include Ordovician rocks (though only the Ashgill Series is seen and it may rest unconformably upon Ingletonian rocks) and rocks of Silurian age. The Ashgillian consists of calcareous shales with trilobites and some graptolites, and is notable for the presence of tuffs and felsites which are approximately contemporaneous with the last evidence of vulcanicity in the Lake District. The Llandovery is shaly and was part of the Stockdale Shales area of sedimentation although thin trilobite-bearing calcareous beds are found. The Wenlock and Ludlow Series comprises flags and grits of generally similar lithology to the contemporaneous strata of the southern Lake District—though the sequence is not the same in detail, and local names are given to the formations found here. The Wenlockian equivalent of the Brathay Flags or Lower Coniston Flags is here known as the Austwick Grits and Flags. The Ludlow age rocks are called the Horton Flags (equivalent to the Coldwell Beds) and the Studfold Sandstone (the equivalent of the Coniston Grits). Higher Silurian beds are not found.

The Silurian rocks of the south of Scotland

The Silurian rocks of the Girvan area are of the same facies as the Ordovician, shallow-water sandstones, conglomerates and mudstones with chiefly neritic faunas. Initially in the Central Belt of the Southern Uplands the conditions which obtained there in Bala times continued, and the lower part of the Llandovery Series comprises a thin group of dark graptolitic shales, the Birkhill Shales. Following the Hartfell Shales of similar lithology no break in sedimentation occurred, although there is the usual major palaeontological break.

Following the deposition of the graptolitic shales, a major change in the kind of deposition took place within Upper Llandovery times. A great thickness of greywackes, the Gala Group, were deposited in the central Southern Uplands and meanwhile, in the Girvan area, sandstones, flagstones and mudstones were laid down with—in some cases—evidence of deposition from turbidity currents. Furthermore, the thicknesses of the sediments of the two areas are of the same order and it appears that by Upper Llandovery times widely uniform depositional conditions had become established. By this time the direction of transport of material seems to have been generally from north-east to south-west, i.e. axial flow occurred in the broad geosynclinal trough (Fig. 6.5).

In the Southern Belt of the Southern Uplands, where predominantly strata of Wenlock age occur, work in the last decade or more has thrown doubt on the age relationship of lithological groups, the Riccarton and Hawick Groups. The former, although chiefly comprising greywackes, includes graptolitic shales which confirm its age as Wenlockian. The Hawick Group was originally believed to lie below the Riccarton Beds and was therefore assigned to the Upper Llandovery Period. Evidence in part of Wigtownshire which suggests that the strata are there inverted, implying that the Hawick rocks are younger than the Riccarton, would necessitate the Hawick Group being also placed in the Wenlock Series. Unfortunately, these unfossiliferous greywackes provide no faunal evidence of age. It has also been suggested from evidence elsewhere that the Hawick rocks are an unfossiliferous facies of the Riccarton Group, while in the extreme west of the Southern Uplands the relationship between the two groups appears faulted and their age relationship therefore indeterminate (Table 6.2). The revision of the succession of these Wenlock strata has also led to a revised idea of the structure in the south of Scotland. Furthermore, the thickness of the Wenlockian strata may amount to about 25,000 feet, or more than ten times earlier estimates. It is interesting to note that this is in one sense a reversion to the pre-Lapworth idea that the Southern Uplands comprised an enormous thickness of strata. Discussion of the Southern Uplands structures is deferred until the next chapter.

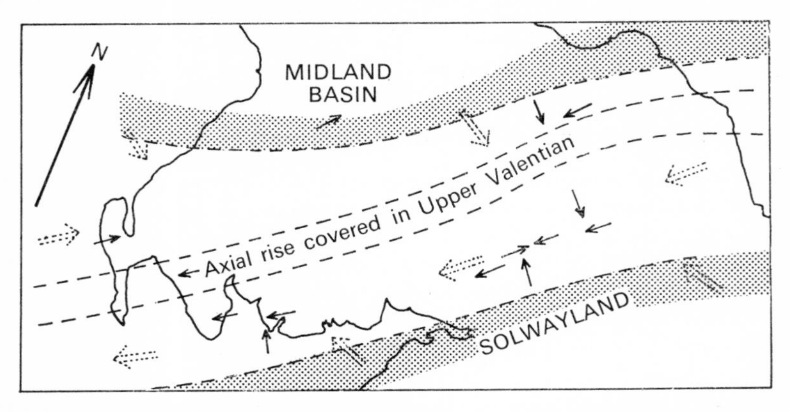

Fig. 6.5 Valentian and Lower Wenlockian palaeogeography in the Southern Uplands region. Solid arrows based on measurements, dotted arrows are inferred current directions. (From E. K. WALTON, 1965, in *Geology of Scotland*, Fig. 6.1b, ed. G. Y. CRAIG. Oliver & Boyd, Edinburgh and London.)

TABLE 6.2

	Girvan	*Moffat*	*Kirkcudbright*
WENLOCK	Dailly Group		? Hawick Rocks Riccarton and Raebury Castle Beds
LLANDOVERY	Newlands Group	Gala Group Birkhill Shales	? Hawick Rocks

In the Girvan area the rocks of the Newlands Group are conglomerates, flags, shales and limestones of generally similar lithology to the Upper Ordovician rocks of the district. They are restricted in outcrop to a narrow belt running inland, a small coastal outcrop at Newlands Point and outcrops in the Craighead Inlier. The fauna is neritic and abundant in some of the sandstones but dark shales, indicating occasional extensions of the deeper water conditions (or recession of the shoreline), permit correlation with the Birkhill Shales of the inliers of the Southern Uplands Central Belt. The Newlands Group is, of course, at about 1,000 feet thick, much thicker than the contemporary graptolitic shales.

The overlying Dailly Group occurs only in a narrow outcrop on the southern side of Girvan Water. Comprising sandstones and shales, the latter includes graptolites which establish the age of the group as Wenlockian. The sandstones are in some cases greywackes and resemble the Gala greywackes—to which they are in part equivalent in age. The Wenlock Series has been divided in the Girvan area into the formations given in Table 6.3, which are of rather local development.

TABLE 6.3

WENLOCKIAN	DAILLY GROUP	Drumyork Flagstones, etc. ≡ Straiton Formation Bargany Formation (flags)
VALENTIAN ≡ LLANDOVERY	NEWLANDS GROUP	Penkill Formation (mudstones) Camregan Formation (sandstones and mudstones) *Monograptus sedgwicki* Shales Sauch Hill Formation (sandstones and conglomerates) Mulloch Hill Formation (sandstones and mudstones)[a]

[a] The lower part of this succession is seen only in the Craighead Inlier.

Where younger rocks overlie the Silurian strata there is strong unconformity. In the south Carboniferous beds rest on rocks now believed to be Wenlock age or possibly Ludlow age. In the Girvan area Old Red Sandstone rests unconformably on the Wenlock age Dailly Group. Clearly, in post-Wenlock times there must have been uplift and erosion to account for the absence of much of the Ludlovian (including the Downtonian) from the south of Scotland.

The Lower Palaeozoic rocks of the Scottish Midland Valley

There are a number of inliers of Lower Palaeozoic rocks along the south side of the Midland Valley of Scotland. Excluding the Girvan area which, lying to the north of the Southern Uplands Boundary Fault, might be considered to be one, the largest is that at Lesmahagow. Other inliers form the Hagshaw Hills and the Pentland Hills and there are small, less well-known outcrops (Fig. 6.6).

On the northern margin of the Midland Valley there are a number of 'wedges' of Lower Palaeozoic rocks known as the Highland Border Series. They are generally bounded by faults. These outcrops lie in line between Stonehaven on the east coast of Scotland and Arran and Bute in the west. Only in Arran can their relationship to older rocks be seen; there they rest unconformably upon the Dalradian Series but are also deformed by Caledonian orogenic movements. Of undoubted Ordovician age they comprise a lower series of black shales and cherts with spilitic lavas which, partly by analogy with the Ballantrae rocks, have been assigned to the Arenig, although re-examination of the fossils from the Aberfoyle outcrop by the Geological Survey in 1962 led to the view that the age could not be given with greater precision than merely Ordovician. Unconformably above these beds occur breccias, grits and limestone (not seen in the Stonehaven section) which were presumed to be post-Arenig and, since elsewhere in Scotland sedimentation did not recommence until Caradoc times, they have been considered to be possibly of Caradoc age. The succession in Arran, belonging entirely to the lower group, is believed to attain a thickness of 1,000 feet.

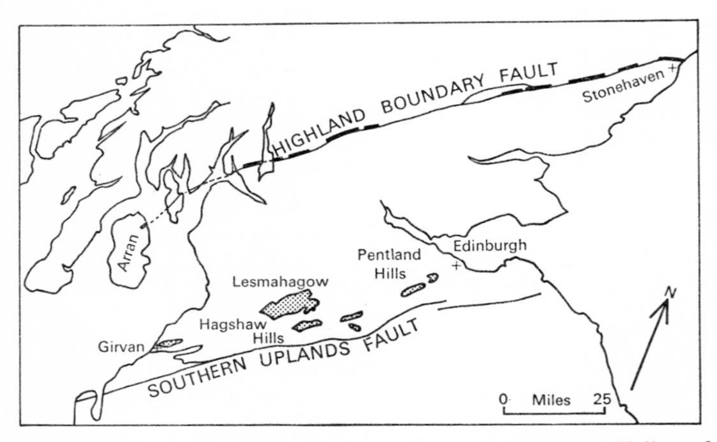

Fig. 6.6 Outcrops of Lower Palaeozoic rocks in the Midland Valley of Scotland. Solid black, the Highland Border Series of ?Arenig age; shaded outcrops, chiefly Silurian.

The sediments of the Highland Border Series are quite unrelated to the Dalradian sedimentation—which had probably ceased by the time of their deposition—but no doubt were laid down contemporaneously with Ordovician rocks in the south of Scotland some 50 miles away. Possibly the lower series of the Highland Border Series are of Arenig age and were contiguous with those of the Girvan area and the whole Midland Valley may have been an area of deposition. However, later, in Silurian times, there is evidence that the area to the south of the Midland Valley became elevated as a geanticline (it has been called Cockburnland). Late Silurian sediments of Downtonian age are present in the east of Scotland north of Stonehaven, where they are unconformable on the Highland Border Series, and comprise over 2,000 feet of sandstones. Some might be more aptly called siltstones, and beds of tuff are also found. Fish horizons and a bed with chiefly *Dictyocaris slimoni*, but from which 8 other genera are known, occur at Cowie Harbour on the Stonehaven Shore.

Downtonian beds were formerly mapped in the Pentlands, Hagshaw and Lesmahagow Inliers but it is now believed that the rocks are older, ranging from Llandovery to Ludlovian (but pre-Downtonian). The rocks of these inliers are greywackes, shales and mudstones and impure limestones, with brachiopods and molluscs in the greywackes, and fish and eurypterids in the siltstones. The latter were probably laid down in lakes or enclosed lagoons. Upwards, sandstones, often red, with suncracks providing evidence of desiccation, follow. They were laid down in shallow water for fish are the chief element of the fauna. A general decrease in the amount of boron up the succession further indicates the gradual transition from marine to non-marine conditions. No such transition from turbidites to freshwater deposits is known from the Silurian of the Southern Uplands. Thus in late Silurian times, while the Midland Valley was a basin of sedimentation, the northern part of the Southern Uplands may already have been emergent (Cockburnland).

The Lower Palaeozoic rocks of Ireland

Ordovician and Silurian rocks outcrop in five main areas in Ireland: (1) in the east, south of Belfast, there is what might be considered to be an extension of the Southern Uplands of Scotland, the Longford-Down Massif. Here is found a northern belt of Ordovician rocks and a broader southern belt of Silurian, both with a Caledonian trend as in the south of Scotland, and bounded chiefly by Carboniferous rocks; (2) the south-eastern outcrops extending from Dublin to Waterford are associated with presumed Cambrian rocks in the Wexford area; (3) south-western County Mayo, recently described in detail by Dewey; (4) a series of inliers in south-central Ireland, where Silurian (and Devonian) strata protrude through the Carboniferous strata of the central plain; and (5) inliers of the 'Midland Valley', chiefly the Pomeroy district.

In the Longford-Down Massif Ordovician and Silurian rocks are closely similar to those of the Scottish Southern Uplands. The rocks are graptolitic shales and greywackes chiefly, but Ordovician rocks of neritic facies also occur —with faunal similarities to those of the Girvan area—in inliers in the southern part of the Silurian belt. The graptolites indicate strata ranging in age from the Caradoc (Glenkiln) age, through the Llandovery (Birkhill and Gala). As in Scotland, graptolites are not common in the Gala greywackes. The sequence in the south of this area appears to more closely resemble the Browgill Beds (Llandovery) of the Lake District than it does the Scottish rocks.

In south-east Ireland lowest Ordovician rocks with early graptolites rest unconformably on the Cambrian age Clara Series of the Bray Anticline. Sedimentation perhaps commenced as early as Tremadoc times but did not continue uninterrupted since Llanvirn age beds are missing. Volcanic rocks attain a great thickness and are chiefly of Bala age. Late Ordovician beds are neritic and include limestones which include faunas like those of Girvan, Scandinavia and Newfoundland. The Lower Palaeozoics of south-eastern

Ireland do not closely resemble those across the Irish Sea in Wales, either in lithology or succession or fauna, and this confirms the presence of a landmass in the region of the present-day southern Irish Sea.

In view of the incomplete information of the Lower Palaeozoic rocks of the Midland Valley of Scotland and the speculation on the nature of the pre-Ordovician floor (and a similar case in the Longford-Down area), the Murrisk, Co. Mayo, area of Ireland is of particular interest. There Ordovician strata rest on Connemara Schists. The latter, assigned to the Dalradian, support the view that Cambrian age sedimentation of Dalradian facies spread southwards beyond the line of the Highland Boundary Fault. There followed a period of uplift and erosion since Arenig Beds follow unconformably—as in Arran. During the interval in sedimentation, deformation and metamorphism of the Connemara Schists took place. The Arenig beds are much thicker than in the Girvan area of Scotland (where pre-Bala erosion may have removed a great thickness) and they consist of spilitic lavas, the Tyrone Volcanic Series, grits and slates. Further grits of Ordovician age were deposited and, later still, post-Caradoc folding affected this area. The Lower and Middle Llandovery are missing in this area but in Upper Llandovery times seas spread again over western Ireland, depositing a thin representative of Upper Llandovery beds (on an irregular floor) which was overlapped by Wenlock rocks of much greater thickness. The changes in palaeogeography in the Murrisk area have been worked out in detail but cannot yet be related to changes in the south of Scotland, since the intervening Longford-Down area is inadequately understood.

Cornwall and north-west Europe

Outcrops of rocks believed to be of Ordovician and Silurian age have long been known from the area immediately to the north of the Lizard Complex, described in Chapter 3. These Lower Palaeozoic rocks occur as discontinuous outcrops or lenses in what has been termed the Meneage Crush Zone, an area of brecciated rocks, chiefly shales of Upper Devonian age. Included lenses may be up to 1 mile in length.

Quartzites have yielded a trilobite-brachiopod fauna including species of '*Calymene*' and '*Phacops*' as well as orthid brachiopods. Their age has in the past been tentatively placed as Llanvirn or Llandeilo, but there is no real confirmation that the beds are necessarily Ordovician in age. Dark blue limestones present contain fossils probably of Wenlock age. The '*Orthoceras*' limestones contain orthocone nautiloids such as *Actinoceras* and *Barrandeoceras*. Brachiopods and bivalvia (lamellibranchs) have also been found.

It has not been possible to correlate these lower Palaeozoic rocks with those to the north in Wales or the Welsh Borders but they are said to resemble the Silurian of Brittany (more precisely the Grès de May). Their presence contributes little to our palaeogeographical knowledge of Silurian times since they form part of what may be a northward thrust mass—although the thrust nature of the northern junction of the Lizard area has recently been disputed.

Lower Palaeozoic rocks are found in infolded inliers in north-west France, chiefly in Brittany, and in the Ardennes of south Belgium. In Brittany they occur preserved in synclines in a group of metamorphosed rocks, the Brioverian, regarded as Precambrian in age. The strata are difficult to interpret, due to the effects of subsequent deformation, and their relationship to the Lower Palaeozoic rocks of British outcrops speculative. Knowledge of the intervening area is derived from a very few boreholes which have penetrated to the Lower Palaeozoic rocks under London and Kent, proving the presence of Ordovician, Llandovery and Wenlockian strata with both graptolites and neritic fossils.

That Bala and Silurian rocks exist in NW France is undisputed and they help to complete the palaeogeographical maps of these periods. The presence of definite Cambrian is unproved, but reddish conglomerates and sandstones succeed the Brioverian though they have yielded few fossils. If they are indeed Cambrian, being of continental facies they differ markedly from the geosynclinal strata laid down to the north in Britain and Scandinavia, as well as to the south in the Mediterranean region. Ordovician and Silurian seas were widespread in northern and western Europe and their extent and palaeogeography can be deduced from the occurrences of rocks of these periods in Scandinavia, in the horst inliers already mentioned and in Bohemia, Thuringia, etc. Wide areas of graptolitic shales and greywacke sedimentation occurred but near the margins of the Baltic Shield shallow-water sediments were deposited, analogous to the marginal facies of south Shropshire and the English Midlands, which were similarly laid down on the margins of the (Midlands) kratogenic block and by shallow seas spreading across it.

Silurian fossils

As we have seen, neritic and graptolitic facies with distinctive assemblages of fossils occurred in the Silurian as in the Ordovician. 'Mixed' facies, really alternations of the two facies, are common. While brachiopods and trilobites were very common, although trilobites had begun to decline, corals, stromatoporoids and bryozoans became abundant, especially in calcareous sediments, and they locally formed reefs. The early vertebrates, especially fish which are found in the later Ludlovian beds, form an important part of the palaeontological record in Silurian times. Elsewhere in the world the first land plants made their appearance in the Silurian.

Many Ordovician forms of brachiopods continued into the Silurian, Orthids and Strophomenids being abundant, but the fauna rapidly became dominated by the rapid rise of the Pentamerids to their acme. Rhynchonellids and spire-bearers such as *Atrypa*, *Meristina* and *Spirifer* became common.

The decline in trilobites was not spectacular and most Ordovician families, except for Trinucleids and Asaphids which died out before the Silurian, continued. Illaenids and Encrinurids were common in the Silurian but did not survive beyond this period. Some bizarre spinous forms, like *Lichas* and *Acidaspis*, and forms losing their trilobation, such as *Homalonotus* [*Trimerus*], occurred. Other arthropods, the ostracods and in late Silurian times the eurypterids (some attaining very large size), became important.

Among echinoderms, Cystids attained a high state of organization, for example *Lepadocrinus*. Crinoids are important as rock-forming organisms in the Wenlockian. *Bothriocidaris*, once considered to be a primitive echinoid, is found and in some beds asterozoans are relatively common.

Many tabulate corals, all of which are of course compound, such as *Heliolites*, *Favosites* and *Halysites* are found with rugose corals of many genera. Two well known genera are *Omphyma* and *Goniophyllum*. Molluscs were common since gastropods were more numerous and varied than formerly, although chiefly Euomphalid and Pleurotomarid forms, and lamellibranchs had become an important element of the fauna—though with little stratigraphical value. Cephalopods are numerous at some horizons. Neritic faunas were, therefore, richer and more varied than those of the Ordovician.

Graptolites are abundant in rocks of shaly facies. In the Llandovery Diplograptids, *Diplograptus*, *Orthograptus* and *Cephalograptus*, are common, but in essence the Silurian is the age of *Monograptus*. This polyphyletic genus exhibits many diverse thecal forms: species with complex thecae arose and died out while some with simple thecae continue through the Silurian Period and in the Ludlow are dominant (Table 6.4). The later Silurian rocks in Britain are devoid of graptolites but this is because suitable marine facies did not exist. Elsewhere, for example in Germany and Poland, graptolites persisted throughout the Silurian and into early Devonian times.

TABLE 6.4

LUDLOW SERIES	Dominance of *Monograptus* with simple thecae	simple thecae
WENLOCK SERIES	Acme of hooked thecae in *Monograptus*. Acme of *Cyrtograptus*	hooked thecae
UPPER LLANDOVERY	Rise of *Monograptus* with hooked thecae. Acme of *Monograptus* with isolate thecae. Disappearance of forms with lobate thecae	isolate thecae
LOWER AND MIDDLE LLANDOVERY	*Monograptus* with lobate and isolate thecae and with simple thecae. *Diplograptus* and *Climacograptus*	lobate thecae

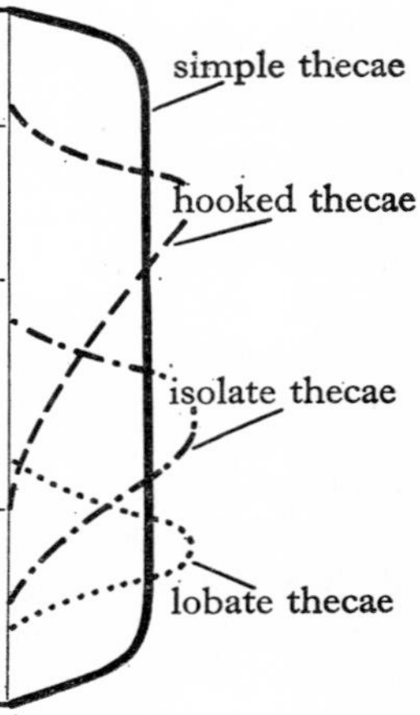

7 The Caledonian Orogeny

The Caledonian Orogeny was as cyclic in nature as each of the Precambrian orogenies discussed in Chapter 3, in that it included geosynclinal, orogenic (i.e. tectonic and metamorphic) and epeirogenic phases. The cycle probably started some time during the late Precambrian and ended during Devonian times, by which time the geosynclinal phase of the Variscan cycle was well developed. Within this long period of time (*c.* 300 m.y.) a general progression from early orogenic, through synorogenic to late-orogenic events is clearly discernible. However, in this, one of the most closely studied of all orogenic belts, no definite conclusions can yet be drawn about the overall sequence of events which formed the Caledonian mountain chain.

It now seems probable that, in Britain, the sedimentation in the geosyncline took place in several separate basins, which were not always elongate in form, and that different areas received their maximum sedimentation at different times. The Scottish Highlands–Donegal area was the chief site of sedimentation in Eocambrian and early Cambrian times, while in Cambrian to Silurian times there were separate basins of sedimentation in western Ireland, in Wales, in south-east Ireland, in the Lake District–Arklow area and in the Southern Uplands of Scotland. Within each of these areas, sedimentation rates varied from place to place throughout this time. The rate of subsidence, therefore, and the earth movements controlling this, must also have been sporadic in both space and time, and the chief interest of this chapter will be the investigation of the periodic nature of the earth movements and metamorphism, during and subsequent to the infilling of the sedimentary basins. Although the metamorphic rocks and slates have been investigated by isotopic age dating, no clear and consistent picture can be made of the results at the present time. It is not certain, for example, that the major fold and metamorphic episodes are even approximately synchronous over the whole of the British Caledonides and, considering the periodic nature of many other aspects of the cycle, it might be suggested that synchronous deformation of the whole region is improbable. The isotopic data neither prove nor disprove this hypothesis. Major periods of orogenic activity are not, in fact, easily dated. Orogenic folding may take place either deep beneath the earth's surface or, when the folding is due to gravity gliding, near or above sea level. In either case the rocks affected by the folding will be older than the stratigraphic age of the event, perhaps considerably so, and the rocks being laid down elsewhere during this time may show no evidence of the diastrophism. Fold movements are thus only datable from the age of the rocks which are affected by the folding. The younger limit may sometimes be found by overlying unconformable strata, or by the dating of undeformed intrusions cutting the folded rocks. Isotopic dates of metamorphic rocks are less reliable due to possibly unrecognized effects such as subsequent heating or a subsequent deformation as well as the fact that an area may take a considerable time after the peak of the 'event' to fall to a temperature at which diffusion of the isotopes ceases.

The relation between the earth movements controlling sedimentation and periods of intense deformation are not clear. In this respect, it is apparent that structural geologists and stratigraphers, by applying different criteria to the study of deformation during the Caledonian, have arrived at quite different views of the intensity of the various earth movements. Movements in the crust during sedimentation can result in the formation of large-scale depressions and areas of uplift, the depressions becoming basins of deposition while the uplifted areas are eroded. Such structures have wavelengths of many miles and amplitudes of only a few thousand feet, and are normally deduced from stratigraphic considerations. They might not be recognizable at all after periods of folding which a structural geologist would call intense. Rarely, the dips on the limbs of these folds are steep enough to be measurable, but correlation of this type of structure with folds produced by penetrative deformation and recrystallization elsewhere in the orogenic belt is not possible at the present time. It is probably best to consider the earth movements which control sedimentation and the intense fold and metamorphic episodes characteristic of the synorogenic episode, as quite separate manifestations of orogenic activity. The correlation of such phases with episodes of synorogenic deformation in the Caledonian chain in northern Europe or New England is obviously highly speculative, although it is often attempted.

The Moine area

The Moinian rocks have been the subject of very detailed structural studies in certain small areas. These have become classic areas for the study of multiple periods of deformation of rocks deep in the crust, and both on a large and a small scale the typical interference patterns of refolded fold structures can be seen (Fig. 7.3 f–j). The north-western boundary of the Moinian is the Moine Thrust Zone, a classic area where thrusting and the rock-type mylonite were both first recognized in Scotland. The relation between the deformation of the Lewisian basement and the overlying Moinian rocks in both marginal and central areas is also clearly displayed.

TABLE 7.1 Structural Sequences in the Moine Area

	Moine Thrust Zone, Lower Nappes	*Moine Thrust Area above Moine Thrust*	*Northern Sutherland and Central Ross*
	Unfoliated alkali pegmatites		Late minettes (372 m.y.) Middle Old Red Sandstone Struie Hill Thrust Lower Old Red Sandstone
F_5	Moine Thrust and conjugate folds	Moine Thrust and conjugate folds	Conjugate folds
	Sole, Glen Coul and Ben More Thrusts		
			Ben Loyal Syenite Lamprophyres
	Assynt Alkaline suite (and Volcanic Cover) (Borolan, Loch Ailsh, Canisp Porphyry, Lamprophyres) (403 m.y.)		Newer Granites (401 m.y.) Appinite Suite (and (?) Volcanic Cover)
	Close of regional metamorphism (415–420 m.y.)		
			Rogart Granite (424 m.y.) (deep-seated Newer Granite)
F_4	Phases of small-scale folding and static porphyroblasts	Large E—W[a] folds (Bac Gobhar). Phases of small-scale folding and static porphyroblasts	
F_3	N—S[a] folds (Tarskavaig, Caradal, (?) Lochalsh)	N—S and NNE[a] folds (Knock, Morar, Loch Hourn, Ben Mhialairidh, Sgurr Mor)	NNE[a] folds
	Tarskavaig, Balmacara and Sgurr Beag Thrusts. Kishorn and Kinlochewe Thrusts		
F_2	ESE[a] isoclinal folds with lineations	Large tight folds (Ben a' Chapuill, Glen Beag, Tarbet)	Large tight folds (Loch Monar, Sgurr na Fearstaig, Glen Orrin, Tarvie)
F_1	Mylonites in axial planes of large isoclines	Mylonites in axial planes of large isoclines, infolding of Lewisian into Moinian	Large isoclinal folds with inthrusting of Lewisian into Moinian

[a] The direction quoted for the various fold phases, throughout this chapter, is the strike of the fold axial plane.

N.B. The successions of events are not closely correlatable nor are many of the positions known with confidence. This table must be regarded, therefore, as a tentative summary of events. (Successions after R. F. CHEENEY and D. H. MATTHEWS, 1965, *Scott. J. Geol.*, **1**, 256; M. R. W. JOHNSON, 1965, in *The Geology of Scotland*, ed. G. Y. CRAIG. Oliver & Boyd, Edinburgh and London; A. B. POOLE, 1966, *Scott. J. Geol.*, **2**, 38; J. G. RAMSAY, 1963, in *The British Caledonides*, eds. M. R. W. JOHNSON and F. H. STEWART. Oliver & Boyd, Edinburgh and London; and N. J. SOPER and P. E. BROWN, 1965, *Geol. Mag.*, **102**, 285.)

The Moinian rocks have been subjected to three major periods of deformation in addition to a late stage phase which may take a variety of forms (Table 7.1). From the intensity of deformation seen in individual localities, it might be imagined that by the end of four periods of folding and recrystallization there would be little evidence of the earliest structures. However, it is fortunate that the deformation during any one phase takes place in restricted areas and varies in intensity, so that in some places only one phase of folding may be recognizable. The metamorphic recrystallization which takes place during the deformation is thus not regional, the areas of intense deformation being the areas of easiest, and therefore of most complete, recrystallization during that period of folding. It may therefore be possible to obtain the isotopic ages of the different phases since the later phases do not everywhere destroy the earlier minerals. (The effect of any later regional heating has, however, to be taken into account in the interpretation of the dates of these earlier phases.)

CENTRAL MOINE AREA

In the central part of the Moine outcrop the second and third folds are the most intense and the interference of these two sets of structures gives rise to the patterns typical of superimposed folds (Fig. 7.3, f–j). In these areas, the first period of folding can only be recognized after detailed mapping of large areas has enabled a reliable stratigraphy to be worked out. Where such work has been done, the Lewisian inliers are seen to form the core of large tight folds (as at Glenelg) or may be thrust into the metasedimentary sequence as huge slices (as with the Scardroy inlier) (Figs. 7.1, 7.2). In the west, the third phase is often absent and the fourth phase was not sufficiently penetrative to destroy the first two phases, so the earlier structures are more readily resolved.

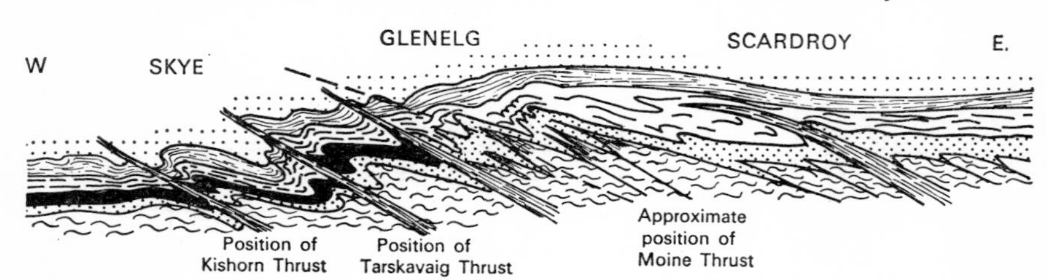

Fig. 7.1 Diagrammatic section through the Moinian area after the first fold movements. The basement has been variously thrust into the cover as a wedge (e.g. Scardroy), folded into tight folds with the cover (e.g. Glenelg) and mylonized in the fold cores (e.g. Skye). The position of the thrusts in the later movements is shown as following these mylonite zones. (Adapted from J. SUTTON and J. WATSON, 1962, *Geol. Mag.*, **99**, 527, and from R. F. CHEENEY and D. H. MATTHEWS, 1965, *Scott. J. Geol.*, **1**, 256.)

In the Lewisian and basal Moinian, the early folding is highly plastic but dies out upwards (Fig. 7.1). Tectonic slides (zones of intense attenuation) occur on the limbs of some of these early structures, although some slides have been proposed on stratigraphic grounds rather than on structural evidence. The second and third folds are developed on a large and a small scale, and where the three sets are all developed the patterns become very complex and the large-scale folds cannot be traced for long distances. The minor folds associated with the larger structures are found over very wide areas, however. The second folds were produced by a strong, penetrative deformation giving very tight folds with strongly attenuated limbs. They often have very well developed lineations parallel to their axes and an axial plane schistosity. The third structures are only strongly penetrative in the central Moines although large open folds of this generation are found near the western margin.

THE MOINE THRUST ZONE; SOUTHERN PART

The Moine Thrust Zone extends from Eriboll, on the north coast, at least as far as Skye and possibly on to Islay (the Loch Skerrols Thrust) but is, at the most, only a few miles wide. It separates the main part of the orogenic belt from its foreland. The Moine Thrust itself represents only the latest of a series of thrust movements and the position of this complex of thrusts was very strongly controlled by the pre-existing structures in this area, i.e. this line marked the boundary between the foreland and the orogen before the thrusting took place. The Moine Thrust carried the gneisses of the orogen completely over the marginal zone for major parts of its length, so that they rest on undisturbed Cambrian and Torridonian, but in other places the underlying thrusts and the earlier folds are exposed, notably between Skye and Kinlochewe and in the Assynt–Eriboll region (Figs. 7.2, 7.4). It is, perhaps, more likely that such structures are formed by underthrusting of the foreland beneath the orogen, and to speak of rocks being 'carried over' others may be incorrect. Only relative movement is therefore implied, but it is simpler to discuss in the traditional terms.

The sequence of events preceding the thrusting is remarkably consistent from Eriboll to Skye and is similar to the main three phases of folding within the orogen (Table 7.1). The first period of deformation resulted in extreme cataclasis of the basement and cover resulting in the formation of zones of very finely granulated rocks—mylonites. They formed in zones, some a few feet wide, others several miles across, and were accompanied by small-scale folds, the large-scale equivalents being recognizable on Skye where it is possible that the mylonite bands form the cores of major folds. It is probable that this period of folding corresponds to the first phase of tight folding of basement and cover in the orogen and to the wedging in of thin bands of basement both near the margin and in the central parts of the orogen (Fig. 7.1).

The next period of folding produced very small isoclinal folds with a very strong ubiquitous, ESE lineation. The first period of thrusting followed. The rocks carried on top of a thrust plane have been termed 'nappes', the Kishorn Thrust, for instance, carrying the Kishorn Nappe over the foreland. There were probably a series of thrusts at this time which brecciated the mylonized rocks but which are themselves folded by later, fairly open, folds. In the south two nappes are recognized. The Kishorn Nappe is the lowest and largest and has four small remnants of a higher nappe carried on the

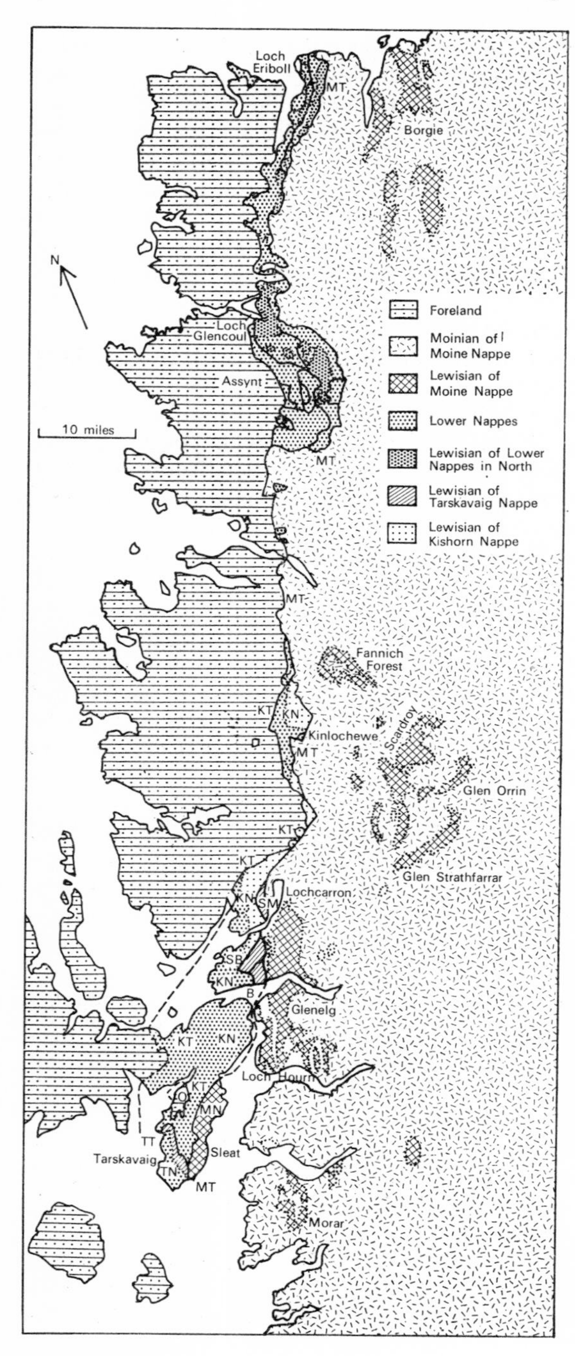

Fig. 7.2 The Moine Thrust Zone of the North-west Highlands. B = Balmacara Thrust, KN = Kishorn Nappe, KT = Kishorn Thrust, MN = Moine Nappe, MT = Moine Thrust, O = Ord window, SB = Sgurr Beag Thrust, SM = Slumbay Mylonite, TN = Tarskavaig Nappe, TT = Tarskavaig Thrust.

Tarskavaig, Balmacara and Sgurr Beag Thrusts; the Slumbay Mylonite at Lochcarron is probably part of the latter nappe. The Tarskavaig thrust plane followed a large mylonite band in the core of an extremely attenuated first generation fold, resulting in an odd, thin zone of mylonized Lewisian at the base of the nappe. One would not normally expect a thrust to develop within the basement just below an unconformity, but this shows the strong structural control which the mylonized zones had on the location of the clean-cut thrust planes. It also explains why the early investigators regarded the thrusting as the cause of the mylonitization since the thrust planes are mostly lying on mylonites, and may mostly, therefore, be following the position of early fold cores.

These first thrusts have been folded by N–S fairly open folds and associated asymmetric small folds have a strain-slip cleavage. The Tarskavaig Nappe is folded by three folds of this generation and the Ord Window (an area of Cambrian rocks underlying the Kishorn Nappe) lies along the continuation of one of these folds where it folds the Kishorn Nappe. It is also probable that the Loch Alsh Synform is of this generation and folds back the Tarskavaig Thrust northwards to outcrop again at Balmacara. There were other small-scale tight fold episodes and a static metamorphism before the last phase of folding, the widespread, brittle, conjugate folds (Fig. 7.3, l–n) accompanying the Moine Thrust.

THE MOINE THRUST; NORTHERN PART

Further to the north, the sub-Moine Thrust nappes are as complex as in the south. At Kinlochewe, at least three thrust slices are seen with imbricate structure. This structure, described in the classic work on the Moine Thrust by Lapworth and Peach and Horne (Fig. 7.4), is probably due to a series of asymmetric folds with steep reverse faults separating them. In Assynt large areas of imbricate structure are exposed, between the lowest, Sole, thrust, and the Glencoul Thrust. The sequence of thrusts in Assynt and Eriboll is similar to that in the south (Fig. 7.4): the Sole Thrust, with imbricate structures above is cut by the Glencoul Thrust plane, which generally brings Lewisian over the Cambro-Ordovician in the imbricate zone. The Glencoul Nappe is also imbricated in places and is cut by the Ben More Thrust. Minor thrusts (some cutting the Assynt Alkaline Suite) lie above this thrust. All these early thrust planes are folded into gentle folds. Tight folding of the Torridonian preceded the thrusting (Fig. 7.4) but the thrusts are almost certainly later than the early thrusts in the south, as they post-date the Assynt Alkaline Suite which is most probably later than the third period of folding. The Moine Thrust cuts across all previous structures and in the Assynt region is warped into a large culmination which exposes the earlier thrust slices beneath.

In the northern part of the Moinian outcrop, there is more definite evidence of the ages of the late events as both the Assynt Alkaline Suite (380 to 400 m.y.) and the Newer Granites (400 to 420 m.y.) are affected by the movements and associated retrogressive metamorphism (Table 7.1). The last move-

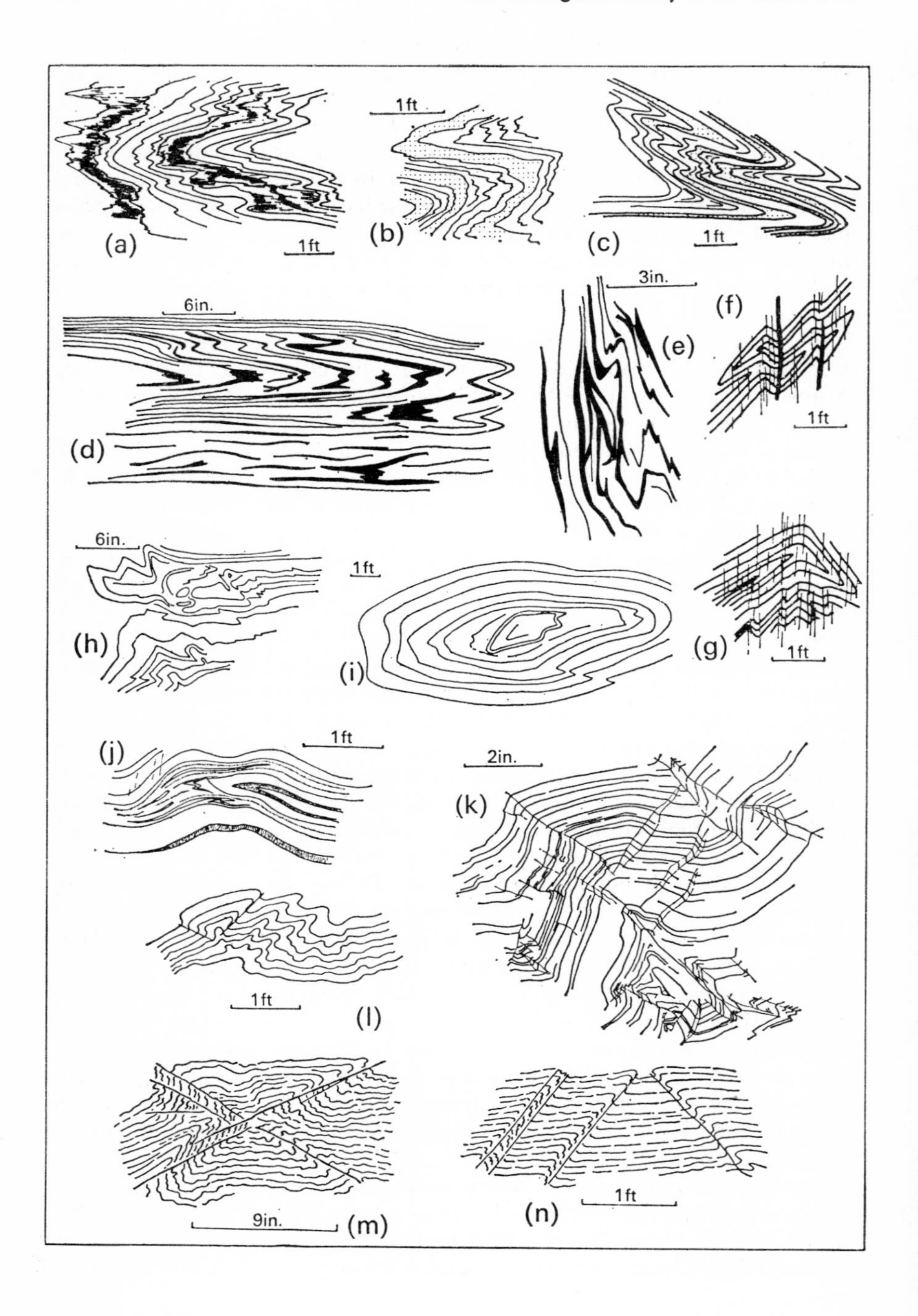
(a)
1ft
1ft
(b)
(c)
1ft
6in.
3in.
(f)
(e)
1ft
(d)
6in.
1ft
(h)
(i)
(g)
1ft
(j)
1ft
2in.
(k)
1ft
(l)
9in.
(m)
(n)
1ft

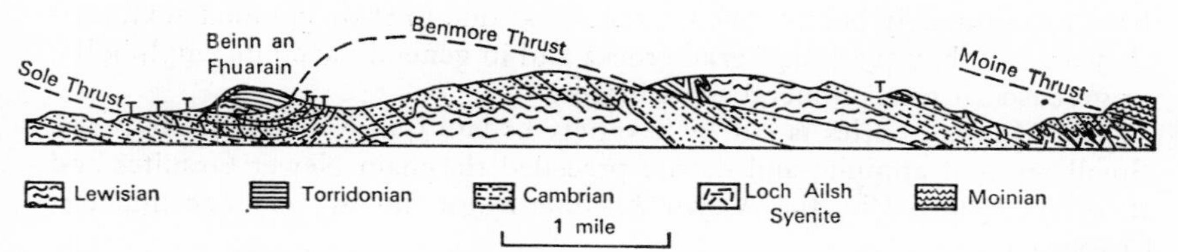

Fig. 7.4 Section across the Moine Thrust Zone in Assynt, showing the folding of the Torridonian before the lower thrusts, the folding of these thrusts and the Moine Thrust cutting across these previous structures. (Based on Crown Copyright Geological Survey map. By permission of the Director, Institute of Geological Sciences.)

ment of the orogeny in this part of the chain was the Struie Hill Thrust which affects Lower Old Red Sandstone strata.

METAMORPHISM OF THE MOINIAN ROCKS

The metamorphic grade of the Moinian varies from low to high. A migmatitic core runs from Ardgour in the south to Bettyhill in the north, and at Ardgour the high grade minerals (chiefly sillimanite) have been deformed by the second phase of folding. High grade zones are also found in the Moinian and Dalradian Series south of the Great Glen but displaced about 65 miles to the north-east. (This is the principal reason for believing in a large sinistral movement on the Great Glen Fault.) The migmatites are surrounded by zones of high grade schists but, as pelitic (aluminous) rocks are rare, kyanite and sillimanite are not sufficiently common to be used as zonal indices and the grade is ascertained by the mineralogy of basic and calcareous rocks. The greatest change in metamorphic grade seen in this area is the retrogression of some of the Lewisian rocks in the fold cores. These have been recrystallized, during the deformation, to chlorite and amphibole schists and their original high grade gneissose character is completely destroyed over wide areas. The metamorphic facies of the Moinian rocks is often difficult to determine as psammitic rocks with little except a quartz-feldspar-mica mineralogy are not very sensitive indicators of grade. They are rarely highly schistose and

---

Fig. 7.3 Fold styles in the Moinian rocks. (*a–c*) First generation folds. Loch Luichart area. (*d, e*) Folding of 'streaky' granulites (rocks produced along slides). Near Loch Frianich, Glen Orrin. (*f, g*) Refolded minor folds, Loch Monar. (*h–j*) Third generation minor folds superimposed on second generation minor folds. Glen Orrin area. (*k*) Conjugate fold in mylonized Lewisian at the base of the Tarskavaig Nappe. Sleat of Skye. (*l*) Conjugate fold in Moinian schist. Coulin. (*m*) Conjugate fold in mylonized Lewisian. Lochcarron. (*n*) Conjugate fold in Torridonian of Kishorn Nappe. Lochcarron. Scales in feet and inches. (*a–c* from P. CLIFFORD, 1960, *Q. Jl geol. Soc. Lond.*, **115**, (1959), 365; *d, e, h–j* from M. J. FLEUTY, 1961, *Q. Jl geol. Soc. Lond.*, **117**, 447; *f, g* from J. G. RAMSAY, 1958, *Q. Jl geol. Soc. Lond.*, **113**, (1957), 271; *k* from J. G. RAMSAY, 1962, *Geol. Mag.*, **99**, 516; *l–n* from M. R. W. JOHNSON, 1956, *Geol. Mag.*, **93**, 345.)

have unfortunately been termed 'granulites' due to their granular texture—they are not, however, high grade rocks and in general have low amphibolite or greenschist facies mineral assemblages.

The Moinian Series is cut by a variety of late Caledonian igneous rocks. Small pipes of appinite and diorite preceded the main Newer Granites and these are followed by the Assynt Alkaline Suite and by very late minettes (Table 7.1).

The Dalradian area

Folding in the Dalradian is again highly complex and two further features are noteworthy: the large overfolds with their associated tectonic slides, and the multiple nature of the metamorphism and its relation to episodes of folding. The classic work of Bailey in the south-west Highlands showed, by detailed mapping of stratigraphic horizons, that the Dalradian rocks have been folded into a series of large recumbent folds (nappes) which are sometimes separated by lines of extreme attenuation where some members of the succession are completely excised. These lines were termed slides by Bailey, and it is probable that they developed at considerable depth as they are accompanied by very strong penetrative deformation. Occasional evidence is seen of the chaotic disruption typical of gravity glides, and many workers suggest that the slides were produced by gravity gliding at an early stage in the tectonic history.

The fact that the slide planes are themselves folded led Bailey to interpret the Dalradian in terms of multiple deformation, following on from the early work of Clough in the Cowal area. More recently, a succession of fold movements has been recognized, with synchronous and intervening phases of metamorphism (Table 7.2).

FIRST FOLDS

The earliest period of folding is generally taken to be that which gave rise to the major structure of the Dalradian. Huge overfolds of alpine dimensions were formed, the major structure being the Iltay Nappe which extends from Deeside to Kintyre and probably forms the area south-east of the Iltay Boundary Slide in northern Ireland also. It is a recumbent fold complex of mushroom-shaped profile, with subsidiary upright anticlines and synclines. It closes, at the Highland Border, in the Aberfoyle 'Anticline', now inverted by refolding into a synform (Fig. 7.5). The large area of horizontal inverted rocks which comprises the lower limb of this fold is usually referred to as the 'Flat Belt' and covers much of the south-west Highlands. The 'Steep Belt' —the area where the fold turns over vertically into its root zone—lies to the north-west. Below the Iltay Nappe lies the recumbent Ben Lui Fold which closes north-westwards. In the Buchan area, a similar complex underlies another nappe which rests on the Boyne Slide. On the north-west side the

Iltay Nappe is separated from the north-west verging* folds of the Ballappel Dalradian by the Iltay Boundary Slide.

The Ballappel Foundation is more complex in its major (early) structures than the Iltay Nappe, being composed of several recumbent folds. The Ballachulish and Appin Nappes are the two major units but each has many slides and the base of the whole unit is the Fort William Slide which separates the Dalradian from much of the Moinian. In Donegal, a similar complex of folds, the Errigal, Aghla and Rosguill Folds are cut by slides, the Horn Head, Derryhassen and Mevagh Slides. In all areas, minor folding of this generation is rare but small folds are found within the foliation in places.

LATER FOLD PHASES

Subsequent fold episodes occurred during periods of metamorphism and give rise to strongly developed axial plane cleavages, schistosity and lineations. These folds are generally fairly open with a north-east trend in Scotland and more easterly in Northern Ireland. Cross-folds (NW and N) occur in both areas and are usually thought to be roughly contemporaneous with the main folds. The later large folds are nearly all third generation folds, e.g. the Ben Lawers Synform, Cowal Anticline, Ben Ledi Antiform, Buchan Antiform and Boyndie Syncline of the Iltay Nappe. Numerous large folds of this generation are also seen in Donegal. The intense movements during this time resulted in further movements on the slide boundaries and may have produced other tectonic slides. Between the second and third fold episodes the major metamorphism took place with the formation of a migmatitic core to the Dalradian—largely in the eastern and northern parts of the outcrop (the 'Older Granites'). In Donegal the similar 'Older Granodiorite' enveloped large areas and obscured both the structure and stratigraphy, particularly near the boundary between the two successions. The rocks were metamorphosed to sillimanite-bearing gneisses at the highest grade, but most of the Dalradian was only raised to chlorite, biotite or garnet grade and the early structures and sedimentary features are thus preserved. The metamorphism in the Buchan area is of a different type (with andalusite and cordierite as typical high grade minerals) and was probably effected at a higher level in the crust.

The third and fourth fold phases deformed the thermal surfaces and a further episode of high grade metamorphism was locally developed. It has even been suggested recently that all the sillimanite is of this generation, being due to a late migmatitic event after the main metamorphism.

* Folds 'verge' in the direction of closure of inclined anticlines; e.g. the folds in Fig. 7.1 verge westwards.

TABLE 7.2

Connemara	*Kilmacrenan and Ox Mountains*	*Creeslough*
	F_{7-8} Brittle structures including steeply plunging conjugate folds	
Later Granites (384–404 m.y.)	Upper Greenschist Facies metamorphism	Cross-cutting granodiorites
		Breccias and minor intrusions
Brittle folding, faulting with Late Thermal Event (445 m.y.)	F_{5-6} Brittle steeply dipping folds and NE trending thrusts	
Oughterard Granite (510 m.y.)		Migmatitic rocks
F_4 E–W Connemara Antiform with metamorphism	F_4 ENE Kink-bands, metamorphism continues	N—S or NW folds with fracture cleavage
F_3 Tight E—W folds with metamorphism (460 m.y.)	F_3 WSW and E—W open to closed upright folds. Fracture cleavage	NE or E—W small folds with flat fracture cleavage. Main retrogression
Migmatization (725 m.y.)	Main metamorphic peak garnet grade	
F_2 Isoclinal folds N—S lineation	F_2 E—W tight similar folds. Axial Plane and fracture cleavage and strong lineation	E—W folds with slaty cleavage. N—S folds.
Migmatites. Basic intrusions. Garnet, staurolite and sillimanite grade metamorphism	Start of regional metamorphism	Garnet grade metamorphism
F_1 Isoclinal folds	F_1 Schistosity and small isoclinal folds	Large recumbent folds and slides
Basic intrusions		Basic intrusions
(P. J. LEGGO, W. COMPSTON and B. E. LEAKE, 1966, *Q. Jl geol. Soc. Lond.*, **122**, 91; and P. J. LEGGO, in press, *Mem. Am. Assoc. Petrol. Geol.*)	(after various authors)	(J. L. KNILL and D. C. KNILL, 1961, *Q. Jl geol. Soc. Lond.*, **117**, 273; and M. J. RICKARD, 1962, *Q. Jl geol. Soc. Lond.*, **118**, 207)

N.B. The successions do not necessarily correlate, and the dates given for the Connemara succession may not apply to the other successions.

Ballappel	*Central Highlands*	*Buchan*
Late Wrench Tectonics	Loch Tay Fault	
Permitted granite intrusions and ring structures (407 m.y.) Forceful granites (423 m.y.) Volcanics, appinites and breccias	Granites	
Steeply plunging folds	F_5 Steeply plunging folds	
	F_4 Gentle NNE upright folds. Metamorphism low grade, high in places	F_4 Asymmetric NNE monoclines verging NNW. Strong local axial plane cleavage. Metamorphism retrogressive
NE large folds, tectonic slides. Slaty cleavage. Main retrogression	F_3 E—W or ENE large folds with fracture cleavage (Ben Lawers Synform, Cowal Anticline, Loch Awe Syncline)	F_3 NNE gentle folds verging ESE (Boyndie Syncline, Buchan Anticline). Small, similar tight folds
	Peak of metamorphism and migmatization (at least 470 m.y.)	Main metamorphism and migmatization Intrusion of Banded Gabbros (490 m.y.)
NW cross-folds Mica lineation	F_2 NW or N—S tight folds, fracture cleavage, strong lineation. Metamorphism to kyanite grade	F_2 Tight folds, tectonic melange, strong lineation. Metamorphism to kyanite grade
Garnet grade metamorphism	Garnet grade regional metamorphism Ben Vuroch Granite (530 m.y.)	Low grade regional metamorphism
Small isoclinal folds NE, recumbent folds and slides	F_1 NE isoclines (Iltay Nappe Complex) slaty cleavage (at least 505 m.y.)	F_1 NNW—N—NNE isoclines. Major recumbent folds
	Metamorphism low grade (540 m.y.)	Metamorphism low grade
(D. R. BOWES and A. E. WRIGHT, 1967, *Trans. Roy. Soc. Edinb.*, **67**, 109; and B. C. KING and N. RAST, 1959, *J. Geol.*, **67**, 264)	(M. R. W. JOHNSON, 1965, in *The Geology of Scotland*, ed. G. Y. CRAIG. Oliver & Boyd, Edinburgh and London; B. A. STURT and A. L. HARRIS, 1961, *Lpool Manchr geol. J.*, **2**, 689; and C. T. HARPER, 1967, *Scott. J. Geol.*, **3**, 46)	(M. R. W. JOHNSON, in *The Geology of Scotland*, ed. G. Y. CRAIG. Oliver & Boyd, Edinburgh and London, 1965)

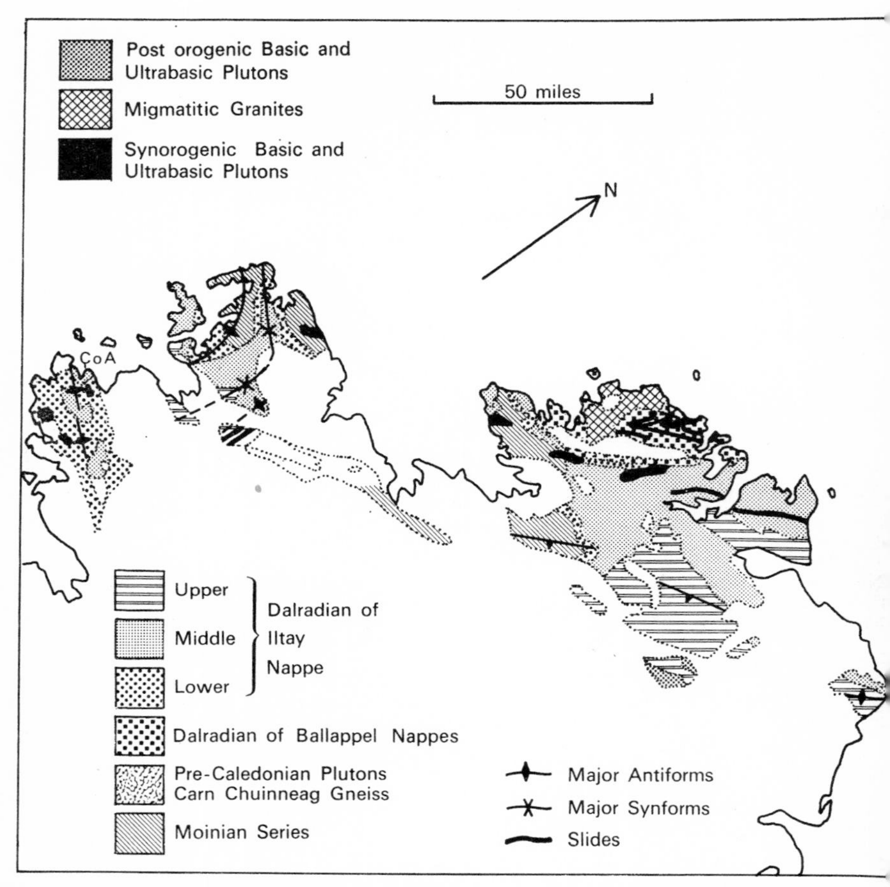

Fig. 7.5 Major folds in the Dalradian, showing the major subdivisions of the Dalradian. AS = Aberfoyle Synform, BLA = Ben Ledi Antiform, BLS = Ben Lawers Synform, BoL = Boyne Lag, BoS = Boyndie Synform, BuA =

LATE-OROGENIC EVENTS

The last effects of the Caledonian Orogeny in the Dalradian area are steeply plunging folds and brittle structures (kink-bands) associated with the

Buchan Antiform, CA = Cowal Antiform, CoA = Connemara Antiform, FW = Fort William Slide, IA = Islay Antiform, IBS = Iltay Boundary Slide, LAS = Loch Awe Synform.

Great Glen Fault. This is the largest of a great number of major NE—SW sinistral wrench faults (Fig. 7.6). Although the sinistral movement on the Great Glen Fault was probably large, there has been no confirmation from more recent studies of Kennedy's suggestion that the Strontian

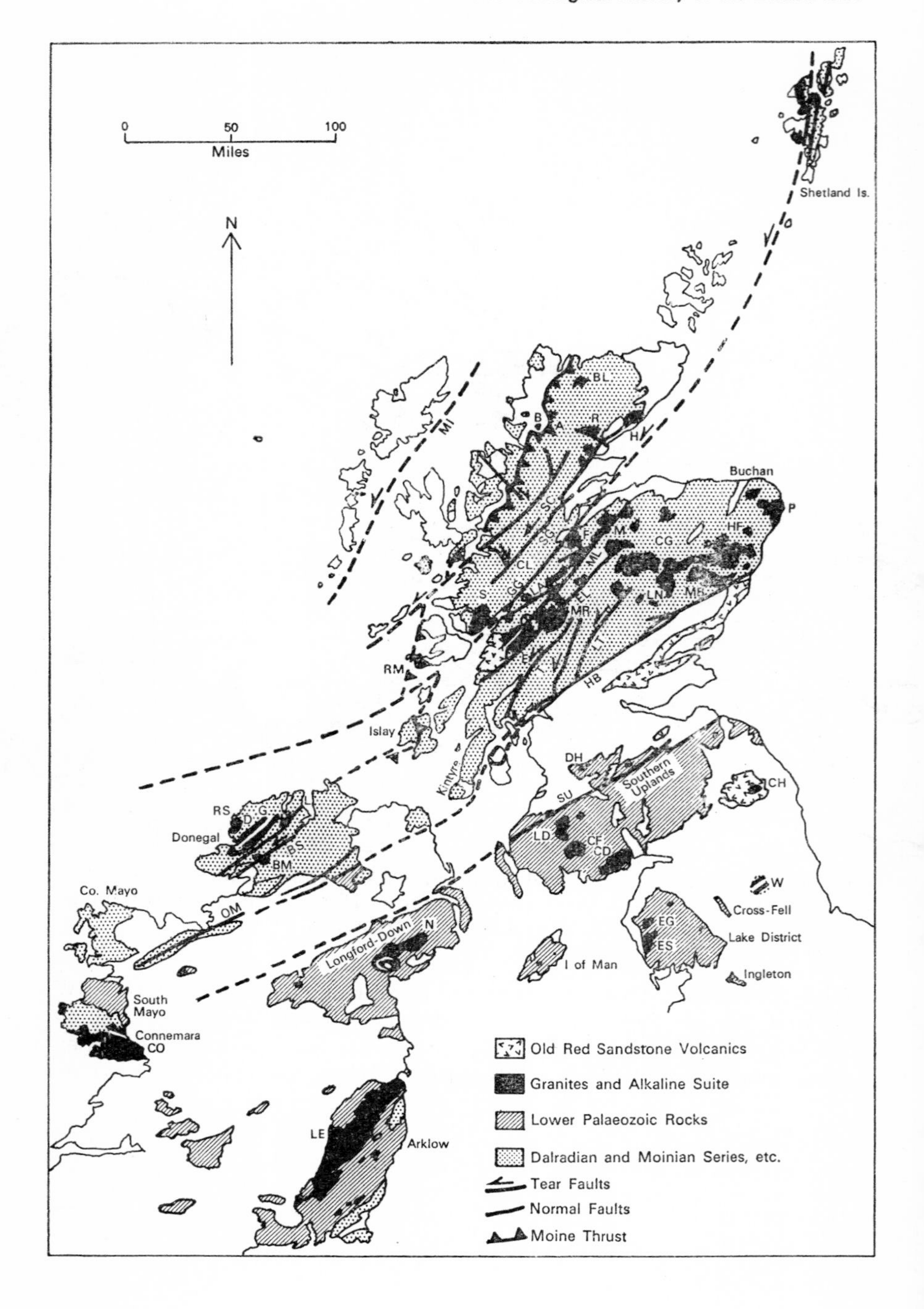
0
50
100
Miles
N
Shetland Is.
MI
BL
B
A
R
H
Buchan
P
HF
CG
SC
F
M
CL
ML
GG
S
LN
MR
LT
HB
RM
Islay
Kintyre
DH
Southern Uplands
CH
SU
RS
D
G
L
Donegal
BS
BM
LD
CF
CD
W
Co. Mayo
OM
Cross-Fell
Longford-Down
N
EG
ES
Lake District
I of Man
Ingleton
South Mayo
Connemara
CO
LE
Arklow
Old Red Sandstone Volcanics
Granites and Alkaline Suite
Lower Palaeozoic Rocks
Dalradian and Moinian Series, etc.
Tear Faults
Normal Faults
Moine Thrust

and Foyers Granites may be equivalent. There is thus no unequivocal evidence to indicate the amount of displacement. The widespread kink-bands are complementary dextral NW—SE structures developed in phyllitic rocks.

The Highland Boundary Fault has probably moved more than once and has largely a normal sense of movement. Some post-Arenig movement was followed by intrusion of serpentinite bodies at several places and the main movement seems to have taken place during Middle Old Red Sandstone times when the Southern Uplands Fault was also initiated, the pair forming the short-lived graben of the Midland Valley of Scotland. These structures continue across into Northern Ireland but are much less strongly developed there and are replaced by series of en-echelon and splay faults or, in the extreme west, by monoclinal structures.

The Great Glen Fault System was preceded and accompanied by the intrusion of many diorite and appinite plugs and the outpouring of andesitic and basic lavas on the surface during a late-orogenic volcanic episode. These probably covered much of the present area of the Scottish Highlands. Subsequently late-orogenic granitic bodies were emplaced, some of which formed cauldron-subsidences (Glen Coe and Ben Nevis) engulfing parts of the lava pile. These granites are of various ages but are much the same age as the Newer Granites which cut the Moinian, i.e. about 400 m.y.

The age of the fold and metamorphic episodes is much less certain. The metamorphic rocks give dates of about 425 m.y., generally being slightly older than those obtained from most Moinian rocks, but this date would seem to be due to the latest of the regional metamorphic events. Dates as old as 505 m.y. have been obtained on slates affected by the First Fold movements, and the Ben Vuroch Granite, which is pre-metamorphic but probably post-F_1, has yielded an age of 530 m.y. In Connemara, at least, it is probable that the main fold episodes are older than the Oughterard Granite (510 m.y.) and

---

Fig. 7.6 The late stage features of the Caledonian Orogeny—the late faulting, including the Great Glen Fault System, the Highland Boundary Faults and the Southern Uplands Faults, the late-orogenic granites and cauldron subsidences and the Devonian late-orogenic volcanics.

Faults: BS, Belshade; EL, Ericht-Laidon; G, Gweebarra; GG, Great Glen; HB, Highland Bounday; L, Leannan; LA, Laggan; LT, Loch Tay; MI, Minch; OM, Ox Mountains; SC, Strathconnon; SG, Strathglass; SU, Southern Uplands; T, Tyndrum.

Granites: BM, Barnesmore; CD, Criffel-Dalbeattie; CF, Cairnsmore of Fleet; CG, Cairngorm; CH, Cheviot; CL, Cluanie; CO, Connemara; D, Main Donegal; DH, Distinkhorn; E, Etive; EG, Ennerdale Granophyre; ES, Eskdale; F, Foyers; H, Helmsdale; HF, Hill of Fare; LD, Loch Doon; LE, Leinster; LN, Lochnagar; M, Moy; MB, Mt. Battock; ML, Monadhliath; MR, Moor of Rannoch; N, Newry; P, Peterhead; R, Rogart; RM, Ross of Mull; RS, Rosses; S, Strontian; W, Weardale (concealed).

Alkali Suite: A, Loch Ailsh; B, Borolan; BL, Ben Loyal.

a date has been obtained from the migmatites there of 725 m.y. However, if the Lower Cambrian age of the Leny Limestone is accepted, it is apparent that in the Dalradian of Scotland all the folding and metamorphic episodes are post-Lower Cambrian. The presence of staurolite in the Arenig rocks of Connemara and of metamorphic fragments in the Bala and Silurian rocks of the Southern Uplands indicates that much of the metamorphism of the Dalradian was pre-Arenig. Similar deformation episodes also affect some of the Highland Border Series rocks of uncertain (? Arenig) age, and it now seems most likely that the folding and metamorphism of the Dalradian was everywhere of late Cambrian-early Ordovician age.

Scotland and Ireland south of the Highland Boundary

In Scotland south of the Highland Boundary Fault, the major outcrop of Lower Palaeozoic rocks is in the Southern Uplands. Inliers in the Midland Valley, however, indicate that probably the whole of the area south of the Highland Boundary was a single unit until after the end of the Caledonian Orogeny both in Scotland and Northern Ireland. Sedimentation started in the area in Arenig times and all the deposits are very similar, none being of near-shore type. The presence of Connemara Schists underlying Arenigian in the Irish continuation of the Midland Valley suggests that Dalradian rocks may underlie the rest of this area.

Folding episodes during deposition in this area are few: at, Girvan the Arenig is folded before the deposition of the Bala; the Upper Valentian of Connemara overlies a large fold with associated thrusts—the Connemara Syncline—in Ordovician rocks; and there is considerable overlap of the Lower Old Red Sandstone in the Midland Valley from Wenlock on to Bala. The most intense deformation, however, seems to be everywhere post-Silurian, during or at the end of Lower Old Red Sandstone times.

The major structure formed in the Southern Uplands during these movements was a large anticlinal structure. This results in an elongate inlier of Ordovician in the core of the structure, which continues into Ireland in the North Down–Monaghan area of the Longford-Down Massif. The major structure of the Southern Uplands is largely controlled by large asymmetrical NE trending folds, with associated thrusts which indicate movements towards the south in the south and towards the north in the north (Fig. 7.7). These form the second of three major periods of folding. The first folds are tight or isoclinal structures but it is now known that these do not form the dominant minor fold style as Lapworth suggested (Fig. 7.7a). They are refolded by the later asymmetrical folds and by E—W steeply plunging tight folds. Refolded folds are particularly well developed on the Berwickshire coast. Wrench faults are the last Caledonian movements in the area and these are contemporaneous with folding in the inliers of the Midland Valley.

In Ireland also, the main post-Silurian folds are of asymmetrical style with associated thrusts and there are also earlier isoclinal folds, some of which are recumbent in the Mweelrea Syncline north of the Connemara Massif.

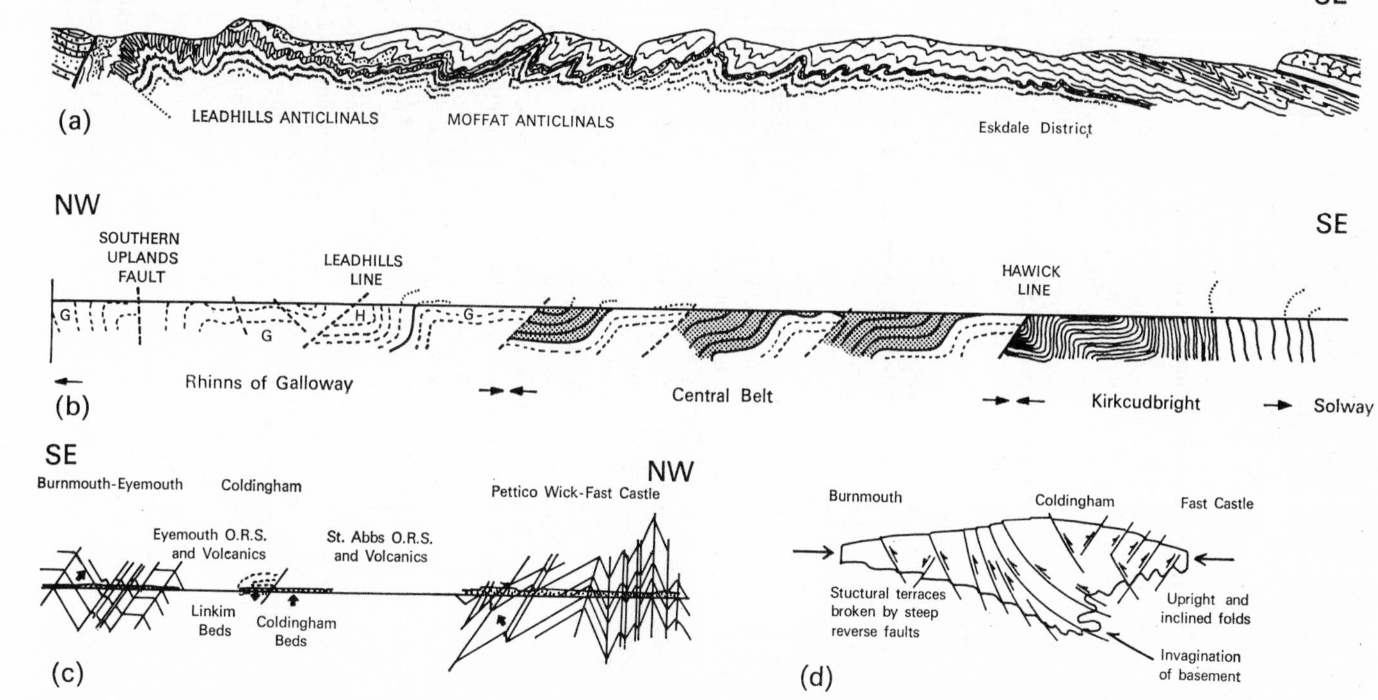

Fig. 7.7 Three interpretations of the structure of the Southern Uplands. (*a*) Anticlinorium and synclinorium. (*b*) A se[illegible] monoclinal folds with faults: G, Glenkiln; H, Hartfell; stipple, Valentian; fine lines, Hawick Rocks; wide-spaced lines, [illegible] (*c*, *d*) Regional syncline (at eastern end of the Southern Uplands; note this section is oriented with NW to the right [illegible] sections; (*d*) interpretation. (*a* from C. LAPWORTH, 1889, *Geol. Mag.*, **36,** 59; *b* from E. K. WALTON, 1963, in M. R. W. J[illegible] F. H. STEWART, *The British Caledonides*. Oliver & Boyd, Edinburgh and London; *c*, *d* from K. A. G. SHIELLS and W. R. DEAR[illegible] *Scott. J. Geol.*, **2,** 231.)

There is no appreciable regional metamorphism in this area—although glaucophane schists are developed near Girvan—but a notable suite of late-orogenic intrusions cut these rocks in the Southern Uplands and the Midland Valley—the Loch Doon, Cairnsmore of Fleet, Criffel-Dalbeattie and Distinkhorn Granites.

The Lake District, Isle of Man and south-east Ireland

The structure of the Isle of Man is known in more detail than that of the Lake District, whose structural pattern is probably the least well known of all the Caledonian areas in the United Kingdom. The structure of the Lake District is dominated by the three stratigraphic divisions, Skiddaw Slates, Borrowdale Volcanics and Silurian Flags and Slates, each of which shows completely different styles of folding (Fig. 7.8). This may be a reflection of their relative competence or they may have different structural histories. The Manx Slates are structurally comparable to the Skiddaw Slates, their stratigraphic continuation. Both formations are largely slates with greywackes and on the Isle of Man three major fold phases can be seen. The first folds are NE—SW upright anticlines and synclines, the dominant fold being the Isle of Man Syncline. A subvertical slide of this generation—the Niarbyl Slide—is also found and other subsidiary folds are abundant. The second folds have shallow-dipping axial planes and there are three major folds again trending NE—SW. The third folds are cross-folds, NW—SE open flexures. All three have planar cleavage structures, the first a slaty cleavage, the later structures fracture cleavage of only local development. The first two phases of this sequence are widely developed in the Skiddaw Slates and the third is locally developed. The boundary with the Borrowdale Volcanics is regarded by some as an unconformity with at least one of these fold episodes being confined to the Skiddaw Slates. Stratigraphic evidence for this is, however, slight and wherever the junction has been examined in detail no unconformity has been proved. The Borrowdale Volcanics are composed of very massive lavas and pyroclastics, which have deformed almost entirely by shearing. Thrust faults are therefore widespread and are aligned NE—SW, denoting movement towards the north-west. Large open folds, the Ullswater Anticline and Scafell–Helvellyn Syncline, preceded the thrusting (Fig. 7.8e).

The structural unit which lies above the Borrowdale Volcanics has the Coniston Limestone unconformity at its base. This transgresses across both Borrowdale Volcanics and Skiddaw Slates, up to 8,000 feet being cut out in 5 miles, the Bala rocks lying on Skiddaw Slates at Millom. There was, therefore, pre-Bala folding with amplitudes of 22,000 feet in the Lake District, which is correlated with the pre-Valentian movements in Connemara by T. N. George. The folds in the Silurian rocks above are a series of upright, symmetrical, NE—SW folds of brittle style, a multitude of medium sized folds lying on the flanks of a large anticline which crosses the southern Lake District in an E—W direction. Thrusting during the folding is a common feature (Fig. 7.8g). Similar structures are also seen in the Lower Palaeozoic

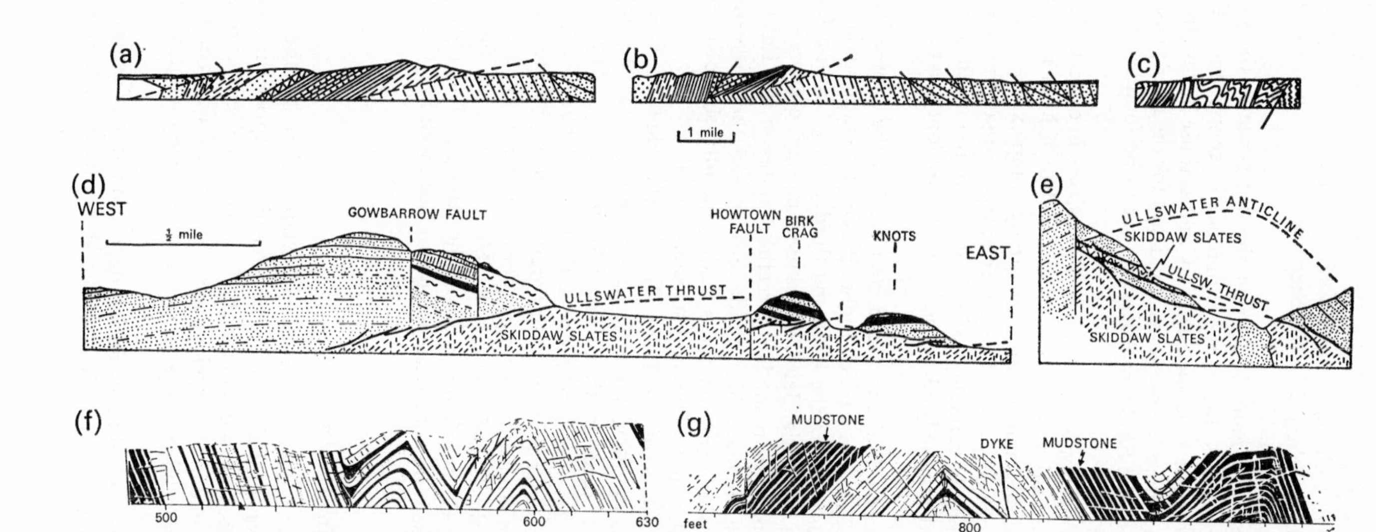

Fig. 7.8 Structural styles in the three main stratigraphic groups of the Lake District and the Isle of Man. (*a-c*) Major folds refolded by low angle folds, Isle of Man. (*d, e*) Major open folds and thrusts in the Borrowdale Volcanics, Ullswater. (*f, g*) Large close folds. Transition beds between the Coniston Grits and the Bannisdale Slates, Shap. (*a–c* after A. SIMPSON, 1963, *Q. Jl geol. Soc. Lond.*, **119,** 367; *d, e* from F. MOSELEY, 1964, *Geol. J.*, **4**, 127; *f, g* from F. MOSELEY, 1968, *Geol. J.*, **6**, 79.)

rocks overlying the Ingletonian in the Craven district, while the Skiddaw Slates of the Cross Fell Inlier are comparable with those of the main outcrop. However, in that region there is also extensive Hercynian thrusting with some associated deformation.

It seems at the moment more likely that the difference in the structures between the three groups is due to the difference in competence, since the rocks above the Borrowdale Volcanic Series, as well as those below, are thrown into much tighter and more intense folds. Evidence for the later phases of the structural sequence in the Skiddaw Slates has yet to be found in the higher beds.

The Arklow and Wicklow areas of Lower Palaeozoics in eastern Eire represent the natural continuation of the Lake District–Isle of Man area although it differs stratigraphically and is also structurally complex, being extremely broken up by faults of many ages. The principal structure of the Arklow area is a syncline with a highly variable profile, isoclinal in the north-east verging north-westwards and gradually changing to a shallow open syncline in the south-west.

Around the Leinster Granite in south-west Ireland the Lower Palaeozoic rocks have been subjected to a much higher grade of metamorphism than in any of the other areas south of the Highland Border. The Leinster Granite is the largest granite in the British Isles, and is more closely comparable to some of the Scottish Granites than the late forceful granites of the Lake District and the Isle of Man. It has a high grade aureole and there is evidence of strong deformation of the country rocks during the intrusion of the granite, giving rise to staurolite and sillimanite schists.

The granitic intrusions of the Lake District, the Skiddaw Granite, the Eskdale Granite and the Shap Granite, are late-orogenic forceful intrusions and have all yielded dates in the range 380 to 400 m.y. The Ennerdale Granophyre gives dates of only 370 m.y. and this may be its true age as there is no proof that it is older than the other granitic rocks (despite earlier suggestions of such a relationship). Younger ages have been obtained from the Foxdale Granite on the Isle of Man (360 to 380 m.y.) and from the Weardale Granite beneath the Carboniferous rocks of the Askrigg Block (364 m.y.). It is apparent that these intrusions were being emplaced well into the Devonian Period.

Wales and the Welsh Borders

South-east of the Lake District–Isle of Man region lay the Irish Sea Geanticline which was probably an area of intermittent uprise throughout the period of Lower Palaeozoic sedimentation. The Lower Palaeozoic rocks of the Arklow area, mentioned in the previous section, were laid down on the north-west side of this ridge while to the south-east lay the geosyncline of north and mid-Wales.

Folding took place several times during deposition and evidence for this is widespread over the whole of Wales and the Welsh borders. Lower Ordovi-

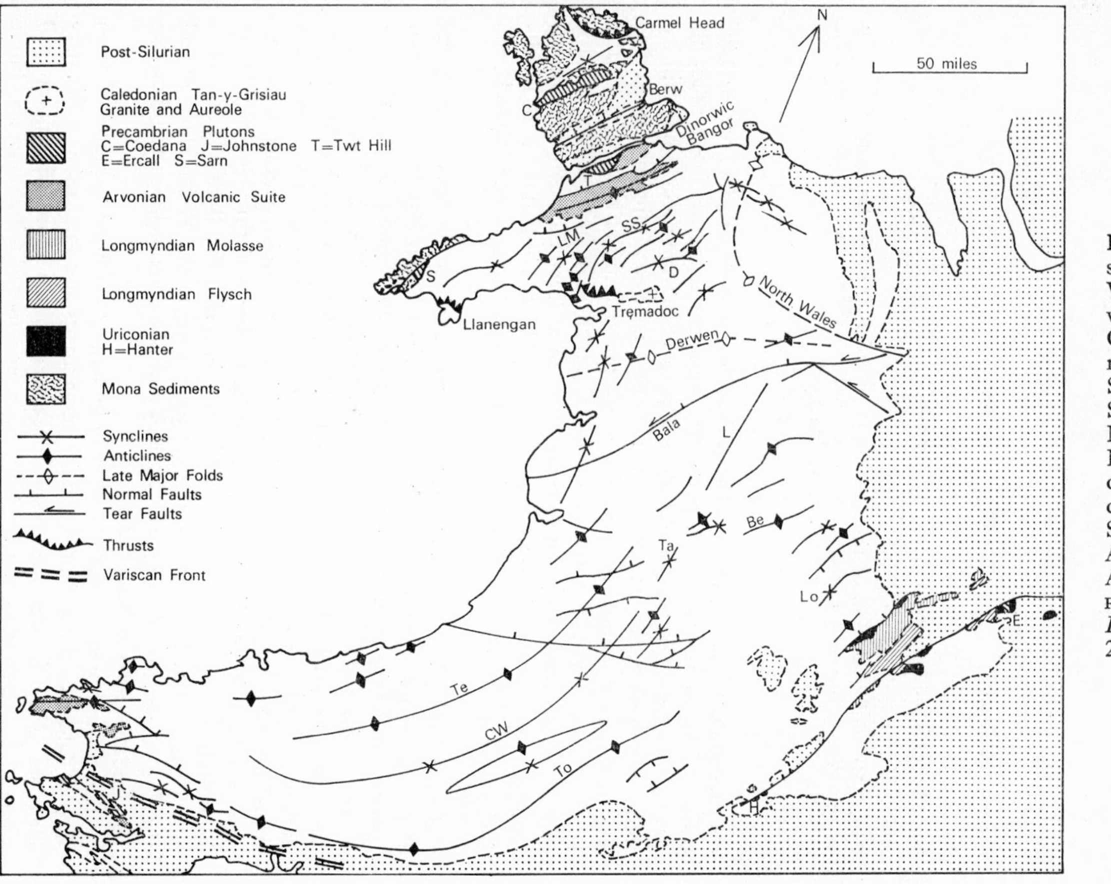

Fig. 7.9 Major folds and structural features of Wales. Folds: Be, Berwyn Anticline; CW, Central Wales Synclinorium; D, Dolwyddelan Syncline; L, Llanderfel Syncline; LM, Llwyd Mawr Syncline; Lo, Long Mountain Syncline; S, Snowdon Syncline; Ta, Tarannon Syncline; Te, Teifi Anticline; To, Towy Anticline. (Partly after R. M. SHACKLETON, 1953, *Lpool Manchr geol. J.*, **1**, 261.)

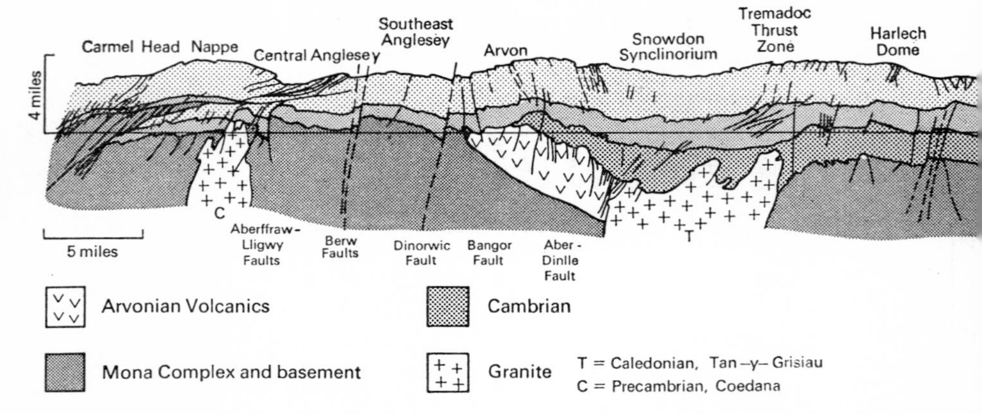

Fig. 7.10 Cross-section of the Caledonian Fold Belt of Wales from Anglesey to the Welsh Borders. (After R. M. SHACKLETON, 1953, *Lpool Manchr geol. J.*, **1**, 261.)

cian movements are apparent in the widespread post-Tremadoc, pre-Arenig, tilting of north Wales and in particular by an anticline at Tyddyndicwm, where a fold of 2,000 feet amplitude of Upper Cambrian rocks is overlain by mid-Ordovician rocks. At the margins of the Irish Sea Geanticline in Anglesey there was a cumulative overstep of all post-Cambrian strata on to the landmass in a series of tilted fault blocks, which have been shown to have moved continuously from late-Precambrian to at least Carboniferous times, allowing deposition in the south-east portions of blocks bounded by the north-east trending Aberffraw-Lligwy, Berw, Dinorwic, Bangor and Aber-Dinlle Faults.

In the Welsh borders, there was some pre-Caradoc movement along the Pontesford-Linley Fault and probably along the Church Stretton Fault also. Then followed the most widespread of the geosynclinal fold episodes, the pre-Valentian folding and thrusting. In south Pembrokeshire the Upper Valentian transgressed both thrusting within the Ordovician and blocks of Precambrian thrust over Ordovician. The Towy Anticline, though mainly a mid-Devonian structure, was probably initiated, not quite parallel to its present direction, in pre-Valentian times, since folding and thrusting of this age is present in its hinge area. The folds of this age in the Builth Inlier trend north-westwards. Later gentle folding took place in the Welsh Borders and the Shelve Anticline during pre-Wenlock and post-Wenlock times and during pre-Ludlow times at May Hill.

The earlier movements thus seem to be largely confined to marginal areas in pre-Valentian times, but the major movements and formation of the slaty cleavage of the Lower Palaeozoic rocks were much more widespread and cannot have taken place until at least the beginning of Devonian times. There

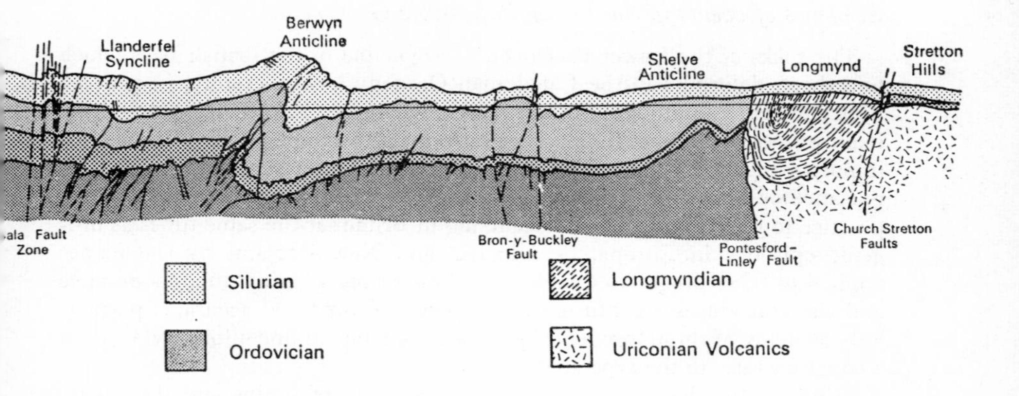

is no justification for calling it a late-Silurian orogenic phase as wherever late-Silurian–early Old Red Sandstone sequences are seen they are invariably uninterrupted and the Old Red Sandstone rocks on Anglesey have been strongly folded, developing an axial plane cleavage.

The sequence of major structures in Wales appears to be similar to that developed in the Isle of Man and the Lake District, and as they affect all the Silurian rocks it seems most likely that the Lake District structures are all of this post-Silurian age. The first structures are again upright folds with well developed slaty cleavage. In the extreme north the folds generally verge northwards and in the south-east they verge in the opposite direction. The area of upright folds (with vertical cleavage) is, however, much closer to the northern than the south-eastern margin.

As in the Isle of Man, the second folds are flat-lying, open flexures with axial planes dipping to the south-east. The third folds deflect the first cleavages, very sharply at Corwen, and have, along the hinge of the fold, a strong strain-slip cleavage and lineation. The axial plane of this fold (the North Wales Antiform) is steeply dipping to the ENE. and the plunge is near vertical.

The general structure can be seen from the accompanying map and cross-section (Figs. 7.9, 7.10). The Carmel Head Thrust, which brings infolded Ordovician and Mona Complex rocks south-eastwards over the main part of the Mona Complex of Anglesey, has analogous structures further south-east in the Tremadoc Thrust Zone and the Berwyn Anticline. The Snowdon Synclinorium and, in particular, the Harlech Dome are gentle structures and it is only in the Cambrian slates, on the north side of the synclinorium that intense deformation has taken place; here compression has been so intense that the vertical dimension has been doubled and the horizontal halved.

The largest acid intrusion in this area is the Tan-y-Grisiau 'granite' which although only small in size has a large aureole, indicating a much larger subsurface pluton.

Sequence of events in the Lower Palaeozoic Geosyncline

The rocks of the Lower Palaeozoic Geosyncline in the British Isles which have been deformed in the Caledonian Orogeny have therefore, in general, been subject to a number of minor earth movements, before the post-Silurian main phases of folding. The earlier movements were generally of only local importance and no one can be shown to have been of such widespread importance or intensity as to warrant a separate phase-name. It has been the practice to refer to movements occurring in Britain at the same times as orogenic episodes in Europe, Scandinavia and New England by the names applied to these orogenies (Table 7.3). This seems a highly dubious practice and also introduces a confusing terminology of unproved accuracy, particularly as many of these 'orogenic' phases are simply an unconformity, or even a conglomerate, in the type area.

Although the deposition took place in a series of basins and the earlier structures were more localized, within the post-Silurian phase a complex series of events can be recognized which is common to most of the area. The separate areas of deposition were still, of course, separate structural entities, but the close correspondence of many of the events and of the sequence as a whole indicates that the whole area south of the Highlands was subjected to a series of fold episodes with intervening periods of stress-release (tension). There is a close correlation between periods of dyke intrusion and periods of tensional faulting and thus a correlation between the orogenic activity in the crust with igneous activity at depth. The last phase of these movements is also evident in the Dalradian–Moinian area as the Great Glen Fault System. This is also preceded and succeeded by volcanic and plutonic igneous activity all pointing to a close association between deep-seated crustal and subcrustal activity with faulting and igneous activity at the surface.

TABLE 7.3 Phases of Orogeny in the lower part of Palaeozoic times

Phase	Age
Acadian	Mid-Devonian—late Devonian
Cheruskic	Early Devonian
Erian	Post-Downtonian
Ardennian	Pre-Downtonian
Caledonian	Late Silurian
Taconic	Late Ordovician
Cow Head	Mid-Ordovician
Tryssil	Early Ordovician
Sardic	Upper Cambrian
Cadoman	Late Precambrian

N.B. Many of these terms are used in a variety of ways by different authors.

The Caledonian pattern

The earth movements which gave rise to the Caledonian mountain chain in the British Isles may, therefore, probably be resolved into at least two main periods of polyphase folding: early Ordovician metamorphism and folding in the Highlands of Scotland and the northern part of Ireland, and post-Silurian folding in the southern part of Great Britain. The late Precambrian Celtic cycle of Anglesey and the English Midlands, with events possibly of a similar age in the Moinian and in the Dalradian (Connemara), are relics of an earlier orogenic belt. The sequence of events which starts with the deposition of the Moinian sediments and ends with the Old Red Sandstone molasse and volcanics, can therefore be regarded as a succession of geosynclinal and orogenic phases of a pair of orogenic belts, the earlier of which has become inextricably involved in the orogenic phase of the latter.

The Torridonian-Moinian sediments were laid down in a large trough with a wide area of foreland deposition; folding and metamorphism of these sediments may then have taken place. The Mona Sediments were being deposited in a geosynclinal trough in the Irish Sea region and were also involved in a Precambrian orogenic episode with subsequent deposition of flysch and molasse at its south-western margin in the Longmynd and Charnwood Forest areas. The Dalradian Series were probably being deposited in Connemara, the north of Ireland and the Grampian Highlands during the deposition and metamorphism of the Mona Complex, although the Connemara rocks may have been involved in this orogenic episode somewhat before the rest of the Dalradian. After this Celtic orogenic cycle in the south of Britain, the Cambrian basins of Harlech, south Wales, Bray and the Isle of Man were initiated and deposition continued here intermittently in a series of basins until post-Silurian times. Uplift of the main part of the Dalradian in the Scottish Highlands during a late Cambrian or early Ordovician orogenic phase preceded sedimentation, in Ordovician and Silurian times, in a series of basins between the Celtic mountains of Anglesey and the newly formed Scottish and Irish highlands. Deposition in these areas, which can be regarded as the synorogenic depositional episode of flysch facies related to the Dalradian orogenic phase, continued until post-Silurian times when the last fold episode of the Caledonian Orogeny gave rise to the Southern Uplands, the Longford-Down massif, the Lake District, the Welsh Mountains and the Wicklow Mountains. Late-orogenic plutonic activity followed and appears to be roughly contemporaneous over the whole area from the north of Scotland to north Wales and Leinster. The late-orogenic deposition of molasse in a series of intermontane basins (the Old Red Sandstone), with accompanying late-orgenic andesitic volcanics, completed the cycle.

The pattern of small basins of deposition being initiated and filling up one after the other is paralleled by the structural isolation of each area from adjacent areas. The Scottish Highlands and Northern Ireland represent the largest unit with the structures facing to the north-west in the north-west and to the south-east in the south-east (the only major part of the Dalradian which verges north-west is the Ballappel Foundation). The Southern Uplands has a similar structure, largely verging southwards but with some structures on the northern margin verging to the north-west. In the Lake District, Isle of Man and Arklow areas the major structures all verge north-westwards away from the Irish Sea Geanticline and, to the south of this structure, the Welsh structures generally verge south-westwards.

There are thus many indications in the Caledonian orogenic belt of multiple phases of folding, sporadic deposition in large and small basins and periodic plutonic activity over long periods of time. The simplified picture of orogenesis, as sedimentation in an elongate trough, compression with folding and metamorphism, culminating in the intrusion of plutons and subsequent uplift must be regarded only as a very generalized sequence. Detailed research, stratigraphical, structural and geochronological, is revealing more and more evidence of the episodic and repetitive nature of practically all features of orogenesis.

IV The Upper Palaeozoic Era

8 The Devonian System

The major orogeny which produced the mountains known as the Caledonides and gave rise to the complex structures discussed in the previous chapter was responsible for important and widespread geographical changes. These were especially evident in Britain. It has been noted that the Caledonian Orogeny did not occur simply at the end of the Silurian Period. Like all orogenies it was prolonged, but it culminated in late Silurian times so that it remains true that throughout much of the Lower Palaeozoic most of Britain was part of an area of marine sedimentation. However, already, by late Silurian (Downtonian) times deposition of continental beds, described in Chapter 6, was widespread. Sedimentation rates exceeded the rate of downwarping of the geosynclines and the Caledonian Orogeny effectively terminated marine geosynclinal sedimentation in Wales, the Lake District and the south of Scotland. The greater part of Britain became part of the Old Red Sandstone Continent which comprised much of northern Europe including the Fenno-Scandinavian Shield and the Russian Platform (Fig. 8.1).

South of the Old Red Sandstone Continent marine sedimentation continued in the Devonian geosyncline which extended across Europe. Rocks of marine facies are now to be found where Palaeozoic rocks protrude through the cover of later strata as Hercynian (= Variscan) massifs. In northern Europe these form the Ardennes and mountains bordering the Rhine, Bohemia and Brittany. The south-west peninsula of England, another Hercynian massif, includes the outcrops of Cornwall and Devon. From the latter the Devonian System takes its name, being first described there by Murchison and Sedgwick, although as a type area it has several shortcomings, notably that the base of the succession is not seen and the succession is not nearly so thick as in the Ardennes and the German (Rhenish) massifs. Further, the full succession of zone fossils has not yet been found.

In Britain there were in Devonian times mountainous areas, areas of erosion, while two very different facies of sedimentary rocks were simultaneously deposited. The marine Devonian beds of south-west England and the continental strata, known as the Old Red Sandstone, naturally differ in rock types and fauna. The term Old Red Sandstone has frequently been used in a stratigraphic sense: formerly deposits of continental facies (often red beds) older than and younger than the Carboniferous were respectively referred to as the Old and the New Red Sandstone. The term is still valuable as a facies name to distinguish beds of Devonian age of nonmarine facies. However, since the Old Red Sandstone beds can generally be dated relatively to the marine sequence, despite problems of correlation, the term should no longer be used as a division of geological time nor should its division into Upper, Middle and Lower be used in that way. The latter, of course, remain useful divisions based on lithology and characterized by different fish faunas. It need hardly be stated that while Old Red Sandstone is essentially a descriptive term, much of the succession is not red nor is it entirely sandstone.

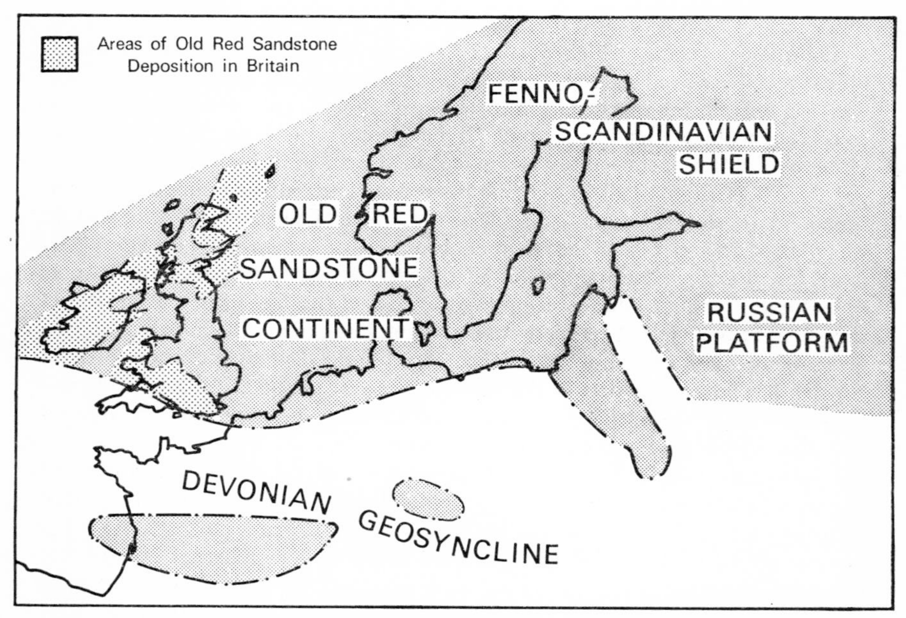

Fig. 8.1 The palaeogeography of Europe in Devonian times, showing the extent of the Old Red Sandstone Continent bordering the geosynclinal areas.

In north Devon marine and non-marine beds occur in the same sequence. There they are interdigitated due to alternating expansions and contractions of the area of marine deposition, that is due to marine transgressions advancing (in a north-north-easterly direction) alternating with recession of the shoreline.

The South-West Peninsula of England

The South-West Peninsula of England consists, essentially, of a broad synclinal structure with an east—west trend. Despite strike faulting of the limbs of the major structure and superimposed folding, sometimes isoclinal, it is responsible for the general disposition of outcrops in Devon and Cornwall. Devonian strata occur to the north and to the south of the broad synclinal structure or synclinorium, the centre of which is occupied largely by Carboniferous rocks (partially overlain by Permian age strata). The Devonian strata of the southern limb (Fig. 8.2) strike east—west occupying south Devon and Cornwall. Of marine origin, they comprise slates, grits and limestones, the latter especially well developed in the Torquay and Plymouth districts. Lateral variation in lithology gives rise to a more argillaceous Devonian succession westwards. The Devonian has been divided into the traditional Upper, Middle and Lower and, in the Ardennes—Eifel region, into stages and zones by means of goniatite ammonoids. Although the faunal sequence has not been completely found in Devon and Cornwall, the succession can be closely related to the continental stages. The succession in south Devon is given in Table 8.1.

TABLE 8.1

	Stages	*Zones*	*Lithology*
UPPER DEVONIAN	FAMENNIAN	*Wocklumeria* *Clymenia* *Platyclymenia* *Cheiloceras*	Purple and green slates Mudstones and shales
	FRASNIAN	*Manticoceras*	
MIDDLE DEVONIAN	GIVETIAN	*Maenioceras*	Limestones passing down into shales and limestones
	EIFELIAN	*Anarcestes*	Couvinian shales
LOWER DEVONIAN	EMSIAN SIEGENIAN	*Mimosphinctes*	Staddon Grits Meadfoot Beds
	GEDINIAN	No ammonoids	Dartmouth Slates

The Devonian System takes its name from Devon through precedence, the succession being first described here (over a century ago). It is only recently that diligent work has produced some of the zone fossils making more detailed corelation with the European succession possible. However, in addition to the goniatites which are rare except at a few horizons, ostracods and conodonts have been found useful for correlation purposes, as well as coral faunas in the rocks of calcareous facies. A number of factors combine to make correlation difficult even within the area of typical marine Devonian strata of south Devon and Cornwall: the succession may not, in fact, be entirely marine for some beds are almost devoid of marine fossils. In addition, there are lateral variations both in lithology and fauna reflecting original differences of sedimentary environment. Subsequent intrusions of granite bosses and basic sills, with accompanying thermal metamorphism, and cleavage imparted to the argillaceous beds during subsequent folding and faulting, make the geology of south-west England far from simple to elucidate.

The rocks ('slates' = Killas) of south Cornwall

The lowest lithological group of the Lower Devonian, the Dartmouth Slates, occurs from Dartmouth westwards to Plymouth Sound and beyond through Looe to Newquay on the north Cornish coast. Occurring in a complex anticlinal structure, the base is not seen and there is no evidence of whether Lowest Devonian is present. It seems probable that the Lowest Devonian (Gedinian) is thinner than contemporaneous strata of the Ardennes which attain 5,000 feet. Overlying the Dartmouth Slates, hence succeeding them northwards, are the Meadfoot Beds and still younger beds. In the west, beyond Looe, due to the anticlinal structure, Meadfoot Beds also occur to the south of the Dartmouth Slates. They are adjacent to rocks of dubious age in south Cornwall known as the Gramscatho Series, generally referred to the Middle and Lower Devonian (Fig. 8.2).

Immediately to the north of the Lizard Complex (p. 70) lies the Meneage Breccia or 'Crush-Zone'. The complex structures in the rocks of this area of conglomerates, breccias, and slates with phacoids of greywacke have been reinterpreted by Lambert as largely of sedimentational origin, although the rocks have suffered some deformation. It also seems probable that their boundary with the rocks of the Lizard Complex is not a thrust of major significance. The limestones in this heterogeneous series, once assigned to the Ordovician on disputable palaeontological evidence, are now accepted as Devonian age as is, by inference, the 'matrix'. Similar rocks also occur at Veryan. The succession of the remaining beds of this area of south Cornwall may be as follows:

Grampound Beds
Portscatho Beds
Falmouth Beds
Mylor Beds

These beds, grits, sandstones and slates, are practically unfossiliferous and they have formerly been assigned to many different geological periods by different authors. It seems probable that all except the Mylor Beds, which may belong to the Lower Devonian, are referable to the Middle Devonian though bearing little resemblance to the presumed contemporaneous strata of north Cornwall and south Devon. However, even the succession is arguable and the lithologically similar Mylor and Portscatho Beds have been equated. Woody plant fossils found at a few localities in the Falmouth and Portscatho Beds confirm the Devonian age, possibly Middle Devonian. Less certain is their mode of origin. Although differing from the typical marine Devonian succession they are, none the less, probably marine despite the occasional occurrence of plant fossils and the lack of corroborative faunal evidence.

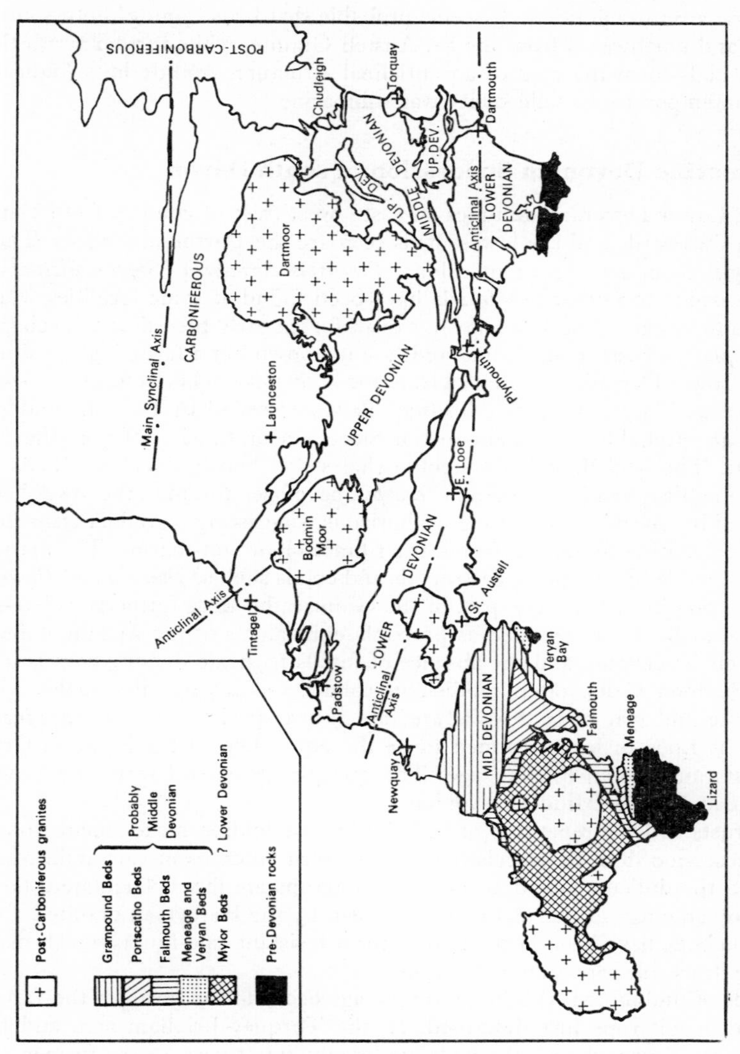

Fig. 8.2 Geological map of the Devonian rocks of South Devon and Cornwall. (Based on several sources, including R.M. BARTON, M.R. HOUSE and published Geological Survey maps.)

Lithologically described as greywackes in earlier literature many of the rocks are quartzose, although clay-shales occur, and are chiefly grits and siltstones. While the stratigraphical divisions are definable on lithological differences they grade into each other. It seems probable that beds 'young' both southwards and northwards from the St. Austell Granite, while further south the Mylor Beds form the core of an anticlinal structure. Spilitic lavas indicate that contemporaneous vulcanicity was submarine.

The marine Devonian Succession of south Devon

The Lower Devonian occupies an east—west tract of country from Dartmouth westwards and the lowest beds seen are the Dartmouth Slates. They are argillaceous and green in colour. *Pteraspis cornubica*, *Steganodictum* (a Cephalaspid) and other fish fossils have been found at some localities while marine fossils are almost unknown, except for an early record of a species of *Bellerophon*, a gastropod. Ammonoids are unknown but a fairly high position in the Lower Devonian is inferred from the known zonal horizons of the overlying beds. These, the grey Meadfoot Beds succeeded by the red Staddon Grits, are probably equivalent to the Siegenian or Emsian of the Rhenish Massifs. The Meadfoot Beds include shales and locally developed beds of limestone. Fish fossils, including *Pteraspis*, have been found in the lower beds followed by marine fossils at higher horizons, suggesting a gradual transition to typical marine conditions from earlier brackish or non-marine. The marine fossils include brachiopods, gastropods and corals such as *Petraia* and *Pleurodictyum problematicum*. The part of the Gramscatho Beds formerly called the Mannacan Beds are regarded as probable equivalents of the Meadfoot Beds but their brecciation and the absence of fossils frustrate definite correlation. The Staddon Grits, often reddish in colour, are largely silty shales with quartzite and grit beds. Fossils are rare but a specimen of *Mimosphinctes* found at Looe in beds assumed to be the equivalent of the Staddon Grits suggests an Emsian-Eifelian age. The group may extend from the Lower Devonian into the Middle Devonian.

Thrusts and the repetition of beds by intense folding make elucidation of the succession difficult especially in the western outcrops in Cornwall. Mapping in the difficult inland areas where outcrops are few is facilitated, however, by an effect of thermal metamorphism by the Hercynian granites. Calcareous beds have been hornfelsed to form resistant calc-flintas which make readily traceable topographic features.

Beds of undoubted Middle Devonian age occur to the north of the Lower Devonian outcrops just described. In the Torquay-Brixham area and the Plymouth-Tavistock area the beds are largely limestones. In the former, the lowest beds of the Middle Devonian are the grey Calceola Shales, with the peculiar shaped rugose coral *Calceola sandalina* and the brachiopod *Atrypa reticularis*, overlain by limestones. At Plymouth, the entire Middle Devonian comprises limestones in which the coral sequence established in Germany has been found. From here westwards a major change in facies takes place, so that in the Padstow area of the north Cornwall coast the succession comprises grey slates, with lenticular limestones only present in some districts. Here, some beds are highly fossiliferous and recently three distinct goniatite faunas have been found, confirming the Middle Devonian age of the strata. The Middle Devonian limestones are highly fossiliferous, the lower limestones are crinoidal and the higher limestones are crowded with compound corals, both tabulate and rugose, and stromatoporoids which together grew as reefs (Fig. 8.3). Pink in colour, these fossiliferous limestones were formerly widely used as ornamental 'marble'. The limestones extend up into the Upper

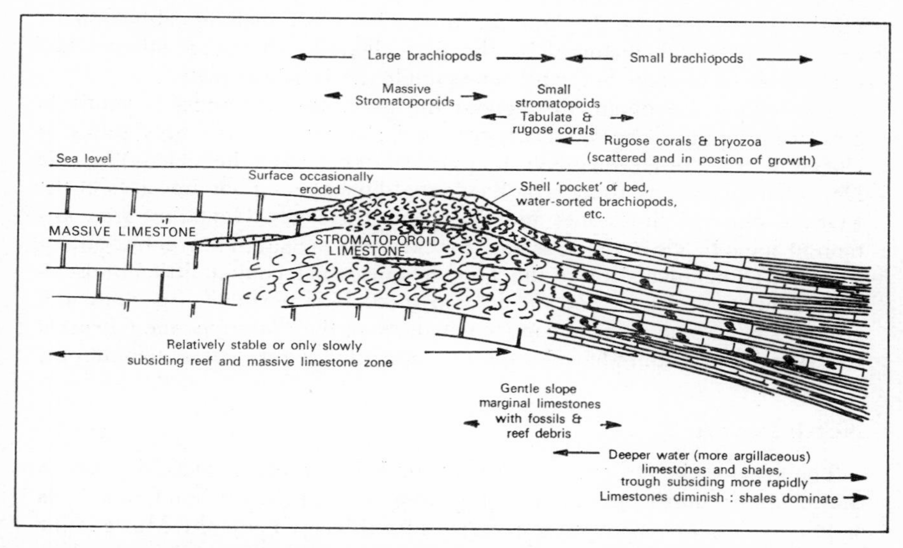

Fig. 8.3 Reef ecology of the Middle Devonian reefs of South Devon. (From D. L. DINELY, 1961, *Field Studies,* **1,** 121.)

Devonian in south Devon where they are succeeded by slates with, in some areas, tuffs and spilitic lavas.

Facies differences in the marine Devonian have been compared with those in Germany and certain analogies drawn. In south Devon, in the Chudleigh area, nodular shales and limestones may have been laid down in shallower water over submarine ridges while ostracod shales were laid down contemporaneously in deeper waters in the Torquay area. This latter facies is widely distributed in Devon and Cornwall but strata of this facies are not exactly of the same time range throughout: deep-water conditions persisted in some areas longer than others. Volcanic rocks occur at numerous horizons in the Devonian, vulcanicity becoming gradually more frequent as the period progressed, although the volcanic rocks are seldom of great thickness. In Cornwall a tuff marks the beginning of the Upper Devonian, and volcanics, chiefly spilitic lavas, are particularly well developed west of Bodmin Moor. They are also exposed in the Cornish coast, for example at Pentire Head. The material between the pillow structures of the spilites is siliceous, argillaceous or calcareous.

Northwards from the development of the ostracod slate facies between the Tamar estuary and Padstow, the Famennian sequence is different. Correlation with rocks to the south is still uncertain and these beds may succeed, or may be in part equivalent to, the ostracod slates. Around Tintagel, west of Bodmin Moor, metamorphism has produced phyllites while pillow lavas have been altered to green schists in an anticlinal structure which plunges to the north-west away from the Bodmin Moor Granite. Fossils, notably *Spirifer verneuili*, are much distorted by the slaty cleavage. Slates are still worked extensively from some horizons, for example the Delabole Slate.

The Upper Devonian, throughout the outcrops, represents a return to argillaceous sedimentation. Dark coloured slates follow the grey slates in Cornwall, while in Devon green and purple slates succeed the (chiefly) Middle Devonian limestones, though dark coloured limestones grade upwards in the Frasnian through limestones and shales into the slates. Goniatites are common in some bands in the Frasnian, for example at Saltern Cove, but absent from the Famennian slates of the Torquay area which contain little but ostracods. Three species of *Entomis* are recorded. It is north of Tavistock and around Launceston that Famennian goniatites of the *Platyclymenia*, *Clymenia* and *Wocklumeria* zones have been found north of the ostracod slate outcrops.

North Devon

The synclinorial structure of Devonshire brings rocks of Devonian age to the surface in north Devon, emerging from beneath the Carboniferous beds of culm facies. The main outcrop extends from Baggy Point and Morte Point along the coast—making spectacular cliff scenery—as far as Minehead in Somerset. Devonian rocks also form the large inlier of the Quantock Hills to the east of the main outcrop.

The succession in north Devon differs markedly from that of south Devon and Cornwall in lithologies and faunas although much of the succession is of marine origin. However, the succession is partly of non-marine origin with marine transgression extending into north Devon and Somerset at intervals and, in late Devonian times, as far north as Pembrokeshire and south-west Ireland. Despite these marine transgressions into north Devon, correlation of north and south Devon is fraught with problems and in fact the greater part of the north Devon succession cannot be related precisely to the goniatite zonal sequence.

Post-Carboniferous folding (generally isoclinal), faulting and the development of cleavage in argillaceous beds combine to complicate the geological picture, but very recently the successions have been deduced in several areas and correlated. Other differences between the north Devon–Somerset succession and that of the type area are the relative unimportance of volcanic rocks and the rarity of limestones, the latter being present but subordinate in the Middle Devonian.

The oldest beds seen are the Foreland Grits outcropping from Foreland Point eastwards. Red and grey quartzose grits and slates, they have yielded no fossils other than plants: *Psilophyton*, generally regarded as a Lower Old Red Sandstone plant, has been recorded. The succession in north Devon commenced as far as is known (the base is not seen) with these non-marine beds. The overlying Lynton Beds, chiefly blue-grey slates with some grits, although containing the fish *Pteraspis*, include lamellibranchs and brachiopods, indicating that the sea swept into this area. The species of *Spirifer* present suggest a Siegenian age, but the fauna which includes corals, both *Pachypora* and *Alveolites*, does not permit reliable correlation with the strata of south Devon. The lowest beds seen in the Quantock Hills, the Little Quantock Beds, may be equivalent to the Lynton Beds. They are slates and have yielded *Chonetes*, *Tentaculites* and crinoid ossicles.

The succeeding Hangman Grits are well seen in the coast as far west as Hangman Point and outcrop in the north-western Quantocks. The lower part, known as the Trentishoe Beds or as the Triscombe Beds, is massive sandstones, siltstones and mudstone pellet conglomerates with some plant remains. The higher Rawn's Beds or Hodder's Combe Beds, shales with flaggy sandstones and conglomerates, have yielded plants—*Calamites cannaeformis* in the Quantocks—and a fish scale of *Coccosteus* which is typical of the Middle Old Red Sandstone. The Hangman Grits represent a return to non-marine conditions and they may be of fluviatile origin. Of approximately Couvinian age, the lowest part of the Hangman Grits may be the equivalent of the Staddon Grits of south Devon.

The transition from the fluviatile Hangman Grits to the marine Ilfracombe Beds which follow, a major marine depositional transgression, has been investigated in detail. The uppermost beds of the Hangman Grits, the Sherry Combe Beds, include sandstones with thick-shelled Myalinid lamellibranchs and gastropods. The '*Myalina*' was long thought to be the brachiopod *Stringocephalus burtini* which does typify the Middle Devonian beds of south Devon. The Hangman Grits may in fact be partly of Middle Devonian age, since there is no apparent break in the succession, and the overlying Ilfracombe Beds are probably Givetian and Frasnian in age. The marine transgression moved in a north-easterly direction, reaching north Devon earlier, then extending into Somerset (the Brendon Hills and still later into the Quantock Hills).

The Ilfracombe Beds have been shown to comprise chiefly slates, probably less than earlier estimates of almost 1,300 feet, with a well developed lower limestone and another 250 feet higher. A thin limestone occurs within the slates between these two. The limestones are not lenticular but are laterally persistent, striking out to sea in the Ilfracombe area. The pre-limestone strata of the Ilfracombe Beds are silty clays and sands and have been interpreted as inter-tidal and delta front deposits. The presence of *Chondrites* and other trace fossils, though not common, supports this belief. Marine beds with a diverse fauna of corals, crinoids, bryozoa, etc., follow. The lowest limestone, the Jenny Start, is probably Givetian in age. The thin Combe Martin Beach Limestone and the highest of the three, the David's Stone Limestone, have similar faunas with abundant corals. Some genera of corals present have been found in the Boulognais area and in Bohemia and it is surprising that they are unknown elsewhere from British outcrops, particularly as limestones in south Devon are contemporaneous. Exact correlation is not possible due to the dissimilarity of the fauna of the two areas, but the Givetian-Frasnian boundary is tentatively placed at the Combe Martin Beach Limestone.

The Morte Slates which succeed the Ilfracombe Beds are grey and green well-cleaved slates with some siltstones in the upper part. Fossils are rare, chiefly being found in thin interbedded sandstones. The lower part of the Morte Slates has yielded *Cyrtospirifer*, other brachiopods and crinoid ossicles, establishing that the lower part of the formation at least is marine. However, the absence of fossils apart from plant fragments and dubious trace fossils from much of the Morte Slates, together with the lithology of the cross-bedded siltstones, suggests a gradual return to non-marine conditions of deposition. The meagre fauna has been said to infer a Lower Devonian age but, if structural relationships have been correctly interpreted, these beds must belong to the Upper Devonian.

A bed of volcanic ash, probably contemporaneous with outpourings of lava in north Cornwall, is regarded by some authorities as marking the base of the Upper Devonian. It preceded the deposition in north Devon of the next formation, the Pickwell Down Sandstone. Purple, brown and green sandstones with some shales, they include fish fossils such as *Bothriolepis*, *Holonema* and *Holoptychius* and fossil wood, clearly being of non-marine facies. The fauna is essentially that of the Upper Old Red Sandstone and the age undoubtedly Famennian.

A further, final, marine incursion took place in Upper Devonian times, depositing sandstones which are now found running inland from Baggy Point to Marwood and beyond. These Baggy and Marwood Beds have a littoral fauna, chiefly of brachiopods. The overlying slates of the Pilton Beds, containing a very badly preserved fauna of trilobites and brachiopods, form a transitional group from Upper Devonian to Lower Carboniferous and represent the maximum spread northwards of Devonian seas—and probably a considerable deepening of the sea in the north Devon area. Only in Devon and Cornwall and in the Tenby area of Pembrokeshire are transitional beds from marine Devonian to marine Carboniferous beds found. They present a particularly difficult problem of defining the boundary between the two systems. The approximate correlation between the north Devon succession of mixed facies with the almost entirely marine succession of south Devon is given in Fig. 8.4.

Before turning to the outcrops of Old Red Sandstone facies, it is relevant to consider the buried Devonian rocks of south-eastern England. Forming part of a sub-Mesozoic ridge of Palaeozoic rocks, a ridge which in places is only 1,000 feet below sea level, an ever growing amount of data is being gained from boreholes. Devonian rocks of Old Red Sandstone facies have long been known from boreholes in north Kent and the margins of the Lower Thames estuary. These rocks, chiefly sandstones with ripple markings and rain pitting, include fossils, and the crustacean *Estheria* confirms a shallow non-marine or brackish environment of deposition. Recently, further south at Brightling (10 miles north-west of Hastings) beds of marine facies of Lower Devonian age have been found. Shales, sandstones and mudstones with *Chonetes* and *Pterinea* occur. Evidence in south-east England is sporadic, since it can be derived only from deep bores—at Brightling Devonian occurs 3,837 feet below sea

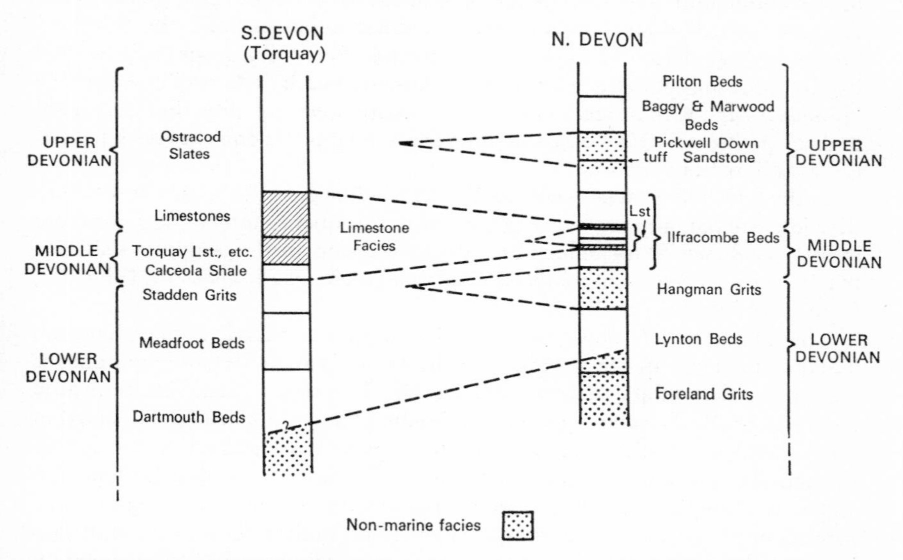

Fig. 8.4 The relationship of the North Devon succession to the marine Devonian succession of South Devon. Limestone beds shaded, beds of non-marine facies stippled.

level—and as yet nothing can be said of possible marine incursions northward or of the relationship of the marine to the non-marine facies, except that the margin of the Devonian geosyncline can be fixed (at least in Lower Devonian times) with a little greater precision.

The Old Red Sandstone

Strata of continental facies, known as the Old Red Sandstone, are widespread in Britain. They occur in northern Scotland, Orkney and Shetland, in the Central Lowlands of Scotland—extending beyond the narrow structural confines of the Midland Valley—and in south Wales and the Welsh Borders. Although virtually absent from the Midlands of England and only sparsely represented in northern England, the Old Red Sandstone may be extensive under the cover of younger rocks in eastern and southern England. Rocks of Devonian age are widespread in Ireland and are entirely of Old Red Sandstone facies except for a late Devonian marine incursion in the south.

The Anglo-Welsh Province

In Shropshire the red marls of the Red Downtonian are followed by Red Marls of Dittonian (Lower Devonian) age. No break in sedimentation took place but evidence of a gradual change in environment is apparent. The base of the Dittonian is regarded as the base of the Devonian in preference to the Ludlow Bone Bed since it is probably contemporaneous with the Silurian-Devonian boundary of the graptolite sequence of central Europe. It is defined by the incoming of ostracoderm fish and is marked by the widely mappable '*Psammosteus*' Limestone. Traced in a south-westerly direction the basal beds of the Lower Old Red Sandstone are overlapped along the north side of the South Wales Coalfield.

In the Old Red Sandstone of the Welsh Borders only the Lower and Upper divisions are faunally proven with a major non-sequence intervening, for there is little evidence of unconformity. The succession in the Anglo-Welsh area has been divided on a lithological and palaeontological basis as in Table 8.2, from work by J. R. L. Allen.

The Downtonian, now generally placed in the Silurian, may have no known equivalent in Devonshire and the Dittonian may be contemporaneous with the oldest strata in north Devon, the Foreland Grits. The transition from the brackish-water lagoonal conditions (with rare marine incursions) of the Grey Downtonian—succeeding the typical marine Ludlovian—to the profoundly different conditions in which the red beds of the Red Downtonian and Dittonian were laid down has been referred to in an earlier chapter. The formation of red beds is no longer taken to be indicative of arid conditions but they do infer weathering of the parent material under hot tropical conditions, perhaps of very intermittent rainfall. Lateritic clays provided some of the material for these micaceous fluviatile deposits of marl. The pellet rocks and sandstones, showing a wide variety of sedimentary features such as ripple-drift as well as trough and planar cross-bedding, were water laid. The conditions of deposition envisaged have been compared with those of the Colorado River subaerial delta, the sandy deposits laid down from braided distributaries and occasionally from widespread sheets of water in times of flood. Such conditions were of considerable extent since similar beds extend as far as Pembrokeshire. There is some evidence that the beds thicken westwards—they are 2,000 feet thick at Knighton compared with 1,200 feet in Brown Clee Hill—so that, although uniformity of sedimentary environment is established, the basin west of the Church Stretton Fault may have continued to subside more rapidly than the 'block' to the east. The whole area must have subsided greatly, sinking generally keeping pace with deposition to maintain similar conditions. During Downtonian and Dittonian times the delta may

TABLE 8.2

			Anglo-Welsh Province			
			Shropshire	*Forest of Dean and Bristol*	*Central S. Wales*	*Pembrokeshire*
FAMMENNIAN	FARLOVIAN	UPPER OLD RED SANDSTONE	Farlow Sandstone Series (400 ft)	Tintern Sandstone Group (400 ft)	Grey Grits (200 ft)	Skrinkle Sandstone (1,000 ft)
FRASNIAN				Quartz Conglomerate Group (100 ft)	Plateau Beds (50 ft)	
GIVETIAN		MIDDLE OLD RED SANDSTONE				
			------- U n c o n f o r m i t y -------			
EIFELIAN						Cosheston Group (? 10,000 ft) Ridgeway Conglomerate (1,200 ft)
EMSIAN	BRECONIAN	LOWER OLD RED SANDSTONE		Brownstones (2,500 ft)	Brownstones (1,400 ft)	
SIEGENIAN			Woodbank Series (770 ft)		Senni Beds (800 ft)	
GEDINIAN	DITTONIAN		Abdon Series (220 ft)	St. Maughans Group (1,400 ft)	Red Marls (4,400 ft)	Red Marls (2,800 ft)
			Ditton Series (1,200 ft)			
	DOWNTONIAN		Downton Series (1,000 ft)	Raglan Marls (1,000 ft)		

(From J. R. ALLEN, 1962, *J. sedim. Petrol.* **62**, 657)

have extended over the greater part of south and central Wales, limited to the north-west by the uplifted Caledonian Mountains of Wales, and extending eastwards across the Midlands as far as Kent. To the south lay the broad geosyncline, already described, extending from south-west England to Poland.

The '*Psammosteus*' Limestones include *Traquairaspis* [*Phialaspis* ≡ *Psammosteus*] and other fish fossils and were deposited in lagoonal conditions which became temporarily established over a wide area from Shropshire to Pembrokeshire in early Devonian (Lower Dittonian) times. In the strata which follow, cornstones (impure limestones) occur more commonly. Erosion surfaces are common, inferring intermittent flooding, and they mark the base of cyclothems in the cyclic sediments. The cyclothems vary greatly in thickness, as might be expected in fluviatile sediments, but they comprise lower arenaceous members followed by more argillaceous members. Towards the end of the Dittonian, lagoonal conditions again became established and the Lower and Upper Abdon Limestones were deposited. The Clee Series, coarse sandstones and grits often with subangular grains, indicate the proximity of a source of material from igneous and metamorphic rocks while Upper Silurian pebbles have been found in the conglomerates. Their presence is enigmatic, for the Silurian rocks of the Welsh Borders must have been covered by Lower Old Red Sandstone at that time. The Clee Series is the equivalent of the Brecon Series of the Brecon Beacons where some 2,000 feet of strata make impressive upland scenery.

In Herefordshire and westwards along the north side of the South Wales Coalfield the Red Marls of Downtonian-Dittonian age are followed by the Senni Beds, micaceous flaggy sandstones and shales with cornstones. *Pteraspis* and plant fragments infer an approximate Siegenian-Emsian age. The generally unfossiliferous Brownstones follow. Chiefly sandstones, they are still used as building stone. Apparently a conformable succession with no major breaks in the geological record above, the Downtonian is present so that the Brownstones would be placed with some confidence in the Lower Devonian. Further, they are overlain unconformably by Upper Old Red Sandstone. However, a small exposure with marine shells has yielded the Upper Devonian *Spirifer verneuili*. The highest beds, the conglomeratic Plateau Beds which form the flat top of the Brecon Beacons, may belong to the Upper Old Red Sandstone.

In Pembrokeshire, south of Milford Haven, thick conglomerates (1,200 feet) occur above the Red Marls, although they are absent to the north of the coalfield where the marls pass upwards into the Senni-Cosheston Beds without apparent break. The fossil evidence and the lithological similarity of the pebbles to the Grès de May of Brittany conflict with clear evidence that the pebbles have not been transported far. It is therefore supposed that they may be derived from a land ridge sited in the position of the present Bristol Channel. In north Pembrokeshire the Cosheston Beds, sandstones, marl and cornstone with breccias in the higher part, attain a thickness of 10,000 feet. Probably equivalent to the flaggy Senni Beds and Brownstones to the north, already described, they are not found south of Milford Haven through the great overstep of the Upper Old Red Sandstone there.

It is in Pembrokeshire, in the Skrinkle Sandstone (as the Upper Old Red Sandstone is known), that marine beds with lamellibranchs, crinoids, fish and brachiopods are found. This fauna includes *Spirifer verneuili.* Representing the last marine incursion of Devonian times, which laid down the Marwood Beds and Lower Pilton Beds of north Devon, it offers clear proof that the transgression extended into south Wales. It extended also into southern Eire where the Coomhola Grits, transitional from the Devonian to the Carboniferous, possess a fauna of lamellibranchs and brachiopods, the latter including *S. verneuili.* The Upper Old Red Sandstone of Shropshire, the Farlow Sandstone, succeeds the Woodbank Series = Brownstones with no apparent break in sedimentation, yet the fauna of the Farlow Sandstone is entirely different. Despite the apparent continuity of deposition, the presence of Middle Old Red Sandstone cannot be demonstrated. Westwards, the unconformity becomes apparent, the Upper Old Red Sandstone—here the Grey Grits—overstepping the Brownstones. However, as fossil evidence is scanty it cannot be ruled out that the Brownstones may be in part equivalent to the Middle Old Red Sandstone. (Similarly, part of the Old Red Sandstone of SW. Ireland may belong to the Middle Old Red Sandstone.)

The Old Red Sandstone of Anglesey, though quite unfossiliferous, is generally regarded as of 'Lower Old Red Sandstone age' on lithological and mineralogical evidence. Resting unconformably upon Ordovician strata and overlain by Lower Carboniferous (Viséan) beds, prior to the deposition of which it was folded, thrust and cleaved, structural relationships give no precise evidence of its age. Similarities to the Old Red Sandstone of both Pembroke and the Welsh Borders support the view that the Old Red Sandstone of Anglesey may belong to the Dittonian. The beds, totalling 1,645 feet, have been divided into four lithological formations of somewhat different facies though all are of continental type of deposition. The different facies have been interpreted as inferring piedmont, playa, stream and lacustrine conditions of deposition. Palaeocurrent directions, dominantly south-east, and lithological similarities suggest that Anglesey was part of the Anglo-Welsh area of Old Red Sandstone deposition, rather than a remote extension of the Caledonian Cuvette, an alternative interpretation.

The Devonian outcrops of the Mendips periclines appear to belong entirely to the Upper Old Red Sandstone, the Lower being eroded in the vicinity of a north—south line which has been called the Bath Axis. Sandstones of Upper Old Red Sandstone with a basal Quartz Conglomerate rest unconformably on Silurian volcanics. Although the Fammennian marine transgression spreading north-eastwards reached the Quantock Hills, there is no evidence that it extended as far as the Mendips. Old Red Sandstone beds occur at a number of places in the Bristol District and are exposed in the banks of the River Severn, but to the west in south Wales and northwards flanking the Forest of Dean the succession is more complete.

The Old Red Sandstone of Scotland

The Caledonian Orogeny which elevated the Highlands of Scotland brought into existence two intermontane basins or cuvettes of deposition. Lying between the Highlands and the Southern Uplands, though sedimentation extended beyond these narrow confines of the Midland Valley, was the Caledonian Cuvette, while to the north and east of the Highlands was the Orcadian Cuvette, bordered by the Northern Highlands and Grampians. Deposition in these two cuvettes was not entirely contemporaneous and considerable differences in lithology and in the successions indicate that it was not until Upper Devonian times that similar environments became established in the two.

THE CALEDONIAN CUVETTE

In the Caledonian Cuvette, Lower Old Red Sandstone was deposited at the foot of mountain areas, extending across the floor of the cuvette. Of the latter little is known, for the cover of Carboniferous rocks is several thousand feet thick and has not been penetrated by boreholes. However, the marginal deposits outcrop extensively—the Old Red Sandstone outcrops over more than half the area of the Midland Valley. The outcrop along the northern side of the Midland Valley is broad due to subsequent folding which produced the broad Mearns Syncline. Volcanic rocks form high hills, the Sidlaw and Ochil Hills in the north and the Renfrew Hills south of the R. Clyde, while the sediments, largely sandstones, form relatively flat, highly agricultural areas. The Old Red Sandstone succeeds beds originally designated Downtonian, but those in the Pentland Hills south of Edinburgh are now known to be somewhat older Silurian, while the Stonehaven Beds are still regarded as belonging to the Downtonian. They contain phyllocarids including the peculiar *Dictyocaris*, eurypterids and, of particular stratigraphical relevance, *Traquairaspis* [*Phialaspis*] similar to that found in the Welsh Borders.

The Lower Old Red Sandstone consists of coarse conglomerates, sandstones and lava flows. The conglomerates, which make up as much as half the succession, with well-rounded pebbles, show evidence of being water transported. They formed as piedmont deposits or scree-fans which coalesced resulting in a variable succession. However, the well exposed succession seen in the eastern coastal sections of Kincardine and Angus (Table 8.3) may be taken as typical—although the lavas there are chiefly basaltic whereas westwards in Perthshire andesites are more common.

The conglomerates thicken towards the north and north-west, the area of supply of material. It has been possible to trace a sequence of events in the sinking of the depositional basin from the somewhat different pebble composition of conglomerates at different horizons. In the lower groups there are many pebbles of jasper, greywacke, chert, etc., of probable Ordovician origin, derived from the Highland Border Series or analogous rocks. However, typical 'Highland' pebbles of gneiss, schist, phyllite, quartzite and vein quartz are common at all horizons where conglomerates occur. The Highland Boundary Fault undoubtedly formed a scarp, the northern limit of a subsiding

TABLE 8.3

Strathmore Group	1,500 ft	Red shales and marls
Garvock Group	4,000 ft	Chiefly sandstones and conglomerates with thin volcanics
Arbuthnott Group	5,000 ft	Chiefly volcanics, andesites and basalts with some conglomerates
Crawton Group	1,600 ft	Alternations of basalt and conglomerate
Dunnottar Group	6,900 ft	Coarse conglomerates and sandstones, with volcanics at four horizons

basin, and was periodically active as the Highlands were from time to time further uplifted and erosion rejuvenated. Meanwhile the subsiding basin accumulated an incredible thickness of sediments—varying from 12,000 to 20,000 feet. The sandstones, purple, reddish or brown, are often micaceous and arkosic including up to 30% of felspar grains, implying a generally arid environment.

However, temporary lakes existed in which fish abounded, perhaps becoming concentrated as the lakes dried up. The contemporaneity of the lavas and sediments is well seen at Crawton on the Kincardine coast, where pebbles of a conglomerate can be seen embedded in the upper surface of a basalt; the spread of conglomerate occurred before the lava was completely consolidated.

The succession in Arran commences with locally derived breccias followed by conglomerates with igneous and quartzite pebbles, the latter possibly from Islay and Jura. Rather similar to the Lower Old Red Sandstone elsewhere along the north side of the Midland Valley, the succession is exceptional in that only one lava flow occurred.

The succession on the south side of the Midland Valley is generally similar to that on the north side although it has not been worked out in the same detail. The lowest part comprises sandstones and conglomerates followed by volcanics with some interbedded sediments and, finally in the Lower Old Red Sandstone, sandstones and conglomerates with only subordinate volcanics were laid down. The pebbles in the conglomerates, being derived from the Southern Uplands, are chiefly of greywacke, chert and igneous rocks with some of quartzite and limestone.

The Caledonian Cuvette was not restricted to the Midland Valley; in Argyllshire (Lorne and Kerrara) and in the Ben Nevis and Glen Coe area there are remnants of formerly extensive sediments and volcanic rocks. The oldest sedimentary rocks of this region may be equivalent to the Stonehaven Beds and fish and trace fossils have been described. In Lorne and Kerrara sedimentary rocks are succeeded by volcanics. In Glen Coe and at Ben Nevis only indurated sediments, conglomerates, sandstones and shales with *Pachytheca*, occur together with a great thickness of volcanic rocks. The Devonian here owes its preservation to downfaulting of several thousand feet, faulting contemporaneous with igneous intrusion forming a 'caldera'. Further north, on the north-west coast of Loch Linnhe, conglomerates outcrop, but these, together with similar rocks found in the Great Glen, are believed to be Middle Old Red Sandstone belonging to the Orcadian Cuvette. One other little-known occurrence of probable Lower Old Red Sandstone occurs north of the Highland Border: one-third of the City of Aberdeen is underlain by it, proved in boreholes and exposed in the banks of the R. Don where it can be seen resting unconformably on Dalradian rocks. South of the Midland Valley, Lower Old Red Sandstone occurs in Berwickshire, while yet further south a great pile of Devonian andesitic lavas forms the Cheviot Hills, penetrated of course by a later granitic intrusion.

The Upper Old Red Sandstone succeeds the Lower in the Caledonian Cuvette with conspicuous unconformity and there is no evidence of deposition of the Middle Old Red Sandstone. On the contrary, several thousand feet of Lower Old Red Sandstone were removed by erosion following faulting and folding. The Midland Valley graben possibly came into existence during mid-Devonian times for, though it had been a basin of deposition of Lower Old Red Sandstone, already described, sedimentation had spread beyond the boundaries of the Midland Valley. The result of downfaulting of the Midland Valley by several thousand feet (the ultimate throw of the Highland Boundary Fault may be as much as 10,000 feet) was to preserve a great thickness of Lower Old Red Sandstone from erosion in the graben while it was almost completely eroded from the adjacent uplifted areas. (It is, of course, probable that in these areas the Lower Old Red Sandstone was initially much thinner. The Upper Old Red Sandstone therefore rests on Lower Old Red Sandstone in the Midland Valley, while to the north, for example on either side of Loch Lomond (the Balmaha area), it rests on Dalradian. Similarly, to the south of the Southern Uplands Fault, the Upper Old Red Sandstone of wide extent in Berwickshire and Roxburghshire rests on the Lower Palaeozoic rocks of the Southern Uplands. This is nowhere better seen than in the coast section at Siccar Point, first described by James Hutton. It is a fact that the largest area of Upper Old Red Sandstone, forming the Lammermuir Hills, lies outside the Midland Valley.

The Upper Old Red Sandstone comprises chiefly sandstones, some red in colour and often cross-bedded. Fossils are locally very abundant and a locality at Dura Den in Fifeshire has yielded many genera of fish.

THE ORCADIAN CUVETTE

Old Red Sandstone sediments laid down in the Orcadian Cuvette make up the entire Orkney Islands, hence the name, and there are extensive outcrops bordering the Moray Firth and in Caithness. Outliers to the west in Sutherland and in the Great Glen and to the south in Aberdeenshire and Banffshire testify to the former great extent of this sedimentary basin (Fig. 8.5). Its northern and eastern boundaries are unknown but it extended beyond the Shetland Isles. Apparently the cuvette did not come into existence essentially until Middle Old Red Sandstone (mid-Devonian) times. Its inception was probably contemporaneous with the renewed downfaulting of the Midland

Valley in post-Lower Old Red Sandstone times already referred to. However, in the south-west of the cuvette, beds below the characteristic basal beds of the Middle Old Red Sandstone, and overlain unconformably by the latter, may be of Lower Old Red Sandstone age. It has also been suggested that the beds of the Rhynie Outlier may be as old as Lower Old Red Sandstone.

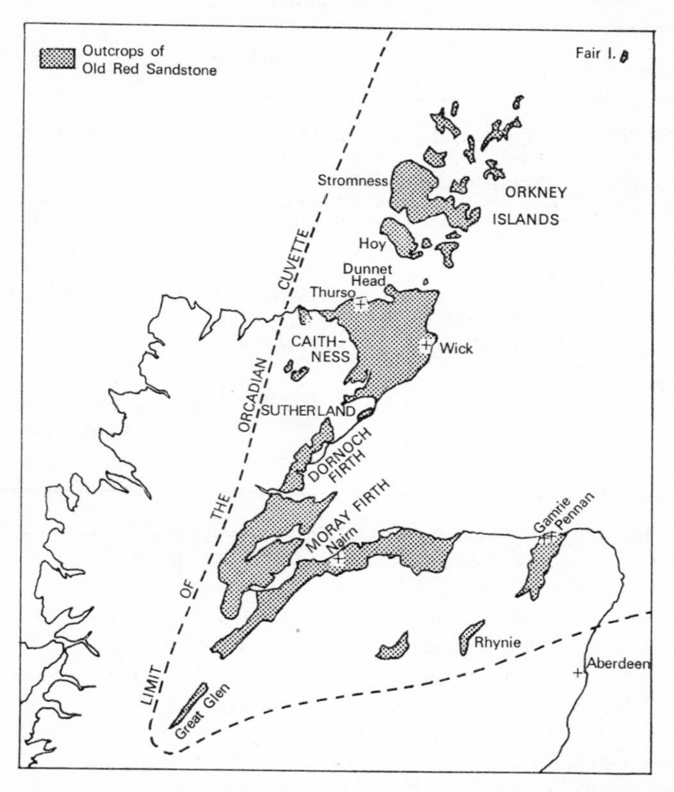

Fig. 8.5 Outcrops of Old Red Sandstone in the north of Scotland, indicating the probable extent of the Orcadian Cuvette, with place names referred to in the text. (Outcrops based on Crown Copyright Geological Survey Maps. By permission of the Director, Institute of Geological Sciences.)

The succession in the Orcadian Cuvette commences with red breccias and arkoses which rest on eroded metamorphic rocks. These are exposed in the coast at Gamrie and Pennan but are also well exposed along the western margin of the main outcrop. These basal beds fill hollows in an old land surface, but, in addition to these variations in thickness, northwards in Caithness there is overlap of higher beds. The basal beds are normally unfossiliferous but a unique locality at Rhynie, an outlier in Aberdeenshire, has provided remarkably preserved early vascular plants and microfossils in chert.

Some lateral variation in succession occurs for the central cuvette deposits thin laterally, while marginal deposits finger out towards the centre of the basin. However, the successions in Caithness and Orkney can be considered as standard (Table 8.4).

TABLE 8.4

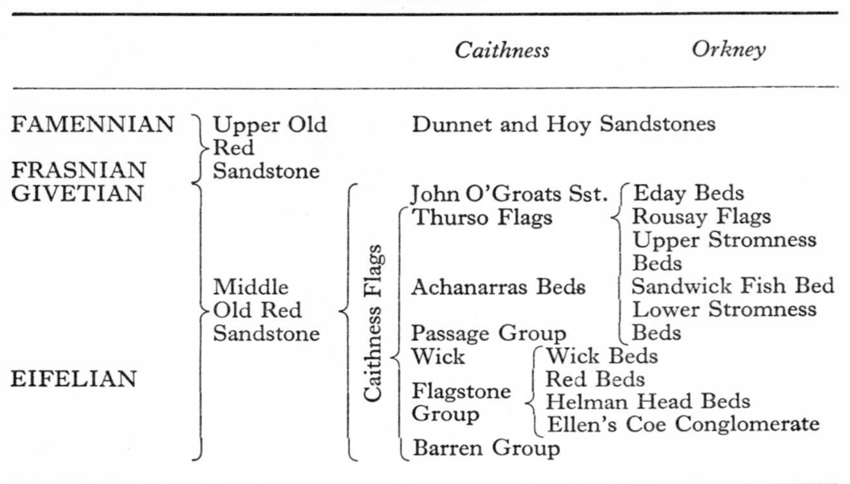

Stage	Series		Caithness	Orkney
FAMENNIAN FRASNIAN	Upper Old Red Sandstone		Dunnet and Hoy Sandstones	
GIVETIAN	Middle Old Red Sandstone	Caithness Flags	John O'Groats Sst.	Eday Beds
			Thurso Flags	Rousay Flags Upper Stromness Beds
			Achanarras Beds	Sandwick Fish Bed
			Passage Group	Lower Stromness Beds
EIFELIAN			Wick Flagstone Group	Wick Beds Red Beds Helman Head Beds Ellen's Coe Conglomerate
			Barren Group	

The Old Red Sandstone is folded into a gentle syncline, the axis of which trending north-east—south-west lies between and parallel to the Cromarty and Moray Firths (the Black Isle area). Higher formations are naturally seen towards the centre of the syncline. Following deposition of the basal breccias and arkoses, the latter with as much as 70% felspar grains indicating formation in arid conditions, the Orcadian Lake was formed. The succeeding beds, of the Wick Flagstone Group, show considerable variation in lithology, approximately 6,000 feet of sandstones, shales, sandy flagstones and limestones being deposited. The 2,500 feet of Passage Beds are grey calcareous flags, and the succeeding Thurso Flags (6,000 feet) are bluer and phosphatic with grey calcareous flags and sandstones. None the less, the whole succession (after the deposition of the Barren Group) shows a cyclic type of sedimentation. A typical cyclothem comprises:

Mudstone with suncracks	10 ft
Sandstone	12 ft
Black slaty flags with sandy beds	15 ft
Limestone	4 ft

The whole succession is essentialy lacustrine and the sediments were deposited in shallow water, cyclothems being commonly between 30 and 70 feet in thickness. Following the deposition of the early scree-formed or piedmont breccias, the lake appears to have been of remarkable persistence—unlike the temporary pools of the Caledonian Cuvette—yet the basin must have continually subsided to accommodate the 20,000 feet of Devonian strata present. The John O'Groats Sandstone and its equivalent appear to be unconformable on the Caithness Flags; certainly this is so in Shetland.

Fish are the only common fossils, as may be expected of this environment, and in some beds they are profuse. The fauna of the John O'Groats Sandstone includes a number of genera restricted to that formation, that is to say quite distinct from the typical fauna of the Caithness Flags.

The Upper Old Red Sandstone rests unconformably on the Middle Old Red Sandstone wherever the relationship can be seen, for example in the island of Hoy. Pre-Upper Old Red Sandstone folding took place followed by erosion, and the Upper Old Red Sandstone is believed to overstep the Middle Old Red Sandstone. However, it has recently been shown that nowhere along the southern margin of the Orcadian basin is there evidence of overstep, the Upper Old Red Sandstone being faulted against Moinian rocks south of Nairn. The Upper Old Red Sandstone commenced with a brief volcanic episode, ashy sediment and a lava flow (or flows) occurring at the base of the succession. The deposition of 3,500 to 4,000 feet of sandstones with some conglomeratic and cornstone beds followed. As much as 1,000 feet of sandstones are exposed in places in the sea cliffs of Foula and some of the Orkney Islands.

In Caithness and Orkney the Upper Old Red Sandstone comprises the Dunnet and Hoy Sandstones with few fossils. Further south, around the Moray Firth, four or five divisions based on a sequence of faunas can be recognized (Table 8.5).

TABLE 8.5

	Moray Firth	*Caithness and Orkney*
FAMENNIAN	Rosebrae Beds Scaat Craig Beds	Dunnet and Hoy Sandstones (Famennian and Frasnian)
FRASNIAN	Alves Beds Boghole and Whitemire Beds	
GIVETIAN	Nairn Sandstone	Bressay Flags Conglomerate

In general the rocks of the Upper Old Red Sandstone are less coarse than those of the Lower or the earlier beds of the Middle Old Red Sandstone, inferring that the relief of the Highlands had been reduced by late Devonian times. They are chiefly yellow and red sandstones, often cross-bedded, with occasional conglomerates, marls and cornstones. Some lateral variation is detectable; for example, piedmont deposits near the margin of the cuvette pass laterally into alluvial and lacustrine deposits. Nevertheless, widespread uniformity of depositional conditions had become established by late Devonian times and the Upper Old Red Sandstone of the Orcadian Cuvette is similar to that of the Caledonian Cuvette. In fact, the beds of the Moray Firth area have been correlated with strata of the Baltic region on the basis of closely similar fish faunas. Since the continental facies deposits are interbedded with marine Devonian beds in the latter area, it is possible to relate the Upper Old Red Sandstone of Scotland to the marine Devonian zonal sequence with some confidence.

It has been suggested that the Upper Old Red Sandstone may not be entirely of Devonian age, perhaps extending into the Carboniferous Period. In Caithness and Orkney the absence of fossils makes it impossible to determine when sedimentation of Old Red Sandstone type ceased. In the Caledonian Cuvette the Lower Carboniferous follows the Upper Old Red Sandstone, usually with apparent conformity, although no early Carboniferous sediments (Tournaisian age) are proved. It seems not unlikely that in Scotland sediments of Old Red Sandstone facies may not belong entirely to the Devonian Period.

The Devonian of Ireland

Rocks of Devonian age are almost entirely of Old Red Sandstone facies and there is evidence that they were laid down in cuvettes like those of Scotland, although at times deposition may have extended over much of Ireland. Late in Devonian times the maximum northwards spread of marine deposition, which reached Pembrokeshire, extended into south-west Ireland.

Present outcrops of Devonian rocks can be arbitrarily grouped into three regions: there are extensive outcrops in the south-west, the Amorican belt where Devonian and Carboniferous beds are folded along east-west axes; in the Central Plain inliers of Old Red Sandstone, together with Silurian rocks in many cases, project through the Carboniferous cover to form prominent hills; in the north are outcrops of Old Red Sandstone analogous to those of the Caledonian Cuvette of Scotland. Devonian age rocks were deposited over much of Ireland and must underlie much of the Central Plain, although they are missing from the Longford-Down Massif and from the Wicklow-Wexford Massif.

In the Armorican belt of south-west Ireland the age of the beds is in dispute through lack of fossil evidence. This is especially true of the strata on the north side of Dingle Bay where some 10,000 feet of conglomerates, grits, sandstones and slates succeed proven Silurian (Wenlock and Ludlow) age beds unconformably, but are themselves overlain unconformably by typical Old Red Sandstone. It seems reasonable to refer these beds to the Downtonian despite the lack of palaeontological confirmation. There is clear evidence that these beds, divided into the Dingle Beds below followed by the Glengariff Grits, are piedmont-type deposits and were formed in a basin flanked to the north by mountains of the Caledonides. They are not found outside south-west Ireland. Perhaps contemporaneous with the deltaic Downtonian of the Anglo-Welsh Cuvette, there is no reason to suppose that the areas of deposition were connected, although they both owe their origin to the uplift of Caledonian mountains to the north. Contemporaneous rhyolitic lava flows occurred in Ireland and can be found in the extreme west of the Dingle Peninsula.

The Old Red Sandstone of south-western Ireland rests on an irregular erosion surface which had not been peneplaned so that the beds are of variable thickness but at least several thousand feet. Piedmont-type breccias, such as the Inch Breccia, and conglomerates pass southwards and upwards into deltaic-fluviatile sandstones. Cross-bedding in the the latter confirms a general direction of transport of material from the north or north-west. The Old Red Sandstone reaches a thickness of as much as 10,000 feet in the Comeragh Mountains of Waterford. There, in the eastern part of the Amorican Belt, the Old Red Sandstone rests on Ordovician and Silurian slates and volcanic rocks, from which many of the pebbles in the conglomerates are derived. Away from the Comeragh Mountains the Old Red Sandstone thins, to only 1,500 feet south of Waterford, while further east it is absent, for the Lower Carboniferous oversteps onto Lower Palaeozoic rocks. Widespread uniform conditions were established by Upper Devonian times and the sandstones and siltstones of the Kiltoran Beds were laid down. This formation is of remarkably consistent lithology and thickness throughout the outcrops of southern Ireland. Both plant and fish fossils confirm the Upper Devonian (Upper Old Red Sandstone) age. Unlike the Lower Old Red Sandstone, the Upper is locally highly fossiliferous and at Kiltoran in Kilkenny includes the earliest non-marine lamellibranch, *Archanodon*. The most important plant fossil is *Archaeopteris hibernica*.

In the periclinal structures of the Central Plain of Ireland the beds belong chiefly to the Lower Old Red Sandstone and may have been laid down in an area of deposition contiguous with the extension of the Caledonian Cuvette.

In northern Ireland there are outcrops of Old Red Sandstone which, palaeographically, belong to the Caledonian Cuvette. The largest outcrop lies chiefly in Co. Tyrone but patches occur to the south-west, forming the Curlew Mountains, and beyond outcrop near Clew Bay. The Old Red Sandstone was deposited beyond the confines of the 'Midland Valley' and outcrops lie to the north of the continuation of the Highland Boundary Fault on the north-east Antrim coast between Cushendall and Cushendun. Resting unconformably on Dalradian rocks, as in Arran, the Old Red Sandstone comprises coarse conglomerates and sandstones with lavas, chiefly andesitic. The pebbles in the conglomerates were locally derived; in the north-east they are largely of Dalradian origin, but in the Fintona area of Co. Tyrone include Dalradian, Ordovician and Silurian rocks, some from the Tyrone Volcanic Series.

Devonian fossils

Although marine Devonian rocks are limited to the South-West Peninsula and the contained fossils are often much distorted by subsequent tectonic deformation of the beds, the faunas are important for correlating from one area to another and do permit correlation with Devonian strata in Europe. In addition, the faunas permit comparison with Lower Palaeozoic fossils and important differences can be seen. By Devonian times Cephalopods had become important and it has been noted that Ammonoids (Goniatites) are used as zone fossils. Nautiloids and Bactritida are also common. This is particularly true of the rocks of south Devon and Cornwall which belong to what has been termed the Hercynian magnafacies, comprising shales and limestones. Common in the limestones are reef-building corals, both tabulates such as *Heliolites* and *Favosites*, and rugose compound corals. Stromatoporoids also form important constituents of the reefs. Simple rugose corals, some of unusual shape (for example *Calceola sandalina*), are also common.

While less common than in the Silurian, trilobites are numerous and varied. *Phacops* and *Cheirurus* are common and in fact most Silurian trilobite families are represented, the exceptions being the Illaenids and Encrinurids. Echinoderms are not generally numerous, with the exception of crinoids at some horizons—the distinctive *Cupressocrinus*. However, blastoids become commoner tending to replace cystids which decline. Very few inarticulate brachiopods survive into the Devonian. Those which do, such as *Lingula* and *Orbicuoidea*, survive to the present day. Orthids and strophomenids as well as rhynchonellids are numerous, as are spire-bearing forms: species of *Spirifer* afford an approximate zoning of the marine Devonian and Athyrids make their appearance.

Of molluscs other than cephalopods, gastropods and bivalvia (lamellibranchs) are common but more especially in the rocks of the quartzitic Rhenish magnafacies which comprises greywackes, quartzites, sandstones and shales. Rocks of this facies include many characteristic genera which are absent from the calcareous Hercynian magnafacies. In addition, the coral *Pleurodictyum*, the trilobite *Homalonotus*, *Tentaculites* and strongly ribbed brachiopods including very broad Spirifers are largely restricted to the quartzitic facies.

The fauna of the Old Red Sandstone facies includes, as well as trace fossils and microfossils, many Eurypterids and vertebrates, namely fish. The fish faunas of the Lower, Middle and Upper Old Red Sandstone are quite distinctive. The fish fauna of the Lower Old Red Sandstone comprises, essentially, three elements. These are the jawless Ostracoderms with dorsal shield and granular armour which had made their appearance in the Upper Ludlow Beds of the Silurian and typify the Lower Old Red Sandstone. Common genera are *Pteraspis*, *Cephalaspis*, *Thelodus* and, in Scotland, *Lanarkia*. The Acanthodians, with paired fins and a dorsal fin supported by a spine, are also common and the Arthrodeirea make their appearance. In the Middle Old Red Sandstone there are several groups of fish. Few Lower Old Red Sandstone genera survived; *Cephalaspis* is one of the few that did, although the Ostracoderms are represented by such forms as *Ptericthys*. Large Arthrodeirans with armoured head and trunk such as *Coccosteus* and *Homosteus* and Dipnoi such as *Dipterus* are common. The slender, fusiform Crossopterygians with rhombic scales and paired fins (e.g. *Osteolepis* and *Diplopterus* and the early Actinopterygii—known as the ray-finned fish—with small rhombic scales) are present. The Upper Old Red Sandstone has a fauna similar to that of the Middle Old Red Sandstone, many of the families continuing although being represented by quite different genera. Crossopterygians are the most important element of the fauna and common genera are *Holoptychius*, *Bothriolepis* and *Asterolepis*.

The Devonian is noteworthy for the record of the earliest land plants of the Northern Hemisphere. Lower Devonian plant remains are scarce in Britain but fragments probably related to *Psilophyton* are known. The remarkably preserved plants of Middle Devonian age were discovered at Rhynie, Aberdeenshire, in 1910 and referred to the genera *Rhynia*, *Hornea* and *Asteroxylon*. New material cannot be found but re-examination of the early material has revealed algae and fungi. Plants are commoner in the Upper Old Red Sandstone and the forerunners of the more abundant Carboniferous flora make their appearance.

9 The Carboniferous System

The Carboniferous System is traditionally divided into the Lower and Upper Carboniferous in Britain and western Europe. The two systems in North America, the Mississippian and the Pennsylvanian, correspond broadly to these divisions. Because of the vast amount of work published on the Carboniferous System it is advantageous to discuss the Lower and the Upper Carboniferous as distinct entities. The division is in many areas readily recognizable due to mid-Carboniferous earth movements which caused important changes in palaeogeography and sedimentation. It is, of course, chiefly in the Upper Carboniferous that workable coals occur and from which the name of the system is derived.

The base of the Carboniferous

The base of the Carboniferous is readily defined where marine Carboniferous sediments conformably succeed Devonian of Old Red Sandstone facies by the appearance of marine fossils. This is so in the South-West Province (Bristol-Somerset area). Of course in the South-West Peninsula, as in the Ardennes and many areas of the Continent, where marine Carboniferous beds follow marine Devonian, transitional marine beds occur and the base of the Carboniferous can only be defined faunally—by the appearance of accepted Carboniferous age fossils. In Scotland non-marine scantily fossiliferous lowest Carboniferous strata pose a problem, while in many British outcrops the base of the system is not exposed. Elsewhere the Lower Carboniferous rests unconformably upon older rocks. The problem of defining the upper limit of the Carboniferous is obscured in Britain generally by the unconformity of the overlying Permian System. Nowhere in Britain is the Carboniferous succession complete (and much of the Lower Permian is missing) so that this unconformity is striking.

The main divisions of the Carboniferous recognized are given in Table 9.1.

TABLE 9.1

		Stages	Lithological Names
UPPER CARBON-IFEROUS	SILESIAN	STEPHANIAN WESTPHALIAN	Coal Measures
		NAMURIAN	Millstone Grit, etc.
LOWER CARBON-IFEROUS	DINANTIAN = AVONIAN	VISÉAN TOURNAISIAN	Carboniferous Limestone and Limestone Shales

Lower Carboniferous

Regional setting of Britain

The Lower Carboniferous saw the continuation and extension of the geosyncline initiated in Devonian times. The history of the Carboniferous is the history of the spread of marine conditions, of the gradual incursion of the former Old Red Sandstone Continent by chiefly shallow epicontinental seas while to the south strata were deposited in the geosynclinal foredeep (sometimes in deep-water conditions). The latter are known as Culm from the presence of inferior coal seams which occur chiefly in the later beds when the foredeep was at times locally silted up though the Culm is, as a whole, marine and partly of deep-water facies with a cephalopod fauna. The distribution of Culm deposits closely corresponds to that of the marine Devonian beds since the Carboniferous geosynclinal deep lay to the south of the site of the Old Red Sandstone Continent. In western Europe geosynclinal Carboniferous sediments are to be found in Armorica (Brittany, chiefly), the Harz Mountains, the Black Forest and the French Massif Central as well as in Devon and Cornwall and south-western Ireland. Further to the south and south-east, in the Alpine region, Carboniferous rocks also occur but they have been metamorphosed to an even greater degree than those of the Franco-German horsts so that their elucidation is difficult.

Most of Britain formed part of the area, to the north of the geosyncline, inundated by generally shallow seas. North of a line through the Bristol Channel to the Ardennes Lower Carboniferous rocks are largely of neritic facies. However, downwarping of the crust must have been far from uniform within this region, for the gradual transgression of the Carboniferous seas was a complex story and the Carboniferous strata show a great variety of succession and facies within this region from one area to another. The main features of Carboniferous geography dated back to the Caledonian Orogeny, despite the effects of earth movements preceding and during early Carboniferous times known as the Bretonic earth movements. Mountain areas created in pre-Devonian or early Devonian times had not been eliminated by erosion, and the Scottish Highlands and the mountains of north and mid-Wales remained. The Welsh Mountains formed part of an upland area known as St. George's Land and the Midland Barrier, a dominant feature of Carboniferous palaeogeography (Figs. 9.1 and 9.2). An area of non-deposition, and frequently of active erosion supplying material to adjacent areas of deposition, it extended westwards to include the Wicklow Mountains of Ireland while eastwards it extended through the south Midlands and East Anglia into Belgium (to include the Brabant Massif). These eastward extensions now lie buried beneath a cover of later rocks, chiefly of Mesozoic age. Despite its

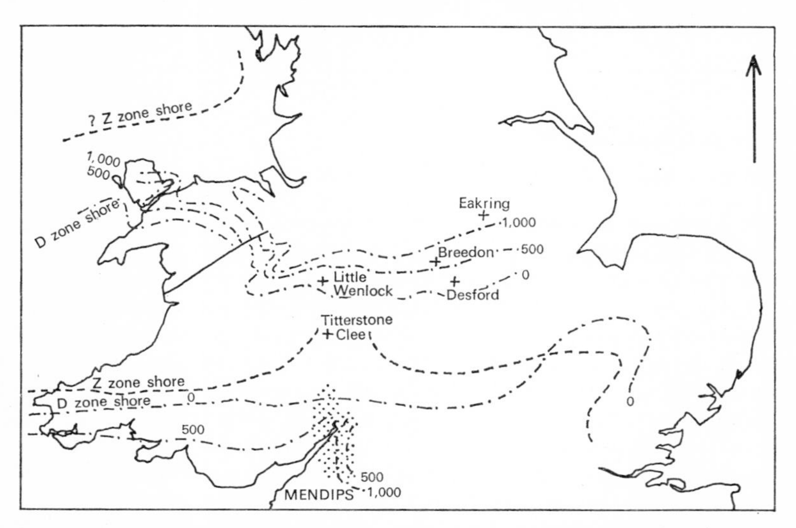

Fig. 9.1 The Midland Barrier—St. George's Land shorelines in early (*Zaphrentis* Zone) and late (*Dibunophyllum* Zone) Lower Carboniferous times. Isopachytes for strata of the *Dibunophyllum* Zone sandy facies stippled. (Based on *British Regional Geology, Central England*, Fig. 6, Crown Copyright Geological Survey Diagram and T. N. GEORGE, 1958, *Proc. Yorks. geol. Soc.*, **31,** 227.)

overall east—west or north-west—south-east trend, the old rocks which formed St. George's Land were arranged in ridges with a Charnoid trend (NE.—SW.). Of the Caledonides, the Southern Uplands appear to have had the least effect upon the facies distribution of Carboniferous sedimentation, although they were possibly never completely covered by Carboniferous seas. New elements controlling sedimentation, notably the North Pennine Fault Block, were important despite a previously little-known history.

The Lower Carboniferous of the South-West Province

The Lower Carboniferous rocks which outcrop in the Bristol area, the Mendips and south Wales belong, collectively, to what is called the South-

West Province. The succession is most complete here, it being one of the areas where the Tournaisian is well developed, and is exposed in a continuous section in the gorge of the River Avon. Some 2,700 feet of Lower Carboniferous are exposed, only slightly less than the greatest development of approximately 3,000 feet in the Mendips. Although a largely limestone succession, it commences with limy shales, which were laid down as muds, with occasional sandstone and conglomerate beds formed when coarser material was brought into the area. These 'Limestone Shales' are widespread in the South-West Province, being best developed in the Mendips and thinning northwards. The shales are succeeded by limestones, some massive and calcitic but others oolitic or dolomitic representing differences in depositional environment. A number of 'Phases' are referred to. The Zaphrentid and Cyathaxonid Phase comprising muddy limestones, characterized by simple corals as well as brachiopods and bryozoa, represents deposition in deeper water too muddy to permit the growth of massive compound corals and often deposited following shales. The Standard Limestone Phase comprises limestones made up of calcite shell fragments bound together by secondary crystals of calcite and is frequently crinoidal but corals and foraminifera may be abundant. The '*Modiola*' or so-called Lagoonal Phase occurs at several horizons in the Carboniferous succession of the South-West Province. It represents deposition in shallow shoals probably *not* in isolated lagoons and the lithology is porcellanous limestone or calcite mudstone (known as 'chinastone') while the fauna is restricted to '*Modiola*', ostracods and other microfossils. No true Reef-Knoll Phase limestones occur in this province. Oolites and pisilotes, often cross-bedded, occur, deposited in shallow tide-scoured seas.

The Lower Carboniferous, known as the Avonian from the type section in the Avon Gorge where it was first described in detail and a zonal sequence given by Vaughan in 1905, is known more generally as the Dinantian from the Dinant Basin of Belgium where strata of similar facies occur. A closely analogous section is there exposed in the gorge of the River Meuse between Dinant and Namur. Tournai and Visé, towns in the Dinant Basin, give their names to the divisions of the Lower Carboniferous. However, the fossil zones based on coral and brachiopod faunas (Table 9.2) are derived, as a matter of priority, from the Avon Gorge succession.

TABLE 9.2

		Lower Carboniferous Zones
DINANTIAN = AVONIAN	VISÉAN	*Dibunophyllum* Zone D (D_3, D_2, D_1)
		Seminula Zone S_2
		Upper *Caninia* Zone C_2S_1
	TOURNAISIAN	Lower *Caninia* Zone C_1
		Zaphrentis Zone Z (Z_2, Z_1)
		Cleistopora Zone K (K_2, K_1 and Passage Beds)

Although the coral *Cleistopora* has been referred to the genus *Vaughania* and the brachiopod *Seminula* to the genus *Composita*, the original stratigraphical nomenclature is retained.

The zonal sequence is largely restricted to sequences of limestones and hence it has geographical limitations, but it permits close correlations with the Lower Carboniferous of the Dinant Basin and with the chiefly limestone succession of north-west England. In these areas some of the characteristic zone fossils are rare or absent and alternative species have been proposed (for NW. England by Garwood in 1918). The main problem of Lower Carboniferous correlation was undoubtedly to relate the typical shelf-sea successions to mainly argillaceous, deeper water, successions. For the latter a zonal sequence of goniatite faunas has been established and approximately equated to the coral-brachiopod zones (see p. 196) despite the fact that corals and goniatites are virtually mutually exclusive of each other.

The uppermost beds of the Old Red Sandstone are called, in Somerset, the Tintern Sandstone (though in the Avon Gorge the Portishead Beds), while in Pembrokeshire the highest Old Red Sandstone is the Skrinkle Sandstone. Sedimentation continued into the Carboniferous without break although the Lower Carboniferous overlaps the Old Red Sandstone to rest on successively lower horizons on the periphery of the South Wales Coalfield. The changes in lithology and fauna were gradual: transitional beds, shaly (with sandstone) passing upwards into limestones, contain a fish fauna which gives way upwards to spirorbids and ostracods which precede the crinoid and bryozoan faunas of the limestones. The latter indicate the eventual establishment of marine conditions. The Limestone Shales which are essentially equivalent to the *Cleistopora* Zone are up to 500 feet thick in the Mendips where they are best developed. They contain few corals apart from *Vaughania* but brachiopods are common. Strata of early Tournaisian age are unknown further north but there is little evidence of the position of the shoreline and an absence of deposits of littoral facies. Certainly, however, the sea was of wide extent, beds of this age occurring from south-west Ireland to Cambridgeshire and Kent (proved in boreholes) and beyond.

In the South-West Province the transition from the Limestone Shales to Main Limestone was relatively rapid and was accompanied by the appearance of *Zaphrentid* corals. The *Zaphrentis* Zone consists of limestones, deeper water conditions giving rise to dark crinoidal limestones in the eastern and central Mendips while in west Wales dolomites are common. Lighter grey limestones and dolomites (such as the *Laminosa* Dolomite) follow almost everywhere in the region except the eastern Mendips where deeper water conditions persisted. The coral *Caninia* makes its appearance (in bed γ), and certain brachiopods such as *Syringothyris* are characteristic. Oolites follow—the *Caninia* Oolite often cross-bedded and laid down in shallow water—completing the Tournaisian succession. The chief characteristic of the Tournaisian is that generally similar conditions were widespread, although shallower water conditions prevailed northwards and, as stated, deeper water conditions were persistent in the eastern Mendips.

In Viséan times, on the other hand, there is much greater variation in facies both laterally and from time to time. At the top of the C_1 Zone there is in general a marked change in lithology and fauna and, at some localities, a physical break due to an interruption in sedimentation resulting from mid-Dinantian earth movements. Among these was the uplift of the Lower Severn Axis which resulted in variation in facies, shallowest water conditions existing over the axis and to the north. South and south-eastwards deeper water conditions persisted. For example, in the eastern Mendips the C_2 limestones are light grey, massive fossiliferous limestones similar to the preceding beds but over much of the Mendips white Burrington Oolite was laid down. This passes northwards into grey limestones with many gastropods; still further north now outcropping in the Avon Gorge were shallower water dolomite chinastones (the *Caninia* Dolomite). West of the R. Severn algal limestones were laid down in very shallow water and they contain few fossils except algae such as *Mitcheldeania*.

TABLE 9.3

VISÉAN	D_3	Sandy Beds (Brandon Hill Grit)
	D_2 D_1	Standard limestone with pseudobreccias
	S_2	*Modiola* 'lagoonal' Phase, concretionary limestone *Seminula* Oolite
	C_2S_1	*Modiola* 'lagoonal' Phase *Caninia* Dolomite
TOURNAISIAN	C_1	*Caninia* Oolite *Laminosa* Dolomite
	Z	Standard Limestone Massive Crinoidal Limestone
	K	Shales with *Modiola* 'lagoonal' Phase

In C_2S_1 times uplift rejuvenated St. George's Land and terrigenous material spread into the South-West Province so that the Lower Drybrook Sandstone was laid down north of Bristol and west of the R. Severn. Meanwhile limestone deposition continued in the Bristol area and in the Mendips, in the latter massive limestones with occasional oolites but shallower water with more varied lithologies near Bristol. Coral limestones, oolites (such as the *Seminula* Oolite) pisolites and algal limestones are found. The limestones of the D Zone are oolitic or rubbly with pseudobreccias and are highly fossiliferous with a wide variety of rugose corals. Minor outbreaks of volcanic activity took place in the South-West Province during Carboniferous times but extensive and continued volcanic activity occurred chiefly in Scotland during the Carboniferous as it had also in the Devonian.

The late Lower Carboniferous arenaceous beds of the South-West Province, at one time referred to the 'Millstone Grit' (Namurian), have been shown to belong largely to the Viséan (but to extend up into the Namurian). However, the southward spread of rocks of sandy facies did not reach Bristol until the end of D_2 times and the Mendips even later. Commencing at an early date in the north of the Province and gradually spreading southwards, the sandy beds form one of the best known examples of diachronic strata, their onset ranging in age from C_2 to post-D_2. They are given different names in different localities such as the Drybrook Sandstone and the Brandon Hill Grit.

Despite the lateral variations in facies, especially of Viséan strata, a typical succession for the South-West Province can be given.

The Lower Carboniferous encircles the South Wales Coalfield, except where it is broken by the sea, and in Gower it is repeated by Armorican folding. It shows variation in thickness and lithology. Thinning generally north-eastwards, it is thinnest north of Pontypool. The effects of mid-Dinantian earth movements are more pronounced and, in addition to the marked lithological change between the C_1 and C_2S_1 Zones, the top of the *Caninia* Oolite is eroded. The Upper Dinantian follows with overstep.

The South-West Peninsula

Outcrops of Carboniferous strata are extensive in Devon (and north-east Cornwall) and only near the axial region of the synclinorial structure is the Carboniferous overlain by a narrow outcrop of younger beds. The Carboniferous strata are of very different lithology and fauna from those of the South-West Province (Bristol-Mendips) which occur as near as 25 miles to the north-east. Between the two areas of Carboniferous rocks occurs the inlier of Cannington where oolitic and dolomitic limestones of C_2S_1 age outcrop. Clearly the latter belong to the South-West Province. The transition to the geosynclinal deposits of the South-West Peninsula must therefore be relatively abrupt but it cannot be investigated, the critical area being concealed by Mesozoic beds.

Known as Culm from the presence of occasional thin seams of impure coaly material, the geosynclinal Carboniferous is dominantly argillaceous with cherty and arenaceous (greywacke and grit) beds. Impure limestones occur but the facies contrasts with the typical calcareous facies found in the Mendips and at Bristol. Fossil evidence of the origin of these beds appears confusing since, while goniatites are the most common kind of fossil, trilobites occur and (especially in the Upper Culm) plant fossils and poorly preserved non-marine lamellibranchs. There seems little doubt from the lithologies that the Culm was deposited in a rapidly downwarping geosynclinal trough or foredeep and that the deposits are analogous to the geologically more recent Alpine 'Flysch'. Occasionally deep-water conditions permitted the establishment of goniatite faunas, but at times the trough was all but silted up. Plant debris was at times washed into the area of deposition.

The Lower Culm, which due to the regional dip occurs to the north and south of the Upper Culm, succeeds the Pilton Beds of late Devonian-Early Carboniferous age. The relationship is complex since major strike faulting of primarily Armorican age is present and there are lateral variations in facies within the Culm. However, a succession has been established in north Devon and it is lithologically characteristic of the Lower Culm.

Instow Beds 1,500 ft	Greywacke and shale
Limekiln Beds 100–200 ft	Siltstones and black shales
Chert Beds 150–300 ft	Central facies all chert: northern facies includes limestone and shales
Pilton Beds	

Volcanic activity was widespread during the deposition of the Lower Culm and spilitic lavas and tuffs occur, though not in the north Devon outcrops. Volcanic rocks are best seen in the north Cornish coast and near Launceston.

The Upper Culm is, in part, similar to the Coal Measures in lithology and perhaps conditions of deposition. Fossil plants and non-marine lamellibranchs confirm an Upper Carboniferous age. It is, however, conveniently dealt with together with the Lower Culm and the problems of Carboniferous stratigraphy in south-west England dealt with all together. Again, the succession is difficult to elucidate due to periclinal folds and strike faulting of Armorican age, but successions have been worked out, for example near Bideford in north Devon.

Shales and sandstones of the Northam and Abbotsham Beds, Coal Measures type strata, succeed the Instow Beds of the Lower Culm and represent an abrupt change in lithology and conditions of deposition. Locally thin bands of culm occur without seatearths. The main culm seam near the top of the formation does have a gannister-like base. Siltstones and sandstones make up the overlying Greencliff Beds with greywackes appearing in the higher part. In the greywackes of the Cockington Beds above, characteristic turbidite sole marks have been described.

Essentially, the Carboniferous history of Devon and Cornwall is of a geosynclinal trough formed by the deepening and extension of the Devonian sea, followed by a southward advance of deltaic 'Coal Measure' type sediments followed by a return to turbidite deposition. Much of the Carboniferous Period appears to be unrepresented by strata and, following the Pilton Beds of undoubted Famennian, *Cleistopora* and *Zaphrentis* Zone age, the Lower Culm goniatite faunas clearly indicate a late Viséan age with P Zone fossils (see p. 196 for lamellibranch-goniatite zones). Again, the sparse fauna of the Upper Culm indicates a Namurian and Lower Westphalian age: goniatites of the *Reticuloceras* Zone and non-marine lamellibranchs of the *Lenisulcata* and *Communis* Zones have been recorded.

The succession in north Devon is summarized in Table 9.4.

TABLE 9.4

Major Stratigraphical Divisions	*Formations*	*Faunal Evidence*	*Unrepresented by Strata (= hiatuses)*
PERMO-TRIAS			Stephanian and much of Westphalian
UPPER CULM (Upper Carboniferous)	Cockington Beds, Greencliff Beds; Northam and Abbotsham Beds	*C. communis* and *A. lenisulcata* Zones; G and R Zones	
LOWER CULM (Upper and Lower Carboniferous)	Instow Beds, Limekiln Beds, Chert Beds	G Zone, R Zone	Lower Namurian
		P_1 and P_{2a}	Much of Viséan and higher Tournaisian
L. Carb.–U. Dev.	Pilton Beds	Z, K and Famennian	

The facies belts in the Lower Carboniferous run approximately east—west so that rocks of limestone facies, similar to those of the English South-West Province, occur in southern Ireland with rocks of Culm facies in the extreme south (Fig. 9.2). The latter are found south-west of Cork. As in the south-west of England, the transition from shelf limestones to culm facies must take place in a very short distance. It has been suggested that in Ireland the terrigenous material of the culm facies, though it accumulated in a downwarp, was perhaps not deposited in a geosynclinal foredeep since the succession includes thin limestones.

The Central Province of England

In early Lower Carboniferous times St. George's Land lay on the north flank of the areas of deposition already discussed, and deposition did not take place generally in the Midlands until later in the Carboniferous Period. However, the northern shore of the South-West Province must have been located as far north as Titterstone Clee where K and Z age limestones succeed the Old Red Sandstone. They are followed unconformably by deltaic sandstones of Westphalian age, called the Cornbrook Sandstone, inferring non-deposition during much of the Carboniferous Period. The transgressive Carboniferous sea did not inundate the area north of St. George's Land until late in Lower Carboniferous times and, although in north Wales there are 3,000 feet of limestones, the greater part belong to the D Zone. The Lower Carboniferous on the north side of St. George's Land rests with conspicuous unconformity on older rocks and thins progressively southwards. From a thickness of about 2,000 feet in the south Pennines the Lower Carboniferous declines to little more than half this thickness at Breedon in North Leicestershire, while at Desford 10 miles west of Leicester the thickness proved in a boring was only 25 feet (Fig. 9.1).

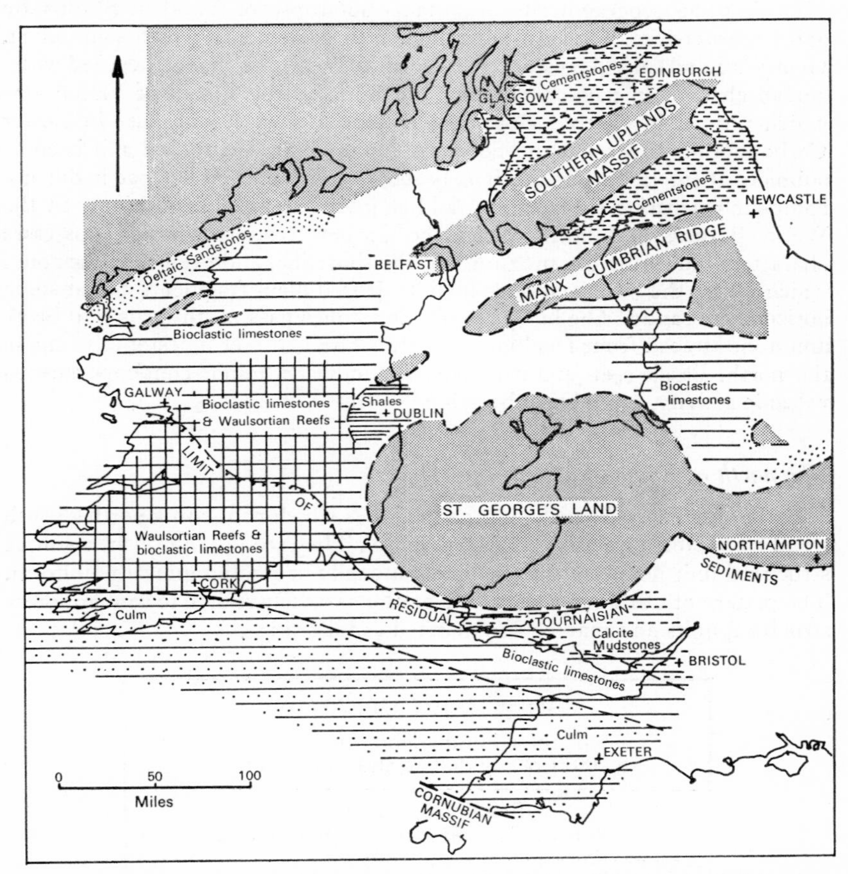

Fig. 9.2 Early Viséan palaeogeography of Britain. (Modified slightly from T. N. GEORGE, 1958, *Proc. Yorks. geol. Soc.*, **31,** 227.)

The Lower Carboniferous shoreline ran in an approximately east—west direction although it was undoubtedly far from straight since the 'grain' of the uplands was north-east—south-west. Embayments may have been present, for example at Llangollen where basal littoral deposits are present. The Lower Carboniferous was deposited from Anglesey, where outcrops occur, eastwards to Lincolnshire where it is known from boreholes. The main outcrops run from Great Orme and along the western side of the Vale of Clwyd, and these are separated by the Clwydian Range (Ludlovian Beds) from outcrops on the west side of the Flint and Denbigh Coalfield. Southwards this latter outcrop narrows due to the overlap of Upper Carboniferous rocks. As a result of this overlap Lower Carboniferous rocks are not seen in the chief coalfields of the west Midlands (S. Staffordshire and Warwickshire) but minor (though palaeogeographically important) outcrops are found in Shropshire and Leicestershire. The mid-Dinantian earth movements, which south of St. George's Land caused erosion, unconformity and facies changes, caused widespread changes in palaeogeography and initiated the spread of Viséan seas which ultimately covered almost all Britain. However, only late in Lower Carboniferous times did the northern flanks of St. George's Land become submerged (in Leicestershire it may have been breached so that it did not completely separate the northern and southern areas of deposition). In the Welsh Borders and north Wales the Carboniferous succession possesses characters indicating the proximity of the shoreline, and even the limestones which follow the basement beds include drifted plant fragments and at some horizons are sandy. The higher sandy limestones pass northwards into black limestones traced from Denbigh into Flintshire and reef limestones occur in the north. Basalt seen at Little Wenlock indicates minor contemporaneous volcanic activity, probably submarine.

The north of England and Pennines

Carboniferous rocks make up the Pennines which run with a north—south trend from Northumberland to Derbyshire. They are not, however, a simple structural unit nor were they sedimentationally or stratigraphically uniform. The pattern of Carboniferous sedimentation is closely related to contemporary structural units and may be summarized as follows:

SOUTHERN UPLANDS Northumberland Trough (6,000 ft partly deltaic and estuarine)
Pennine Fault Block (1,500 ft cyclic sediments succeed limestone)
Mid-Pennine Trough (+ 10,000 ft argillaceous marine beds, limestones subordinate
South Pennine Massif (2,000 ft chiefly limestones) ST. GEORGE'S LAND—MIDLAND BARRIER

Unfortunately Lower Carboniferous rocks cannot be traced along the whole length of the Pennines (they are overlain by Namurian (Millstone Grit) which outcrops in the high Pennine Moors of west Yorkshire), but changes in thickness and facies can be deduced.

Sedimentation did not commence over the Pennine Fault Block until S_2 times when it was submerged. As a result the Carboniferous succession there is much attenuated. To the south of the block lay the subsiding mid-Pennine Trough in which a great thickness of largely argillaceous sediments was accumulated. Here there is little similarity to the Avon Gorge succession in either lithology or fauna. The Pennine Fault Block, until its mid-Viséan submergence, was an effective barrier or threshold preventing the southward spread of deltaic material so that the succession of the Northumberland Trough has little in common with the mid-Pennine Trough, the former being largely deltaic and estuarine. On the other hand the Lower Carboniferous of the Northumberland Trough and the Scottish Midland Valley have general similarities despite the fact that the Southern Uplands remained emergent through much of Lower Carboniferous time.

West of the Northern Pennines, in the North-West Province, the Lower Carboniferous succession is yet again different. Westwards and southwards in this area the succession is largely limestone and more closely resembles the South-West Province succession than elsewhere, although there are important faunal differences. Because of the wide variety of facies it is not surprising that the zonal sequence of the Avon Gorge section is not readily applicable to the Lower Carboniferous of the north of England. However, in the North-West Province the zonal equivalents can be identified, but by diagnostic fossils more readily available in that province.

Dibunophyllum Zone	= D Zone
Productus corrugato-hemisphaericus Zone	= S_2 Zone and S_1 Zone
Michelina grandis Zone	= C_2 Zone
Athyris glabristria Zone	= C_1 Zone

In addition to the zonal forms *Saccaminopsis* (a foraminifer), found in the D_2 Zone, and calcareous algae at three horizons form recognizable and useful index horizons. The *Girvanella* band marks the base of D_2, and *Ortonella* and *Solenopora* occur in the C_1 Zone.

In the Mid-Pennine Province, bounded to the north by the Craven Fault system which must have been contemporaneous (though it has post-Carboniferous displacement, too), lay a basin or trough of continuing subsidence in which a vast thickness of Lower Carboniferous strata accumulated. Sedimentation commenced later than Z times, though the base of the succession is not seen and no Tournaisian age strata are proved to be present. Limestones (? C_1 age) occur but the greater part of the succession is more like the Culm deposits of south-west England. Hence, corals are rare but a sequence of goniatite-lamellibranch zones is applicable.

The P Zone, sometimes known as the Bollandian, is further subdivided on a faunal basis into P_{1a}, P_{1b}, P_{1c} and P_{1d} and P_{2a} to P_{2c}.

Limestones, of perhaps C_1 Zone, occur in the Craven Lowlands in the Skipton-Clitheroe-Bowland area and extended as far south as Dovedale.

	Lamellibranch/Goniatite Zones		*Coral/Brachiopod Zones —Approximate Equivalents*
NAMURIAN	Eumorphoceras E Stage		
DINANTIAN	Posidonia P Zone	P_2	D_3
		P_1	D_2
	Beyrichoceras B Zone	B_2	D_1
		B_1	S_2 and C_2S_1
	Pericyclus Pe Zone		
	Protocanites Pr Zone		
DEVONIAN	Wocklumeria		

Though undoubtedly flanked to the north by the Pennine Fault Block and to the south by the Derbyshire Massif (Figs. 9.3 and 9.4), no marginal deposits can be seen because of the overlap of higher beds. In the Viséan the margins of the trough were the site of the development of reef limestones. South of the Mid-Craven Fault unbedded reef limestones make conspicuous hills and are reef knolls which were situated close to the margin of the trough while apron reefs occurred on the basin's slopes. Sometimes of unfossiliferous calcite mudstone, the reefs include shelly limestones with a diverse characteristic fauna. The lower beds of the Mid-Pennine Trough possess sufficient faunal evidence to establish the presence of C, S and D_1 Zones while the higher shales are shown to belong to the P Zone by the goniatite-lamellibranch fauna. The basin facies laid down in the subsiding trough with continuous supplies of terrigenous material and with a sparse fauna, chiefly molluscan, bears a certain similarity to the Culm. Similar conditions of deposition persisted into the Upper Carboniferous although there is an unconformity between the Upper Bowland Shales (E_1) and the Lower Bowland Shales (P_{1-2}). Due to differential subsidence of the basin the boundary between basin and shelf facies fluctuated so that the succession is variable and a representative succession is difficult to define. Nowhere can a complete typical succession be found exposed in a continuous section. Lateral variations are summarized in Fig. 9.3.

The southern margin of the Mid-Pennine Trough cannot be fully explored due to the cover of younger Carboniferous rocks, but in north Derbyshire there has been found the transition from basin to shelf deposits of the Derbyshire Massif, a standard limestone succession. Apron reefs appear to have been banked up against the standard limestone submarine scarp of the massif.

The main outcrop of Lower Carboniferous rocks in the south Pennines occurs in the Derbyshire Dome and minor outcrops occur on its flanks in inliers at Ashover, Crich and Kniveton. Some 1,500 feet of Viséan limestones are seen and a further 900 feet are known to exist from boreholes. The Tournaisian is absent, through non-deposition, Viséan seas spreading across

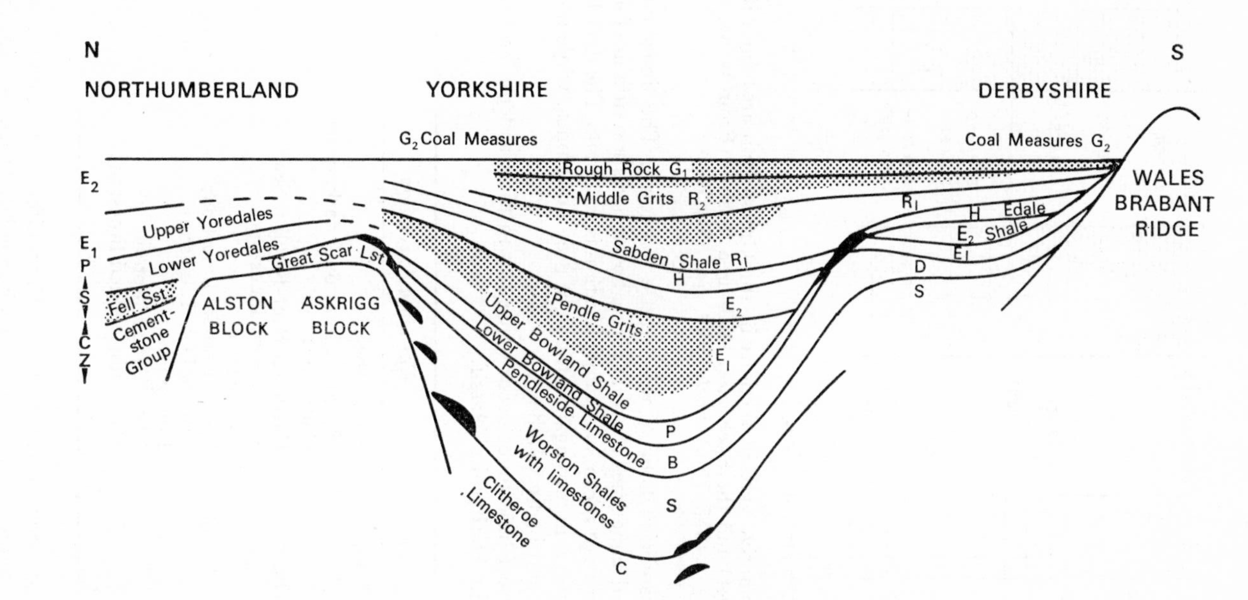

Fig. 9.3 Lateral variations in the Viséan and Namurian rocks of Northern England. Reefs shaded, chief developments of sandy facies stippled.

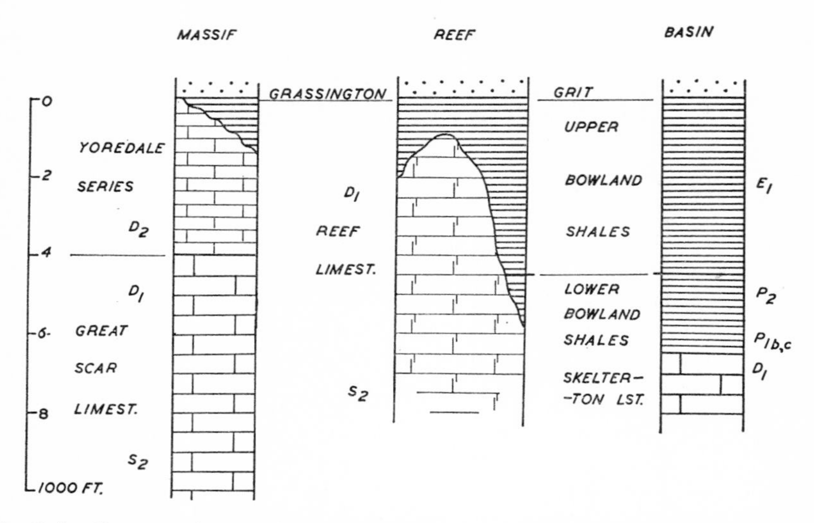

Fig. 9.4 Comparative vertical sections in the Grassington area (massif), the Thorpe area (reef) and the Skelterton Beck area (basin). (From W. W. BLACK, 1958, *Proc. Yorks. geol. Soc.*, **31**, 391.)

a possibly Precambrian floor of which little is known. The limestones are thick-bedded and sometimes cherty. Olivine basalt lavas and tuffs are found as well as later intrusive rocks, and mineralization is common. The dark basalts are conspicuous in outcrops against the light grey limestones, while the tuffs are often green and badly weathered.

In the west, reef limestones occur as well as in the Castleton area on the north flank of the Derbyshire Massif.

Zone	*Castleton Succession*
E and H	Edale Shales
P	Nunlow Limestone, dark, cherty
	Calcareous tuff
B_2	Castleton Reef-Knoll facies
	Grey cherty limestone
	Upper lava
	Grey crinoidal limestone
	Millers Dale Limestone
	Lower lava
	Chee Tor Limestone
	Dark thin-bedded limestone

The east—west extensions of the Lower Carboniferous beneath the bordering coalfields are known from relatively few boreholes. At Eakring, in Nottinghamshire, an oil borehole penetrated 3,000 feet of limestones which rest on red sandstones and conglomerates with igneous rocks at some horizons. The Viséan limestones of massif facies undoubtedly are widespread, the facies belt running east—west. The underlying red beds may be, at least in part, of Devonian age.

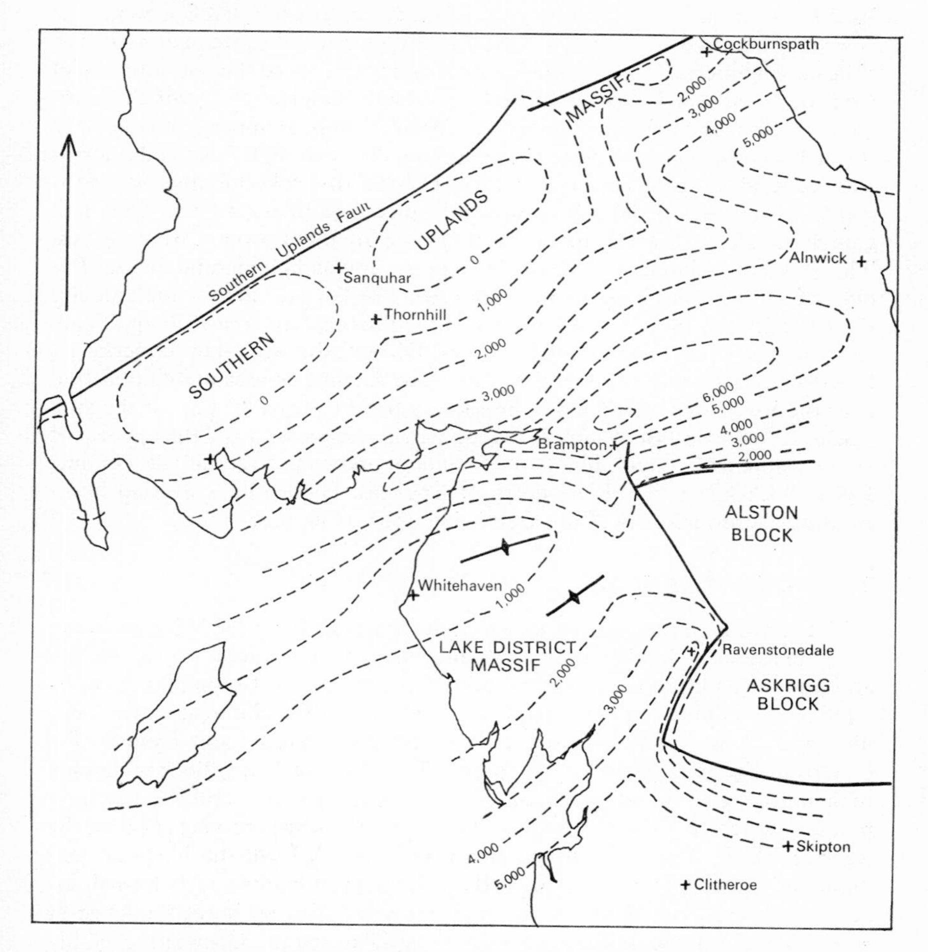

Fig. 9.5 Viséan isopachytes in the north of England and southern Scotland. (From T. N. GEORGE, 1958, *Proc. Yorks. geol. Soc.*, **31**, 227, Figs. 21 and 23)

One important downwarp, at Mold in north Wales, though not far removed from the shoreline which lay near Corwen, led to the accumulation of over 2,000 feet of Lower Carboniferous. It is remarkable for being a limestone sequence despite downwarping of this magnitude.

West of the Pennine Fault Block thick Carboniferous sediments were laid down. The Ravenstonedale area may be considered as a gulf of the Mid-Pennine Province though the succession is largely limestones and is not of basin facies. (It is often referred to as the North-West Province.) On the east side of the Lake District thick conglomerates rest unconformably on Lower Palaeozoic strata. At the western end of Ullswater the Mell Fell Conglomerate is several hundred feet thick. It is remarkable for the absence of Borrowdale Volcanic pebbles, although pebbles from the (now) more distant outcrops of Coniston Grit and Coniston Limestone occur. Evidence of age of this conglomerate is lacking, but it may be pre-Carboniferous. It appears to have been formed as a scree-fan deposit thinning laterally, although it extends northwards round the periphery of the Lake District to Cockermouth and southwards to Ravenstonedale. It is probably incorrect to regard the Mell Fell Conglomerate as the basement conglomerate of the Lower Carboniferous. The lowest Carboniferous beds include several conglomeratic beds. The oldest Carboniferous beds are the grey-green Pinskey Gill Beds which underly the typically red 'basal' conglomerate of the Carboniferous near Shap. There they rest on the Silurian Brathay Flags. Of doubtful age, they underlie C_2 limestones. The limestones of the Shap-Ravenstonedale area are dolomitized near the base. South of Penrith, limestones form the greater part of the succession. Northwards limestones, though persistent, are thin and the upper part of the sequence is much like that of contemporaneous strata of the Pennine Fault Block and most of the north of England. Higher beds overlap lower northwards towards the Manx-Cumbrian uplift (Fig. 9.5).

The Pennine Fault Block

The north Pennines consist, structurally, of a fault-bounded block, or horst, tilted to the east by Miocene earth movements. The faults are of greater antiquity and played an important part in the control of Carboniferous sedimentation. Bounded to the north by the Stublick Fault, the west by the Pennine and Dent Faults and to the south by the Craven Fault System, the eastern margin is masked by later strata. The Pennine Fault Block is divided by a shallow east—west syncline, the Stainmore Syncline, and the northern part is frequently called the Alston Block while the southern part is called the Askrigg Block. The pre-Carboniferous rocks which form the block are exposed only in deeply cut valleys north of the Craven Faults and in a small inlier in Teesdale, but they have been encountered in deep bores the latest of which was put down to investigate the Weardale Granite. Comprising highly folded Silurian, Ordovician and Precambrian rocks, the form of the pre-Carboniferous surface can only be inferred. Certainly the block was not submerged by Carboniferous seas until much later than the Mid-Pennine Trough to the south or the Northumberland Trough to the north. Remaining above sea level until late in S times, except for its southern margin, the total thickness of Lower Carboniferous strata over the block is less than 2,000 feet. The sediments are of shelf type and the succession differs in both thickness and lithology from the successions found in these troughs. Recent borehole evidence and geophysical work indicate the presence of a gulf to the east of the Askrigg Block in east Yorkshire and the Ravenstonedale Gulf to the west of the block has already been described.

On the Pennine Fault Block, Carboniferous sedimentation commenced with the deposition of basal conglomeratic and arenaceous beds which can be seen at a number of places along the western escarpment of the northern Pennines. Wherever Lower Carboniferous strata rest unconformably upon older rocks the succession commences with such a basal series. However, despite lithological similarities they are far from being contemporaneous, since Carboniferous sedimentation commenced at very different times at different places. Following the basal series the lowest, and much the thickest, limestones of the succession occur. The Great Scar Limestone of the southern part of the block (the Melmerby Scar Limestone is equivalent to the lower part of it) is laterally uniform, indicating that the pre-Carboniferous rocks had been considerably peneplaned permitting the spread of shelf-sea conditions.

The Long Fell–Roman Fell area lying between the Outer Pennine Fault and the Swindale Beck Fault provides an interesting contrast in succession with that of the Fault Block. West of the Swindale Beck Fault the more complete succession (beginning with C_1 age *Seminula gregaria* subzone) infers that the fault was contemporaneous, in earlier Lower Carboniferous times forming a fault coastline—the block supplying detritus to the Ravenstonedale gulf (e.g. the Ashfell Sandstone). Later uniform conditions became widespread. Several conglomerates occur in this area. In the thrust block of Roman Fell, the Roman Fell Beds, sandstones and shales with a quartz conglomerate, rest on the Polygenetic Conglomerate of unknown age but with pebbles including andesites probably derived from the Old Red Sandstone.

Following the Great Scar Limestone, cyclic sedimentation became established giving rise to the Yoredale Series. Characteristically developed in Wensleydale, they are found over the whole of the Pennine Fault Block and are closely analogous to contemporaneous strata occurring over most of northern England and the Scottish Midland Valley. Thus the Yoredales link the Pennine Fault Block with areas further north, rather than with the Mid-Pennine Trough where strata of equivalent age are the Bowland Shales and Pendleside Series. The base of the Yoredales is conveniently taken as the *Girvanella* Band of P_1 age but the onset of cyclic sedimentation was not everywhere contemporaneous, commencing earlier in Wensleydale than in the Ingleborough area. Cyclic conditions of deposition persisted into the Upper Carboniferous, and the base of the Namurian can be fixed by the appearance of Namurian goniatites but not by lithological changes in northern England and Scotland. A typical cyclothem, though many are incomplete and workable coals occur only in some cycles, is given on p. 202.

Individual limestones rarely attain a thickness of more than 20 to 30 feet, while cyclothems are characteristically about 60 to 90 feet, so that limestones do not comprise the bulk of the succession—although they are the most constant members. The different lithological units (beds) represent changing conditions from clear shallow seas, in which limestones were deposited, silting

Lithology	Inferred Conditions of Deposition
Coal Fireclay	Forest and soil—land
Sandstone Sandy shale Black shale	Delta
Calcareous shale Limestone	Marine

up with the spread of terrigenous material so that emergence occurred and led to the development of a soil supporting a flora. These changes suggest oscillatory subsidence. Limestones have been given descriptive names such as the Five Yard or Cockleshell or local names such as the Corbridge. Contemporaneous limestones may have several names when traced from the Yorkshire Dales through Durham to Northumberland. The succession of limestones of the Pennine Fault Block, Yoredale Series and their probable equivalents elsewhere in northern England are given in Table 9.5. By late Viséan times similar cyclic sedimentation was widespread.

Northumberland Trough (including NE Cumberland)

North of the Pennine Fault Block, north of the Stublick Faults, a subsiding basin accommodated some 6,000 feet of Lower Carboniferous strata. Submergence commenced early, possibly in Tournaisian times, and the oldest beds resting unconformably on the Old Red Sandstone lavas of the Cheviots may belong to the C_1 or even Z_2 zone. A twofold division of the Lower Carboniferous into Bernician and Tuedian has long been applied to the succession of this region.

	Zone	Lithological Groups		
BERNICIAN	D_1–E	Upper Limestone Group Middle Limestone Group Lower Limestone Group		
	S	Scremerston Coal Group	≡	Birdoswald Limestone Group Craighill Sandstone Group
TUEDIAN	C_2	Fell Sandstone Cementstone Group Conglomerate		

The Cementstone Group comprises chiefly grey or red shales with thin impure limestones alternating with the shales in a rhythmic pattern. Up to 3,000 feet are exposed in the Tweed Valley but the group thins towards the

TABLE 9.5

	Zones	*Cockermouth*	*Wensleydale*	*Teesdale/Brampton*		*N. Northumberland*		*Scotland*	
NAMURIAN			Stonesdale	Upper Fell Top					
			Crow	Lower Fell Top	UPPER LIMESTONE GROUP		UPPER LIMESTONE GROUP		LIMESTONE COAL GROUP
			Upper Little	Corbridge					
		Little		Little		Lickar			
	E_1	1st or Great =	Main =	Great	=	Dryburn			
VISÉAN	P_2	2nd or Four Fathom	Undersett	Four Fathom		Sandbanks		Top Hosie	
		3rd	Three Yard	Three Yard					
		4th: Five Yard	Five Yard	Five Yard	MIDDLE LIMESTONE GROUP	Acre	MIDDLE LIMESTONE GROUP		LOWER LIMESTONE GROUP
		4th: Scar	Middle = Scar, Cockleshell, Single Post	Scar					
		4th: Single Post		Cockleshell		Eelwell			
				Single Post					
		4th: Tyne Bottom	Simonstone	Tyne Bottom		Budle			
		4th: Jew	Hardraw	Jew					
			Gayle	Lower Little					
		4th: Rough =	Hawes =	Smiddy	=	Bankhouses			
	D_2					or Oxford	=	Hurlet	
	D_1	5th or White				Watchlaw	LOWER LIMESTONE GROUP		
		6th	Great Scar	Melmerby Scar		Woodend			
		7th				Dun			

[See Table 9.6 for details of Scottish succession]

Cheviots. At the base of the group the Birrenswalk Volcanic Group is seen in Liddesdale while in the Tweed Valley the well known Kelso Traps occur. Fossils are few in the Cementstone Group but such as occur (*Spirorbis*, *Naiadites*, fish scales and plant fragments) indicate a non-marine environment usually termed lagoonal. However, westwards into Cumberland, limestones with a crinoid-brachiopod fauna indicate that marine conditions became established there before the end of Cementstone Group times.

The Fell Sandstone which follows, varying in thickness from 600 to 1,000 feet, makes a pronounced topographical feature in the Border Country. It represents the incoming of vast amounts of detritus. Sparsely fossiliferous, only plant fossils occur except in the west where fossiliferous limestones give the age as C_2 Zone. The Scremerston Coal Group, representing a return to cyclic sedimentation, consists of limestones, shales and coals. These are the oldest worked coals in Britain, and in the north near Berwick there are 10 workable seams and the group is about 1,000 feet thick. It thins southwards and coals are less numerous. Westwards the group thickens but this is accompanied by a facies change and the deltaic Craighill Sandstone is followed by the cyclic Birdoswald Limestone Group. The Arbigland Beds of the Solway area continue this facies change further and are highly fossiliferous.

The Limestone Groups, Lower, Middle and Upper, are cyclic sediments of Yoredale type deposited during D Zone times, sedimentation continuing into the Namurian. Individual limestones are generally thin but some in the Middle Limestone Group are as much as 60 feet thick. The Lower Carboniferous, as a whole, becomes more calcareous south or south-westwards inferring derivation of terrigenous material from the north-east.

Scottish Midland Valley

The Scottish Lower Carboniferous was laid down in a subsiding trough in which some 6,000 feet of sediments accumulated. Non-marine, deltaic and terrigenous cyclic sediments occur and limestones make up only a small percentage of the total thickness of Carboniferous rocks. They show certain similarities to the Carboniferous rocks of the Northumberland region but there are significant differences which imply that the basins were not unified. The Southern Uplands appear to have been covered, perhaps only partially and intermittently, by late Carboniferous seas. Patches of Lower Carboniferous, for example at Thornhill, are thin and infer long periods of uplift and erosion with only occasional periods of deposition. In other words, the Southern Uplands were effectively the southern margin of the Scottish Carboniferous Trough. The position of the northern margin and the northern extent of Carboniferous deposition is in doubt, though deposition occurred north of the Highland Boundary Fault. The eastern extent of the Scottish Trough is unknown: there is no evidence of the succession thinning eastwards; on the contrary, most groups of the Carboniferous thicken eastwards. The western extension of the trough into Northern Ireland is confirmed by relic outcrops found there.

The lowest beds of Carboniferous age in Scotland appear to be conformable upon Upper Old Red Sandstone Beds but they lack both coral-brachiopod or goniatite faunas which would permit their relative dating. However, by implication they must be of Tournaisian age. Professor George dissents, disputing the normal unbroken sequence and believes that no Carboniferous strata older than Viséan were deposited in this trough. (It cannot be ruled out that the Upper Old Red Sandstone may be in part of Tournaisian age.) Certainly the parts of the Lower Carboniferous which have been dated with reference to the goniatite zonal sequence are high in the Viséan and it seems unlikely that the strata below these could represent deposition through the whole of Tournaisian and Lower Viséan times.

The base of the Carboniferous is taken on lithological grounds as the lowest cementstone. The cementstones are widespread in western and central parts of the Midland Valley, but in Kinross Carboniferous sandstones follow Old Red Sandstone and it is difficult to place the boundary. Even the contemporaneity of the lowest cementstone is dubious since the base of the Cementstone Group is probably diachronic. The general succession in Scotland is:

	Stage or Zone	*Lithological Group*	
NAMURIAN	H–G_2	Passage Group (formerly called Millstone Grit)	
	E_2	Upper Limestone Group	(formerly called Carboniferous Limestone Series)
	E_1	Limestone Coal Group	
VISÉAN	P_2	Lower Limestone Group	
	B_2P_1	Oil Shale Group	Calciferous Sandstone Series
? TOURNAISIAN	B_1	Cementstone Group	

The Cementstone Group of the Calciferous Sandstone Series comprises repeated cycles of argillaceous limestone and limy shale. Best seen in Ballagan Burn on the south side of the Campsie Fells, where over a hundred repetitions occur and they are more than 200 feet thick, they are often known as the Ballagan Beds. However, the greater part of the Calciferous Sandstone Series comprises volcanic rocks, the Clyde Plateau Basalts. Lava was poured out upon lava to pile up a thickness of over 2,000 feet. Lava flow tops are scoriaceous but often show evidence of being subaerially weathered, to form a red bole, before a succeeding flow occurred. They are chiefly basalts but more acid lavas are encountered further south. Relatively resistant to erosion, the Clyde Plateau Lavas make up the Kilpatrick and Campsie Hills north of the R. Clyde and the hills of Renfrewshire and north Ayrshire south of the river. The lavas are followed by an upper sedimentary group of varied lithologies including volcanic debris with some coal seams.

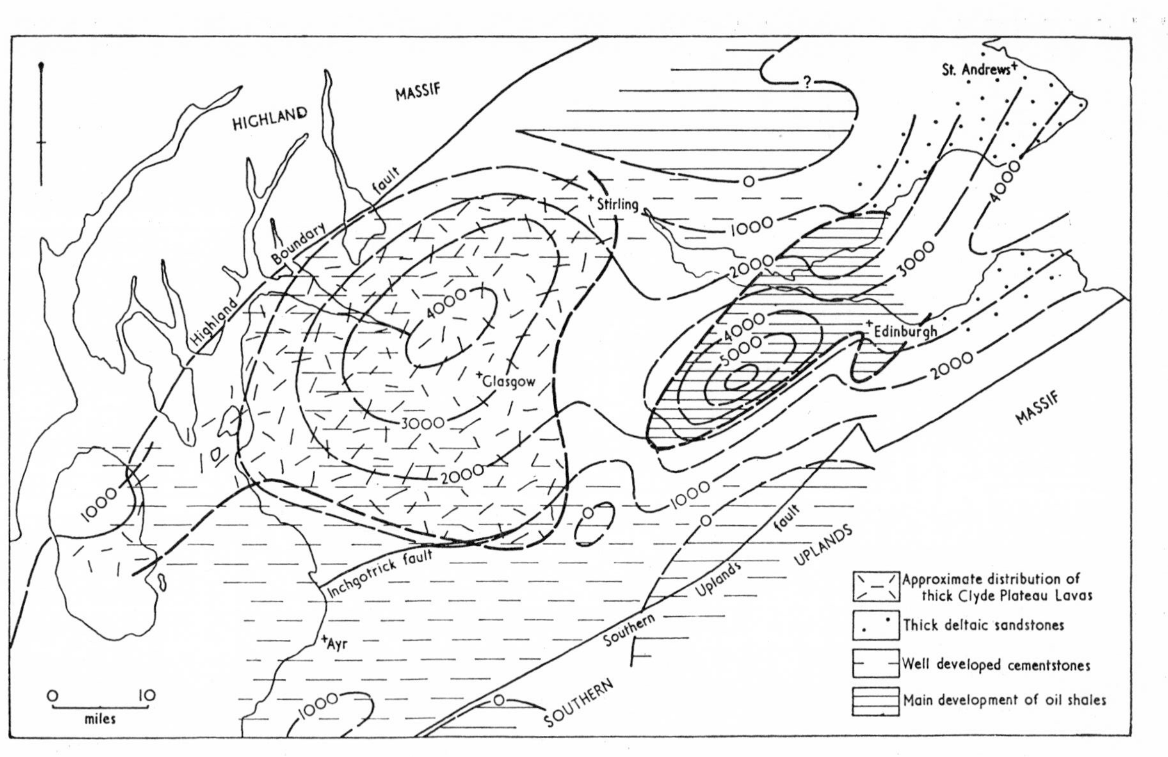

Fig. 9.6 Isopachytes showing the relation of facies to structure in the Calciferous Sandstone Series of Scotland. (Isopachytes in the heart of the Central Coalfield are conjectural.) (From T. N. GEORGE, 1958, *Proc. Yorks. geol. Soc.*, **31**, 227.)

The Calciferous Sandstone Series is without volcanic rocks in south Ayrshire and again in eastern Fife. In the latter it attains its greatest thickness of about 4,000 feet. The stratigraphy is still imperfectly known but the strata are rudely cyclic comprising some limestones, shales, sandstones and coals. Palaeocurrent directions indicate transport of sediment from the north-east, possibly axial flow along a trough since the deposits thin rapidly to only a few hundred feet in west Fife. Separated from the Cementstones by about 1,000 feet of lavas with some tuffs which together with volcanic necks make up the spectacular outcrops in Queen's Park (Arthur's Seat, etc.) Edinburgh, are the Oil Shales. The volcanic history of this area has been worked out in great detail; basalts of several types occur and, in addition to the volcanic vents which penetrate the flows, there are post-Carboniferous sills. It is possible that these lavas are continuous with the Clyde Plateau lavas at depth beneath the cover of younger rocks in the intervening area. The Oil Shales were laid down in a basin (Fig. 9.6). The oil shale (or torbanite) occurs as thin seams in a shale-sandstone succession and most of the workable seams are near the top of the sequence. They have been worked for over two centuries west of Edinburgh and the tips of red shale produced during the distillation of the shale are a characteristic feature of the Midlothian scenery. The strata of this facies are cyclic, but essentially non-marine. Of restricted area of deposition, they may be truly lagoonal. Fossils are rare but the Lower Oil Shales are separated from the Upper by the freshwater Burdiehouse Limestone. The sparse non-marine fauna consists of ostracods, fish remains and some non-marine lamellibranchs, but in the Wardie Shales division marine horizons are present.

North of the Forth volcanic rocks occur in the Upper Oil Shale Group, outcropping south of Kirkcaldy. A thickness of 1,400 feet has been proved in a colliery and the volcanics extending under the Forth may link up with those of Garleton Hill in East Lothian (16 miles east of Edinburgh).

The volcanic history of the Midland Valley of Scotland is extremely complex. In addition to the Dinantian occurrences of vulcanicity discussed, vulcanicity continued into the Namurian, in West Lothian without interruption, and mostly tuffaceous volcanic rocks occurred at three main centres. These were in east Fife, west Fife and in the west of the Midland Valley. They give rise to topographical features (such as Largo Law and the Saline Hills), as do post-Carboniferous intrusions discussed in Chapter 10.

The Carboniferous Limestone Series, to give it its former name, comprised the Lower Limestone Group, the Limestone Coal Group and the Upper Limestone Group. All three consist of rather similar cyclic sediments. Once referred to the Lower Carboniferous, goniatites have shown that only the Lower Limestone Group belongs to the Viséan, the other groups being Namurian. The Lower Limestone Group is defined at the base by the persistent limestone, the Hurlet. It has been given several different names in various parts of the Midland Valley (see Table 9.6). The top of the group is placed to include the Hosie Limestones. Like the other groups, the Lower Limestone Group shows considerable variation in thickness. In general it thickens eastwards, attaining a maximum in the Lothians, and it shows variation from north to south. The latter can be related to contemporaneous faulting, a mid-Ayrshire shelf sinking slowly. Deposition was not, apparently, continuous over the whole of the Scottish basin and strata were not laid down in part of Lanarkshire. On the other hand, although the Calciferous Sandstone volcanics show evidence of having been subaerially weathered, almost all were submerged by the Hurlet Limestone sea. Subsidence was not great (the limestone may have been laid down in shallow water) but was none the less widespread since the Hurlet Limestone correlates with the Oxford Limestone of Northumberland.

TABLE 9.6

		N. Ayrshire	*Lanarkshire*	*Midlothian*	*Fifeshire*
NAMURIAN	UPPER LIMESTONE GROUP E_2–H	Castlecary =	Levenseat	Castlecary	Castlecary
		Upper Linn or Calmy	Arden or Calmy	Calmy	Jenny Pate
		Lower Linn or Linn Brig	Orchard	Orchard	Capledrae Marine Band
		Third Post	Lyoncross	Lyoncross	Marine shales
		Index or Highfield	Index	Index	Index
	? E_1 LIMESTONE COAL GROUP				
VISÉAN	LOWER LIMESTONE GROUP P_2	Hosie D =	Top Hosie =	Top Hosie =	Upper Kinniny
		Hosie C	2nd Hosie	Bilston Burn	Middle Kinniny
		Hosie B	Mid Hosie	Upper Vexhim	Lower Kinniny
		Hosie A	Main Hosie	Lower Vexhim	Seafield Marine Band
			Blackhall	North Greens	Charlestown Main
			Craigenhill	Dryden	Charlestown Green
		Dockra =	Hurlet =	Gilmerton =	Charlestown Station

The Lower Carboniferous in Ireland

Rocks of Carboniferous age occur in every county except Wicklow. However, although so widespread and not extensively covered by younger formations, outcrops are widely obscured by drift in the Central Plain and other areas. In the south of Ireland Carboniferous slates occur in synclinal structures from Cork westwards to Bantry Bay. The rocks which are of Culm facies have been subjected to post-Carboniferous (Armorican) compression, producing cleavage. (The Armorican front lies parallel to, but north of, the northern limit of the Culm facies as in Pembrokeshire.) Sedimentation continued from the Devonian into the Carboniferous without break. The deltaic Coomhola Grits of late Devonian to *Cleistopora* Zone age are approximately the equivalent of the Skrinkle Sandstones of Pembrokeshire. Some 3,000 to

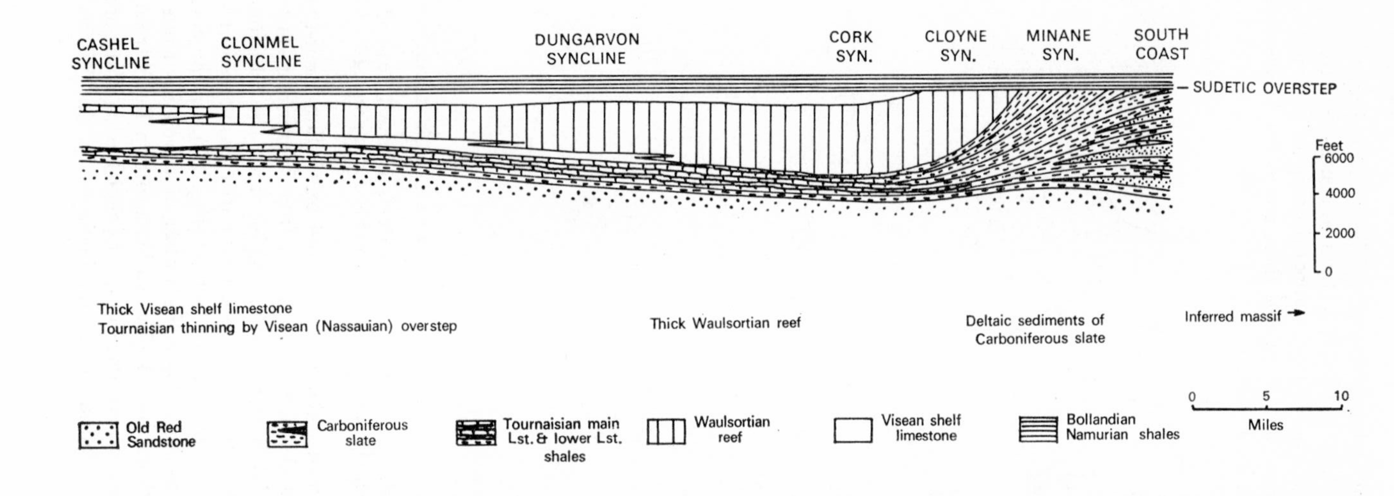

Fig. 9.7 Variations in the Lower Carboniferous rocks of Ireland. (From T. N. GEORGE, 1962, in *Some Aspects of the Variscan Fold Belt*, ed. K. COE. Manchester Univ. Press, Manchester.)

5,000 feet of typical Culm facies rocks follow ranging up to the P_2 Zone or later. Palaeogeographical evidence is slight in south-west Ireland but palaeocurrent directions generally imply transport of material from the south-west, although the nature of the strata and their contained fossils suggest a basin flanked to the north by a shelf.

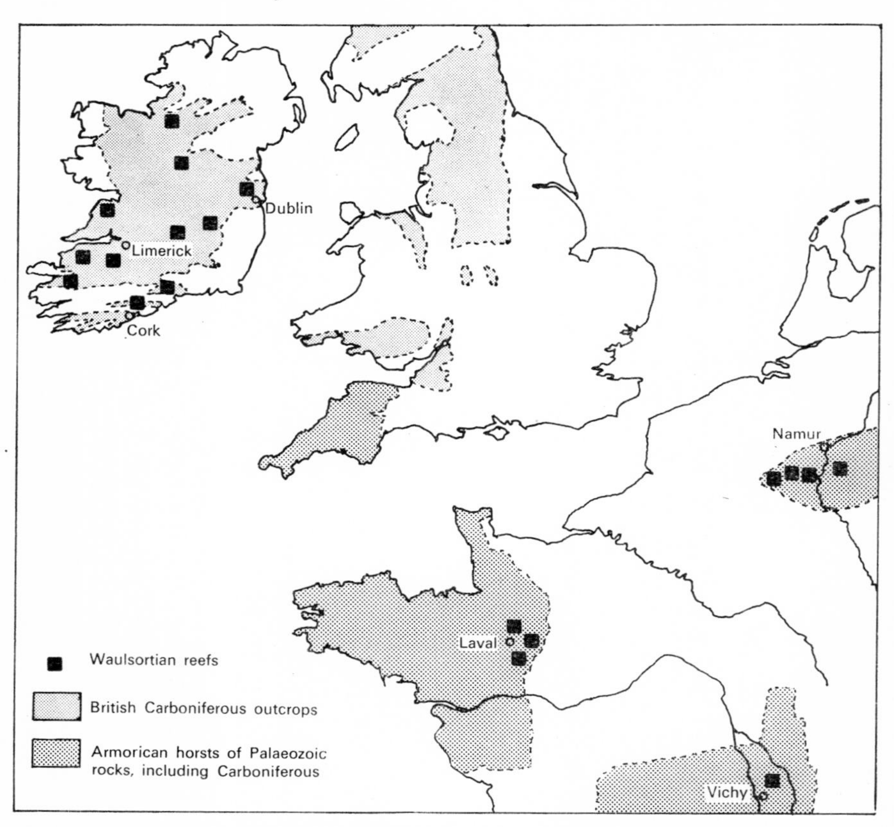

Fig. 9.8 The distribution of Waulsortian Reefs in the Lower Carboniferous of NW. Europe. (From G. DELÉPINE, 1951, *Proc. geol. Ass.*, **62,** 140.)

The Central Plain of Ireland was covered by shallow shelf-seas in Lower Carboniferous times. As far north as a line from Waterford Harbour to Galway the Lower Carboniferous is conformable upon the Old Red Sandstone and Tournaisian strata are present. Further north seas spread in Viséan times to cover almost all the remainder of Ireland (Fig. 9.7). On the northern flanks of the Culm basin, probably covering almost half Ireland at one time, great thicknesses of reef limestones, known as Waulsortian Reefs from their similarity to those of the southern flank of the Dinant Basin (Waulsort lies south-west of Dinant), are found at several horizons in the Viséan (Fig. 9.8).

Extensive sheet reefs, not necessarily primarily organic in origin, as well as reef-knolls are widespread, the reef-knolls occurring on the flanks of massifs.

In Ireland, with the exception of the extension of the St. George's Land

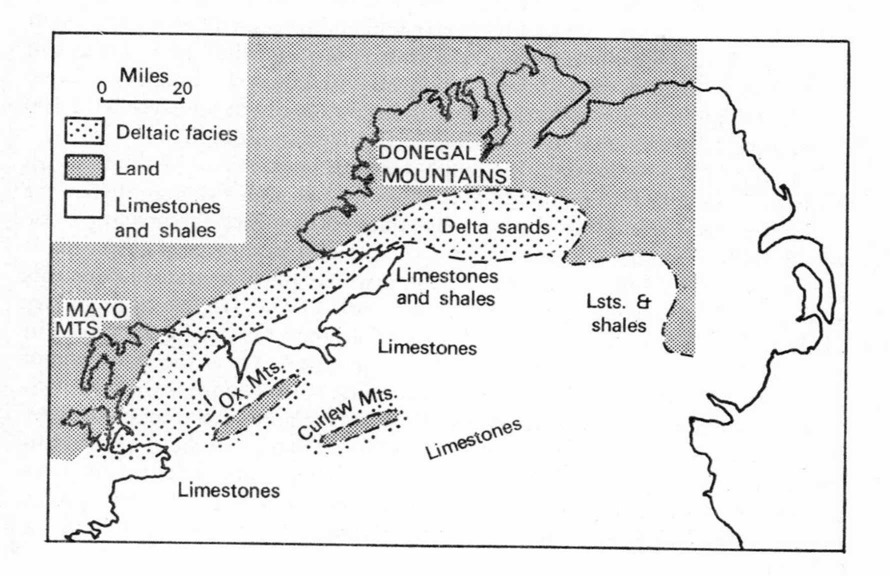

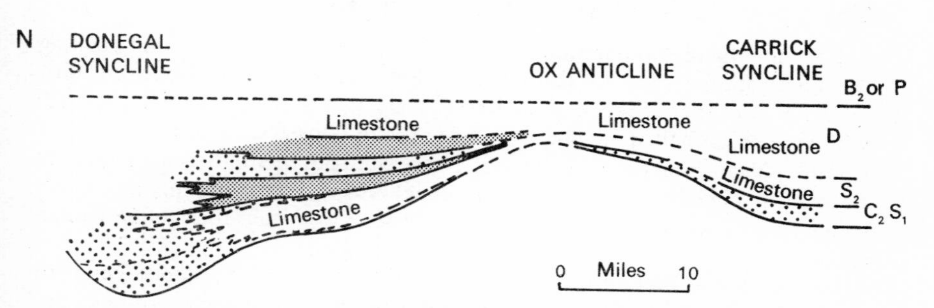

Fig. 9.9 Palaeogeographic reconstruction of north-western Ireland in early Viséan times with a section showing lateral changes in the Viséan rocks. (On the section, sandy facies stippled, shales shaded, limestones unshaded.) (Both simplified from T. N. GEORGE, 1958, *Proc. Yorks. geol. Soc.*, **31**, 227.)

massif into Leinster, no major emergent barriers appear to have existed comparable with those of Britain, though the Longford-Down Massif may have been an extremely shallow-covered shield with only D Zone strata being deposited over its eastern end. It follows that the transition from one facies belt to another is uninterrupted. Work carried out over the last two decades has done much to elucidate the palaeogeography, especially of the northern half of Ireland. The pattern of regional sedimentation can be related to the Caledonian tectonic framework. North of St. George's Land thick deposits accumulated in a downwarping basin. Here the Rush Slates, Z–C_1, and Rush Conglomerate, C_2, are followed by cherty and argillaceous limestones. Northwards the Lower Carboniferous strata thin rapidly and little now remains of the former cover of Carboniferous which extended over the Longford-Down Massif. Of Viséan age, the Carboniferous here probably never exceeded 1,500 feet compared with 4,000 feet of strata in the Dublin Trough.

The Slieve Beagh Trough, the south-westwards extension of the Scottish Midland Valley, bears similarities to the Scottish Trough although the Lower Carboniferous strata deposited were less than half as thick. Westwards these significant Caledonian structures lose their effect but a unique feature is present, the Ox Mountains arch, north of which thick arenaceous deposits accumulated in a subsiding trough. In Scotland the Highland Boundary Fault did not form the northern limit of Carboniferous sedimentation but evidence is restricted to outcrops in Kintyre, Arran and near the south end of Loch Lomond. In Ireland the evidence is much more complete that the continuation of the line of the Highland Boundary Fault was transgressed by early Viséan seas. Downwarping was considerable in a basin south of a mountainous area of which we have little direct knowledge, though deltaic and conglomeratic beds, comprising detrital material including strained quartz, schists, mica and felspar, infer a shoreline north of the Donegal Syncline (Fig. 9.9).

The Upper Carboniferous—the Namurian

Forming the high moorland between Sheffield and Manchester and flanking the Derbyshire Dome, sandstones and shales known as the Millstone Grit occur. However, the range of facies of strata laid down contemporaneously with the typical Millstone Grit is so wide that it is preferable to use the time-division Namurian for these strata, restricting the term Millstone Grit to rocks of that age but of a particular (deltaic) facies. This is broadly analogous to the restriction of the term Old Red Sandstone to certain rocks of Devonian age.

In the South-West Peninsula, and the continuation of that facies belt into south-west Ireland and into the European Continent, Culm conditions of deposition persisted at least through some parts of the Namurian. In the north of England and Scotland and Northern Ireland where cyclic Yoredale-type sediments were deposited in late Dinantian times this kind of sedimentation persisted for some part of the Namurian. The repeated spreads of deltaic material forming coarse sandstones interbedded with marine shales which we think of as typical Millstone Grit are not therefore everywhere representative of the Namurian. They occur, especially in Britain, in the central and southern Pennines and in south Wales, and of course in the Dinant-Namur Basin of Belgium—from which the Namurian takes its name.

In the mid-Pennines the Millstone Grit attains a maximum of 6,000 feet and thins steadily southwards towards St. George's Land. Locally sedimentation was continuous from Lower to Upper Carboniferous, the base of the Namurian being defined by the replacement of Viséan goniatites by other genera such as *Eumorphoceras* and *Cravenoceras* which typify the Lower Namurian. The Upper Bowland Shales thus follow the Lower Bowland Shales without a break. This conformable relationship is exceptional and largely confined to the basin areas of Carboniferous sedimentation. On the massifs a pre-Namurian topography of Viséan reefs was buried by Namurian sediments so that Millstone Grit rests unconformably on Lower Carboniferous. This kind of relationship can be seen at Ingleborough and more markedly southward onto the edge of the Derbyshire Massif. Again in the Peak District, Millstone Grit succeeds Lower Carboniferous conformably but it overlapped to the south, against St. George's Land, and was itself overlapped by Coal Measures. The mid-Carboniferous movements, the Sudetic earth movements, responsible for the unconformable relationship of the Namurian were much more marked in the South-West Province. In the South Wales Coalfield the evidence of unconformity is apparent from the marked overstep of Namurian beds, on to beds as old as Ordovician, while in the Bristol-Mendips area Namurian is almost completely absent. Beds of sandy facies there are chiefly of Viséan age and the Coal Measures rest unconformably on beds older than the Namurian, as they do further east in this same facies belt in the Kent Coalfield. Further south-east, at Namur, the Namurian is well developed and closely similar to the British Namurian strata of Millstone Grit facies. In Scotland rocks of Millstone Grit facies were laid down, following cyclic sedimentation, rather later than in the mid-Pennines.

The Mid-Pennine Province

Following the culm-like Bowland Shales, there were widespread depositions of deltaic sands. While the actual form of the delta cannot be defined, the areal spread of successive sheets of deltaic material is known. Cyclic sedimentation then occurred but, unlike the Yoredale cyclothems, limestones are absent and the succession comprises marine shales or mudstones, succeeded usually by barren mudstones after which coarse thick sandstones were deposited followed by seatearth and coal. Fossils, especially in the upper half of the Millstone Grit, are chiefly goniatites and marine lamellibranchs. The goniatites are often restricted to relatively thin but persistent marine bands. Fossils occur typically in the mudstones and limy shales, where they are crushed, and in concretions known as 'bullions'. The Namurian has been zoned by means of goniatites, the sequence of faunas being worked out by Bisat as long ago as 1924 in the Pennine area. It is only in the mid-Pennines that the old lithological classification can be applied but the sequence of goniatite stages is of wide application. The 6 goniatite stages called after goniatite genera, e.g. Lower *Eumorphoceras* Stage (E_1), were formerly called zones but they are stages further divided into 16 zones (and many subzones).

Since they are stages, names derived from typical localities have been given to them, in line with general stratigraphical practice, and these are listed in Table 9.7.

TABLE 9.7

Stage	*Stage Name*	*Zone*
Lower *Gastrioceras* G_1	YEADONIAN	*G. cumbriense* *G. rurae*
Upper *Reticuloceras* R_2	MARSDENIAN	*R. superbilingue* *R. bilingue* *R. gracile*
Lower *Reticuloceras* R_1	KINDERSCOUTIAN	*R. reticulatum* *R. eoreticulatum* *R. circumplicatile* [*inconstans*]
Homoceras H	SABDENIAN	*H. eostriolatum* *H. beyrichianum*
Upper *Eumorphoceras* E_2	ARNSBERGIAN	*Nuculoceras nuculum* E_{2c} *Cravenoceratoides nitidus* E_{2b} *E. bisulcatum* E_{2a}
Lower *Eumorphoceras* E_1	PENDELIAN	*Eumorphoceras* sp. E_{1c-d} *E. pseudobilingue* E_{1b} *Cravenoceras leion* E_{1a}

The Namurian, in the type area in Britain, begins with the Bowland-Shales forming a transition from the Viséan. It is followed by the Pendle Grit, a great thickness of almost unfossiliferous grit, and further sandstones and shales make up the remainder of the Lower *Eumorphoceras* Stage. Among the best known of these higher sandstones is the Grassington Grit, widespread in the mid-Pennines and possibly equivalent to the Warley Wise Grit which succeeds the Pendle Grit. Further south, in Derbyshire, Edale Shales represent both the Lower and Upper *Eumorphoceras* Stages.

The Upper *Eumorphoceras* Stage is dominantly shaly, the Sabden Shales, but locally sandstones are more important and coarse grit occurs at the top. The *Homoceras* Stage, some 500 feet of sandstones and shales but thinning southwards, occupies the upper part of the Sabden Shales and in Derbyshire the upper part of the Edale Shales.

The beds of the Lower *Reticuloceras* Stage, the Kinderscoutian called after the thick Kinderscout Grit at the top, are 1,500 feet thick and make bleak moorland in Derbyshire. Where dissected it makes 'edges', steep scarps or valley sides. On the east of the Derbyshire Dome, both in outcrops near Ashover and in Nottinghamshire boreholes, beds with R_1 fossils lie above Viséan beds suggesting overstep of earlier Namurian strata. Though the Upper *Reticuloceras* Stage comprises largely sandstones or grits, the strata are cyclic. Coals occur and widespread marine bands. Non-marine lamellibranchs occur in other shell beds. The uppermost grit, the Holcombe Brook Grit of Yorkshire and the Chatsworth Grit of Derbyshire, is of wide extent.

The Lower *Gastrioceras* Stage consists of further cyclic sediments of coal measure type with, in fact, workable coals which have been worked by opencast methods. The most important sandstone is the Rough Rock which, in parts of Lancashire, is interrupted by a coal seam, the Sandrock Mine.

Vulcanicity had apparently ceased by Namurian times in England from the evidence of outcrops but Nottinghamshire boreholes have proved volcanics of this age.

Southwards from the area of maximum deposition—Lancashire—the Namurian rocks thin but are present in Derbyshire, Staffordshire and Flintshire. In the coalfields of the west Midlands, which lie close to the site of St. George's Land, the southward thinning is further demonstrable. In Staffordshire and Leicestershire only the R_2 and G_1 Stages (Rough Rock and Third Grit Groups) are well developed, while in South Staffordshire and Warwickshire the Millstone Grit is overlapped by Coal Measures—which rest on Silurian beds at Dudley and on Cambrian beds at Nuneaton. Southwards from Flintshire to Denbigh and into Shropshire the Millstone Grit outcrop narrows and a change in facies from the Holywell Shale in the north (cf. the Edale Shale) to sandy beds in the south takes place.

The South-West Province

In the Bristol-Somerset region the Namurian is not fully represented and everywhere Coal Measures rest with unconformity upon early Namurian or pre-Namurian beds. (The beds of sandy facies are largely of Viséan age.) It is not clear if a more complete Namurian succession was laid down and subsequently removed. Certainly in south Wales folding and uplift was of pre-Namurian age, the Millstone Grit resting unconformably on older beds—on Old Red Sandstone and Lower Palaeozoic rocks along the north side of the coalfields where the elevation was greatest. The Millstone Grit outcrop surrounds the South Wales Coalfield (except where broken by sea) but only in the west is the succession complete. The lowest beds are found there but are overlapped northwards and north-eastwards on to the southern flank of St. George's Land. Erosion of Lower Carboniferous strata took place over the line of the Usk Anticline, an axis to be rejuvenated in post-Carboniferous times, and the sub-Millstone Grit floor is shown in Fig. 9.10.

The Millstone Grit is at its thickest in Gower, attaining 2,000 feet, but it thins northwards and eastwards to only 300 feet at Pontypool and Merthyr due both to the attenuation of stages and the overlap of the earlier stages. A threefold lithological division of the Millstone Grit has been recognized in south Wales, although the sediments are essentially cyclic as in the Mid-Pennine Province.

Naturally, lateral variations occur in south Wales for the Millstone Grit was deposited on the flanks of St. George's Land with the Culm trough to the south in Devon. The thick grits of the northern outcrops die out and the rocks are of a generally finer grain in the south. It is clear that the material was de-

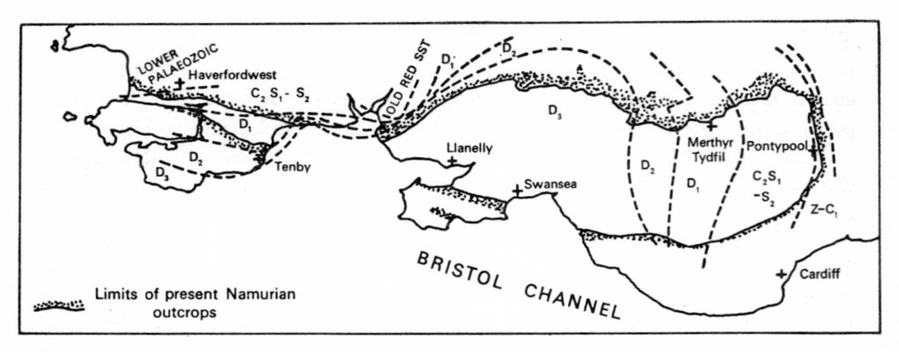

Fig. 9.10 The pre-Namurian surface in South Wales. (After J. PRINGLE and T. N. GEORGE, *British Regional Geology, South Wales.*)

rived from the north, from the Precambrian and Lower Palaeozoic rocks of Wales. Variations in thickness can be related to post-Dinantian folding which continued to operate causing differential subsidence of the area.

Stage	*Lithological Division*	*Characteristic Lithology*
G_2	Farewell Rock	Sandstones
G_1	Middle Shales	Shales with sandstones (Twelve Foot Sandstone near base)
R_2, R_1, H	Basal Grits	Sandstones, conglomerates, shales especially in higher part
E	Plastic Clay Beds	

South-West Peninsula

The Upper Culm, described on p. 191, is undoubtedly partly of Namurian age since R Stage goniatites have long been known from it. However, the combined evidence of goniatites, non-marine lamellibranchs and plants indicates that much of the Upper Culm is of Westphalian age and is contemporaneous with the Lower Coal Measures. Whether early Namurian deposition took place in the South-West Peninsula is unknown.

Northern England

The earliest Namurian goniatites have been recorded from just above the Great Limestone of the Yoredale Series. It occurs at the base of the Upper Limestone Group which indicates that cyclic sediments of the Yoredale type continued into Upper Carboniferous times. The presence of *Tylonautilus nodiferus* in the Upper Limestone Group of northern England (and Scotland) confirms the Namurian age of this group. Early attempts to define the base of the Namurian singularly failed in the northern part of the Alston Block where no rocks of Millstone Grit facies have been recognized. There Coal Measures succeed the Upper Limestone Group and comprise similar cyclic strata differing chiefly in the absence of limestones. In the Brampton area goniatites have not been found, but since limestones can be widely correlated it is clear that the Upper Limestone Group is Namurian. The proven Lower Westphalian age of the Coal Measures confirms that this is so. North-eastwards in Durham and Northumberland cyclic measures in which sandstones are more important occur but there is little comparable with the Millstone Grit facies of the mid-Pennines. Possibly the uplift through the Isle of Man–Lake District, which separated areas of different successions during the Lower Carboniferous, was renewed, preventing the spread of goniatite faunas northwards during the Namurian. The *Homoceras* and *Reticuloceras* Stages cannot be palaeontologically confirmed in the Northumberland area.

Scotland

In Scotland the cyclic sediments of the Lower Limestone Group (Viséan) are followed by similar cyclic sediments of Namurian age (Fig. 9.11). The base of the Namurian is taken as the base of the Limestone Coal Group, i.e. above the Hosie Limestones. The three groups formerly classed together as the Carboniferous Limestone Series differ little, but limestones are absent from the Limestone Coal Group. Its former name, the Coal and Ironstone Group (and it is also called the Edge Coals in Midlothian), was more apposite since workable coals occur. More than 50% of Scottish coal comes from this group while clay ironstones were at one time worked commercially. All the groups thicken generally eastwards, attaining a maximum north of the Forth estuary. In addition there is, especially in the west of Scotland, variation in thickness from north to south. Changes in thickness are relatively abrupt, coinciding with faults which must have acted contemporaneously, although they have a post-Carboniferous throw too. Thickest in the north of Ayrshire, the groups are thin in mid-Ayrshire which formed a slowly subsiding shelf; they are thick again in south Ayrshire except near the Southern Uplands Fault. The fault was not, however, the shoreline since sediments of this age overstep it and the precise position of the shore is unknown. Thinning of the Limestone Coal Group in certain areas appears to be due to two factors: fewer cyclothems are present, and individual cyclothems are not so fully developed. The amount of workable coal present in an area is not closely related to the thickness of the group since in some cyclothems the coal is thin or absent. The presence of numerous workable coals, as for example in south Ayrshire, reflects a rate of subsidence suited to the frequent production of coal-forming conditions (land developing and vegetation becoming established) followed by conditions of optimum conditions for the preservation of the coal-forming peat.

The limestones of these groups include a brachiopod-coral fauna not closely diagnostic of age. Non-marine lamellibranchs are present at a few

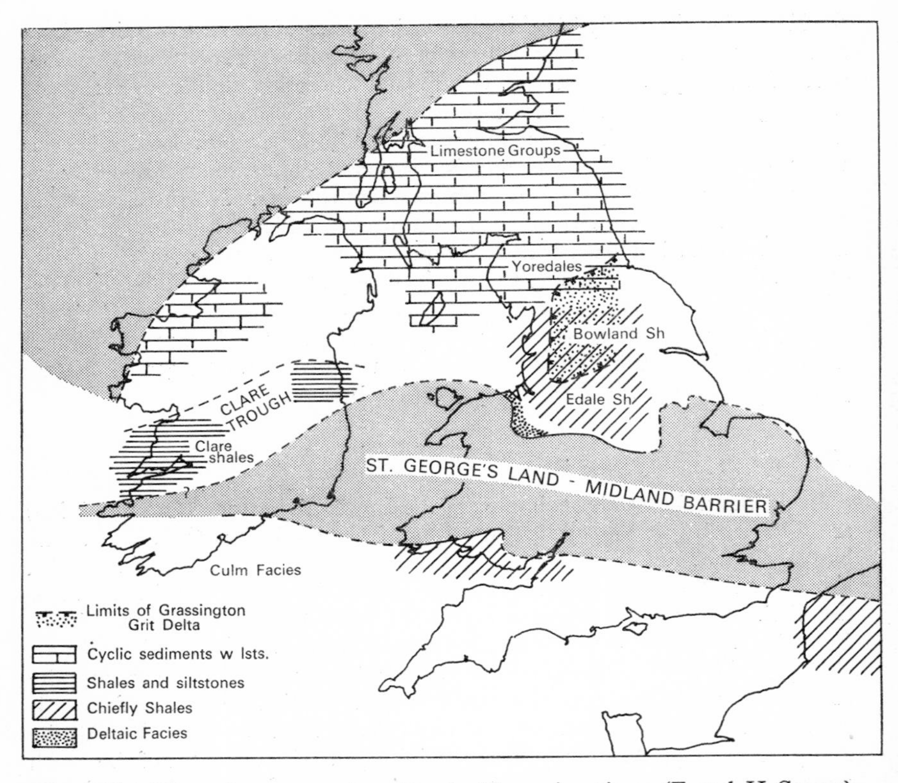

Fig. 9.11 The palaeogeography of early Namurian times (E and H Stages). (Based on L. J. WILLS, 1951, *A Palaeogeographical Atlas*, Plate VII, Blackie, London, and on J. K. CHARLESWORTH, 1963, *Historical Geology of Ireland*, Oliver and Boyd, Edinburgh.)

horizons but are of little stratigraphical consequence. Dating is dependent on goniatite evidence. The Lower Limestone Group has been shown to belong to the P_2 Zone and the Upper Limestone Group to the E_2 and H Stages. By inference the Limestone Coal Group must be also Namurian in age E_1. Correlation becomes of primary importance in a group such as the Limestone Coal Group because of the economically important seams. It is facilitated by the presence of *Lingula* bands but rests partly on lithological similarities of successions.

Above the Castlecary Limestone which marks the top of the Upper Limestone Group occurs the Passage Group, formerly called the Scottish Millstone Grit. It has been renamed because of the stratigraphical implications of the term 'Millstone Grit' and since the Passage Group is partly Namurian and partly Westphalian in age. A dominantly arenaceous series, it includes shales and important commercially exploited fireclays. Coal seams are only locally workable, although the thickest known seams in Britain occur in the Passage Group deposited in a subsiding basin at Westfield in Fife. The sandstones are spreads of deltaic material but they are of variegated colour, sometimes reddish.

In the west of Scotland there was renewed volcanic activity and volcanic rocks are to be found in Kintyre, Arran and Ayrshire. In the latter, lavas were poured out to pile up a thickness of 500 feet. Subtropically weathered they gave rise to the deposition of the Ayrshire Bauxitic Clay which was mined in north Ayrshire. In general the variation in thickness of the Passage Group reflects those of the underlying groups and it is thin in the west except where volcanics occur. It thickens eastwards to a maximum of over 1,000 feet in Clackmannan while east of the Pentlands it is about 500 feet, chiefly of sandstone called the Roslin Sandstone.

Naturally, goniatites are not found in deltaic sediments such as these so that the stages between E_2 and G_2 have not been confirmed. Further, a 'plant break' (a change in the characteristic fossil plants) takes place about one-third of the way up the Passage Group, but recent work on spores shows that at least some of the formerly unproven stages are present. The period of non-deposition may have been short.

The Namurian in Ireland

Little remains of the vast extent of Namurian strata which must have overlain the Viséan rocks over much of Ireland. Beds of Namurian age are now found in three areas, always overlying Lower Carboniferous and generally surrounding Coal Measure Basins. They occur in the west from Co. Clare to Kerry, in the Kilkenny-Carlow area and there are several minor outcrops in the north.

In broad outline the Namurian of Ireland is analogous to that of Britain. In the south-west culm facies conditions persisted, while in the north—the westward extension of the Scottish Midland Valley—cyclic sediments of Yoredale type persisted from the Viséan into the Namurian. However, the cycles of Upper Viséan and Lower Namurian are shale-sandstone cycles without limestones.

In County Clare the Foynes (= Clare) Shales were laid down initially in a downwarping trough centred in the region of the Shannon Estuary. An unconformity due to the Sudetic earth movements accounts for the absence of varying amounts of strata. P_2 and E_1 beds are missing and, towards the flanks of the basin, higher beds overlap the earlier ones. This basin of deposition was of Caledonian trend and can be deduced in the region of Dublin too. The Clare Shales, which include siltstones, vary from only about 50 feet to several hundred feet (600 feet at the Shannon Estuary) and they are generally unconformable upon Lower Carboniferous, as are the Edale Shales on the flank of the Derbyshire Massif. Despite the overlap of the lower stages away from the centre of the basin, beds as high as the lower part of the *Homoceras* Stage are overlapped, the succession of Namurian is very thick due to the great development of beds of the Upper *Reticuloceras* Stage. Thus, even in northern County Clare it totals 1,300 feet, though this is less than half the thickness exposed along the River Shannon in central Clare. In parts of north-west Clare there are nodules and extensive thin seams of phosphorite deposited during periods of lack of terrigenous material, possibly lagoonal. It occurs in commercial quantities. The coastal sections of County Clare provide

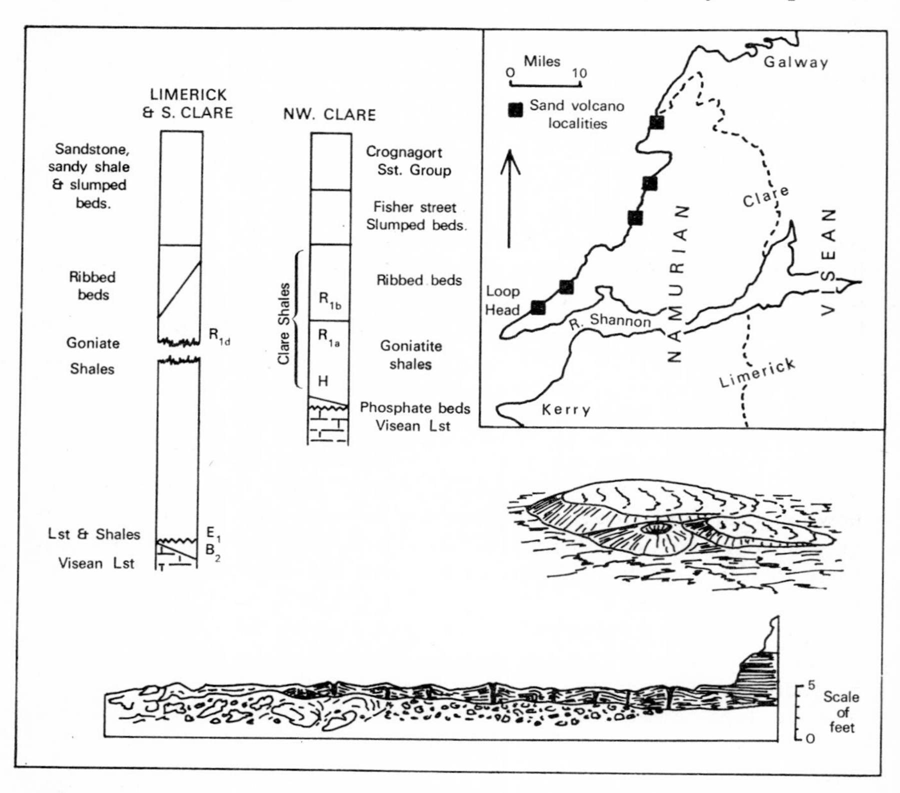

Fig. 9.12 The occurrence of sand volcanoes in the Namurian of Co. Clare, Ireland. (From W. D. GILL and P. H. KUENEN, 1957, *Q. Jl geol. Soc. Lond.*, **113**, 441, with stratigraphical data from F. HODSON and G. C. LEWARNE, 1961, *Q. Jl geol. Soc. Lond.*, **117**, 30.)

excellent examples of slumping, sediments having slid while still unconsolidated down the slope towards the axis of the trough in a southwards direction. Comparable slumps in the Tipperary-Carlow outcrops occur but the direction of sliding was northwards towards the trough's axis. Slump structures occur in sheets and in broad channels gouged to a depth of up to 150 feet. On the eroded surface of some slumped sheets occur small sand volcanoes (Fig. 9.12). These were caused by water-charged sand (quicksand) forcing its way through a cover of slumped material, the weight of which caused it to be under pressure.

North-west of Dublin thick argillaceous Namurian beds up to the *Homoceras* Stage succeed the Viséan conformably in the Summerhill Outlier, a succession comparable to that of the mid-Pennines. The Namurian of the several outcrops which lie to the south-south-west of Lough Neagh comprises cyclic sediments with only thin limestones but workable coal seams. Lacking goniatites, which elsewhere in the Namurian of Ireland are common, the presence of *Tylonautilus* confirms their Namurian age.

The Upper Carboniferous—The Westphalian and Stephanian

Problems of correlation

In Britain the greater part of the Coal Measures belongs to the Westphalian and only the highest beds in some coalfields, usually red beds, belong to the

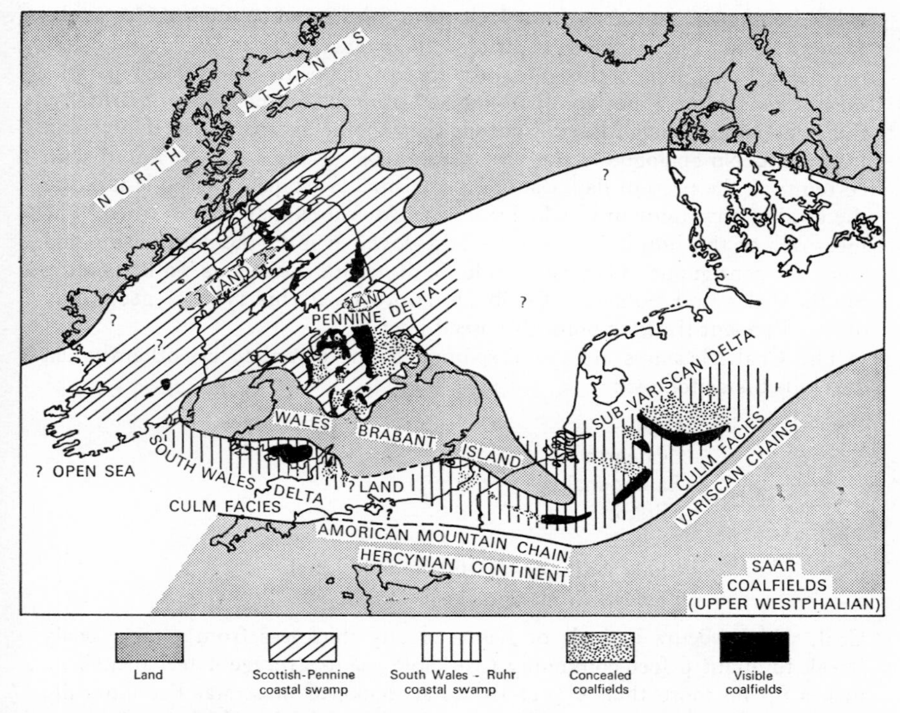

Fig. 9.13 Palaeogeography of north-western Europe in Westphalian times. (Note: recent work by P. E. KENT, 1966, revealing new information on the buried Carboniferous rocks of northern England necessitates minor changes to the northern shore of the Wales-Brabant Island.) (From L. J. WILLS, 1951, *A Palaeogeographical Atlas*, Plate IX, Blackie, London.)

Stephanian. The terms Productive Coal Measures and Westphalian are nearly synonymous for this country therefore. Originally divided on lithological bases into Lower, Middle and Upper Coal Measures, these divisions have been re-defined in relation to the zonal sequences (see p. 225).

The general palaeogeography of Westphalian times resembled that of the Namurian. The important feature was still the Wales-Brabant Island, while the northernmost parts of Britain remained elevated and the Hercynian Continent lay to the south (Fig. 9.13). Elsewhere erosion had produced vast flat areas which were at times inundated by shallow sea and at other times were the site of spreads of deltaic material. Periodically the sedimentation exceeded the rate of subsidence and swamp forests grew and soil formed. Close to sea level at all times, small depression or elevation of the crust caused widespread changes in the kind of deposit laid down.

Post-Carboniferous folding has resulted in the Coal Measures being preserved, as a rule, in synclinal structures now separated but showing close similarities of succession. The remarkable feature of Westphalian times is the extremely wide development of similar conditions. Coal-forming conditions existed contemporaneously from Ireland to Silesia and at times as far as the Donetz Basin. The coalfields of Britain, the Franco-Belgian, Ruhr and Silesian fields belonging to this one vast area of deposition are called paralic, meaning maritime. They are distinguished from the limnic basins of Brittany, the Saar and French Massif Central, that is, the basins of the Hercynian Continent. No analogue of the vast paralic coal swamps can be found with certainty at the present day, though the Dismal Swamps of Virginia indicate the kind of environment in which the Coal Measures may have formed. The measures of the limnic basins were laid down in inter-montain basins and were not contiguous. They lack evidence of marine incursions. In Britain the South Wales and Somerset Coalfields approached this type of geography during Pennant times, a point discussed later.

The Coal Measures are cyclic sediments, the complete cyclothem when developed is as follows:

Coal
Seatearth
Sandy shale
Siltstone
Shale
Shale with non-marine lamellibranchs
Shale with marine fossils

Coal, which occurs in beds or seams of any thickness from a mere coaly streak to about 6 feet commonly (workable seams average 4 feet 6 inches), makes up no more than 4% of the total thickness of strata. Economically important, as well as coal, are the seatearths found beneath the coal seams. Where a clay soil existed, leaching out of minerals has left a grey blocky fireclay (often with roots) sometimes thick enough to be worked economically. Sandy soils gave rise to very pure siliceous sandstones called gannister, commercially important, though sandy clay soils produced a useless impure ganister. Ironstones, argillaceous material cemented by iron carbonate, may have an iron content of over 25% and were formerly widely worked. The proximity of coal and ironstone accounts for the distribution of heavy industry in Britain although the pattern has changed with the development of other sources of ironstone. The commonest rocks of the Coal Measures are siltstones and shales, though sandstones are conspicuous. The different lithologies represent different conditions of deposition but the Coal Measures may be described, as a whole, as estuarine.

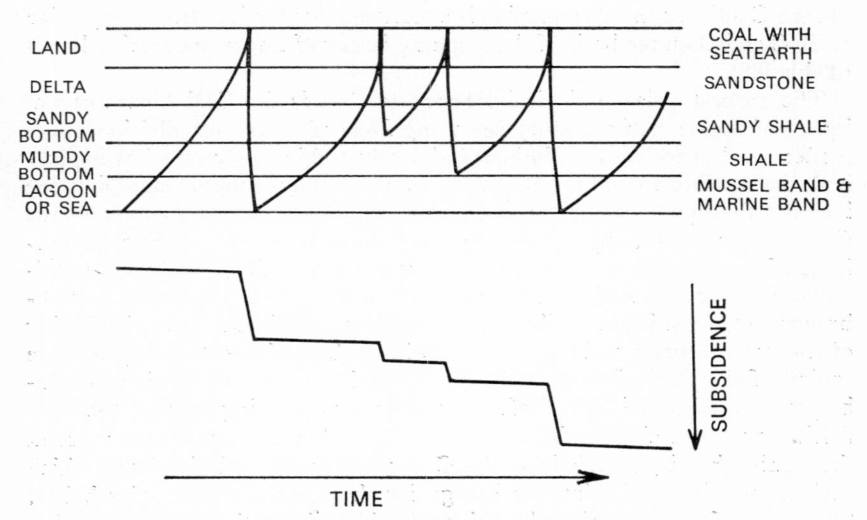

Fig. 9.14 Cyclic sedimentation illustrated diagrammatically. Many cycles, as shown, are incomplete. (Subsidence was intermittently accelerated to initiate further cyclothems.)

Coal-bearing strata are not confined to the Carboniferous System and are known from every geological system since terrestrial vegetation became abundant at the close of the Devonian period. However, two factors were necessary for the preservation of coal seams: firstly, the development of a flourishing land flora, requiring periods of emergence; secondly, the covering by sediment of the peat thus formed to prevent its rapid bacterial decay. Hence rapid submergence terminating coal-forming conditions must have occurred (Fig. 9.14). The kind of flora and its abundance infer a subtropical climate, and this conveniently relates to palaeomagnetic evidence of the position of the earth's poles in Carboniferous times.

Because of the importance of coal, still our primary raw material, more attention has been given to the detailed correlation of the Coal Measures than perhaps any other part of the stratigraphical record. However, the cyclic nature of the sediments introduces a difficulty, for the strata of one cyclothem are similar to those of any other. Furthermore, marine fossils, often used in correlation, occur too infrequently in the record to permit correlation of the order of exactitude required. However, the marine horizons (called marine bands) are sometimes of great lateral extent and act as vitally important index horizons. The more important can be recognized throughout the British Coalfields and in Belgium and Germany. Although marine fossils occur at infrequent intervals in the succession, for few cyclothems include marine shales, goniatites also utilized in zoning the Viséan and Namurian serve to divide the Coal Measures into the *Gastrioceras* and *Anthracoceras* stages. The base of the Westphalian is defined by the *Gastrioceras sub-crenatum* Marine Band. On the Continent the Westphalian is commonly divided into the Westphalian A, B, C and D on the basis of the goniatite faunas of important marine bands (Table 9.8).

The earliest palaeontological attempt to classify the Coal Measures was by Kidston over half a century ago using fossil plants. Four divisions were made: the Lanarkian, the Yorkian (originally ambiguously called Westphalian), the Staffordian and the Radstockian. The lowest of these Plant Divisions includes the upper part of the Namurian. Much more precise zoning of the Coal Measures (and upper Namurian) was effected by Dix utilizing assemblages of fossil plants to define zones designated A to I. This gives approximately as precise a zoning as the more commonly used non-marine lamellibranchs. Non-marine lamellibranchs occur in 'mussel bands' at the beginning of many cyclothems and less commonly in the beds above. Seven genera are typical of the Coal Measures. The zones are based on characteristic assemblages of species and named after a commonly occurring species, the zone fossil. Important changes in non-marine lamellibranch faunas take place at the marine bands, especially at the Cwmgorse or Top Marine Band which divides the Westphalian into the Morganian above and the Ammanian below. The range of genera is given in Table 9.9 which shows that *Carbonicola* dies out to be replaced by *Anthracosia* at the Mid-*Modiolaris* Marine Band and that other genera die out at the Cefn Coed and Cwmgorse Marine Bands. The widespread recognition of the mussel bands from coalfield to coalfield (especially in the Ammanian) confirms that the Coal Measures were formerly continuous.

In addition to the zoning of the Coal Measures by means of non-marine lamellibranchs and macroscopic plant fossils, correlation of seams has been effected using microspore assemblages and by means of detailed lithological comparisons. The latter are still invaluable for correlation over short distances where the lateral changes in a seam are not great and can be discounted. Correlation of seams is complicated by the historical fact that many seams were named before modern methods of correlation were known. As a result a laterally persistent seam may have several different names in different parts of the same coalfield. The corollary of this, that the same name may be given to different seams, is also found. Names of coal seams are themselves an interesting study. Many are purely descriptive, such as the Top Hard, the Two Foot or the Wee, while the majority are named after localities where they out-

TABLE 9.8

Europe	*Lithological Divisions*	*Plant Divisions*	*Floral Zones*	*Goniatite Stages*	*Non-Marine Lamellibranch Zones*	*Chief Marine Bands*
STEPHANIAN		RADSTOCKIAN	I		*Prolifera*	
WESTPHALIAN D	UPPER COAL MEASURES	STAFFORDIAN	H		*Tenuis*	MORGANIAN
WESTPHALIAN C			G		*Phillipsi*	
						—Top M.B.
	MIDDLE COAL MEASURES	YORKIAN ['Westphalian']	F	*Anthracoceras* = A	Upper *Similis-pulchra*	AMMANIAN
WESTPHALIAN B			E		Lower *Similis-pulchra*	—Mansfield M.B.
			D		*Modiolaris*	—Clay Cross M.B.
WESTPHALIAN A	LOWER COAL MEASURES	LANARKIAN	C	*Gastrioceras* = G	*Communis*	
					Lenisulcata	
NAMURIAN			B			
			A			

TABLE 9.9

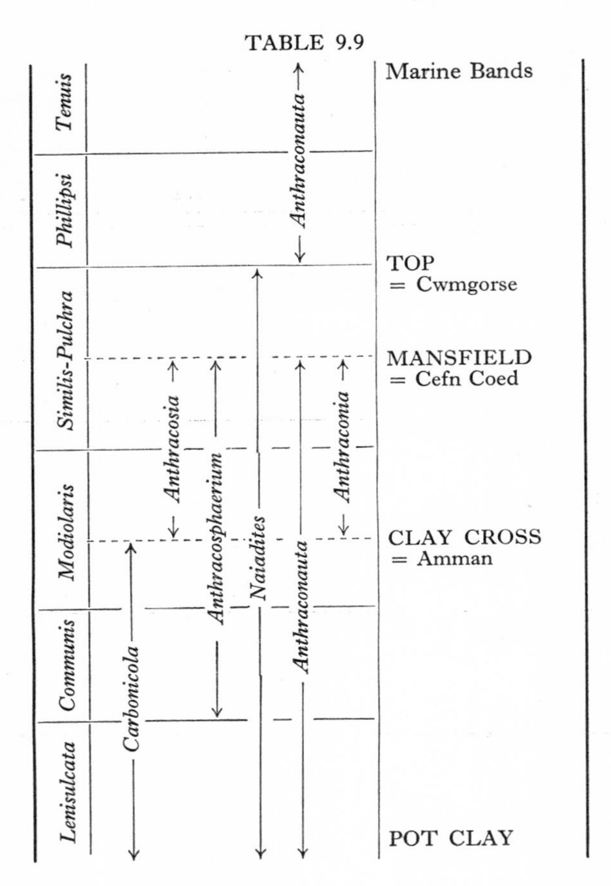

crop or have been mined, for example the Tupton Coal, the Alton Coal, etc. Common names such as the Main Coal and the Cannel Coal have led to confusion in correlation.

A major problem of correlation of seams is consequent upon the palaeogeography of Westphalian times. Near to land such as the Wales-Brabant Island where subsidence was slow, coal-forming conditions persisted for a long period and thick coal accumulated. However, traced away from the flanks of such a stable area the thick seams split into a number of seams separated by shales and sandstones, coal-forming conditions occurring only intermittently (Fig. 9.15). The best known example of a thick coal is that of South Staffordshire and Warwickshire where the Thick Coal is up to 30 feet. Northwards it splits into several 'leaves'. Similar examples are known from almost every field. The Nine Foot Coal south of Barnsley has a 1-inch dirt parting which northwards increases to 50 feet separating the Top Barnsley or Warren House Seam from the Bottom Barnsley. The presence of a thick coal often, but not always, implies the proximity of a stable area where coal-forming conditions did not become established.

Interruptions in coal seams, called washouts, are valleys filled with detritus. Many washouts were contemporaneous or penecontemporaneous cut through the peat of the coal swamp by meandering streams. The sandy infilling of these washouts, often cross-bedded, may be conglomeratic including pebbles of ironstone. Washouts formed after the compaction of the peat result in a seam being thickened at its edges (Fig. 9.15).

Coalfields are most conveniently described in regions related both to their palaeogeographical setting and structural pattern.

Fig. 9.15 (*a*) A washout in a coal seam with an infilling of cross-bedded sandstone. (*b*) Seam splitting in the South Staffordshire Coalfield with the development of thick coal in the proximity of the Wales-Brabant Island.

Coalfields of the Variscan Foredeep

South of the Wales-Brabant Island, coalfields are now found in south Wales, Somerset and Kent. The latter is completely concealed; that is the Coal Measures do not outcrop, being covered by younger rocks. The successions of all these fields have a lower shaly series with coals followed by a sandy series, deltaic or limnic, but with some workable seams. Except in Kent an upper shaly series with coals follows.

SOUTH WALES

The South Wales Coalfield is a large synclinal structure, plunging at the eastern end and compressed at the western end in Pembroke where thrusting has taken place. There are many post-Carboniferous faults, chiefly north—south, and minor folds. The succession is given in Table 9.10.

The Lower Coal Series outcrops peripherally and along the eroded Maesteg

TABLE 9.10

			Former Lithological Units	*Flora*	*Zones*	*Marine Bands*
MORGANIAN	Pennant Measures (≡ Upper Coal Measures)	Upper Pennant	Upper Coal Series	West. D	*Tenuis*	
		(Hughes Vein)				
		Lower Pennant	Pennant Series	West. C	*Phillipsi*	
						Upper Cwmgorse
AMMANIAN	Middle Coal Measures				*Similis-pulchra*	
			Lower Coal Series		*Modiolaris*	Amman
	Lower Coal Measures				*Communis*	
			Farewell Rock		*Lenisulcata*	
						G. subcrenatum

Anticline. In the Rhondda it outcrops in the valley bottoms, the Pennant Sandstone making the high moors between the valleys. The series thins from 3,000 feet at Swansea to 1,400 feet at Merthyr Tydfil and only 600 feet at Pontypool. The Pennant Series follows the Upper Cwmgorse Marine Band and comprises thick felspathic micaceous sandstones much used for building. Mineralogical evidence shows its derivation, in part, from the Old Red Sandstone. Paleocurrent directions indicate considerable variation in direction of transport of material but dominantly from the south or south-east with some supply from the north. It seems probable that the Pennant Sandstone was deposited in a basin of deposition of similar form to the present structural basin but larger, and that transport was towards the axis. However, the Pennant Series shows a similar pattern of variation in thickness to that of the Lower Coal Series, thinning to the north and east from a maximum in the Swansea district. In the western parts of the coalfield it includes workable coal seams. The Upper Coal Series is preserved in synclines and downfaulted troughs. It has been removed by erosion over the Maesteg Anticline. The lower beds are normal grey measures with coal seams but the uppermost 200 to 300 feet are red beds—marls, sandstones and shales—similar to the red beds which characterize the uppermost beds of every coalfield in Britain. These are discussed later.

The South Wales Coalfield is remarkable for the presence of seams of anthracite. These extremely valuable low-volatile coals (carbon = approx. 94%) occur in the Lower Coal Series of the western and north-western parts of the coalfield (Fig. 9.16). The distribution of anthracite seams bears no relationship to the present depth of the seams nor does it appear to be related to the former depth of burial (as Hilt's Law supposed), although it must be admitted that the evidence for estimating the former distribution and thickness of Mesozoic rocks in south Wales is necessarily inadequate. The rank of coal increases generally to the north-west but no complete explanation can be given.

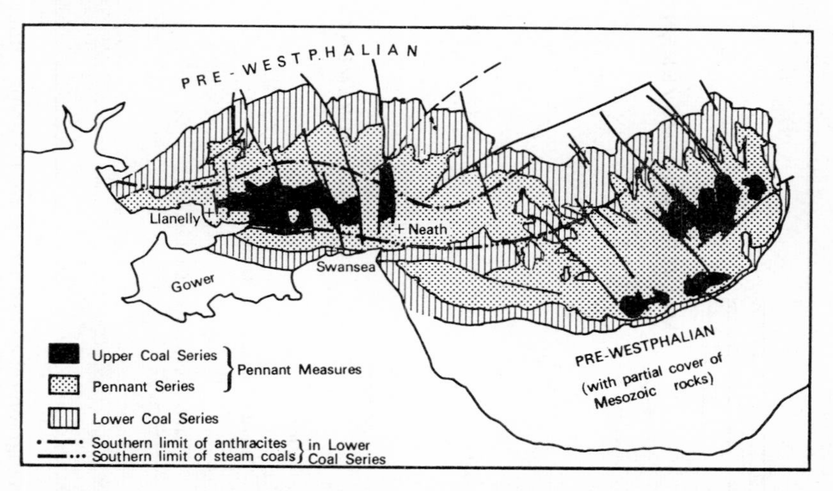

Fig. 9.16 Simplified geological map of the South Wales Coalfield with the area of occurrence of anthracite seams shown. (Based on L. R. MOORE, 1954, in *The Coalfields of Great Britain*, ed. A. E. TRUEMAN. Edward Arnold, London.)

FOREST OF DEAN

To the east of the South Wales Coalfield occurs the Forest of Dean field, a small basin with an essentially similar succession to that of south Wales but less disturbed by post-Carboniferous tectonics. Namurian strata are largely absent and the lowest group, here called the Trenchard Group, rests on beds from Upper Drybrook Sandstone to Lower Old Red Sandstone (Fig. 9.17). The chief seam worked marks the base of the Pennant Series which here comprises thick sandstones with little shale or coal. Workable seams occur in the Supra-Pennant Series found in the centre of the basin.

SOMERSET-GLOUCESTER COALFIELDS

These coalfields have been worked since mediaeval times and were probably worked as early as Roman times. Essentially a main basin or Central Basin is separated from the Gloucester Basin by the Kingswood Anticline.

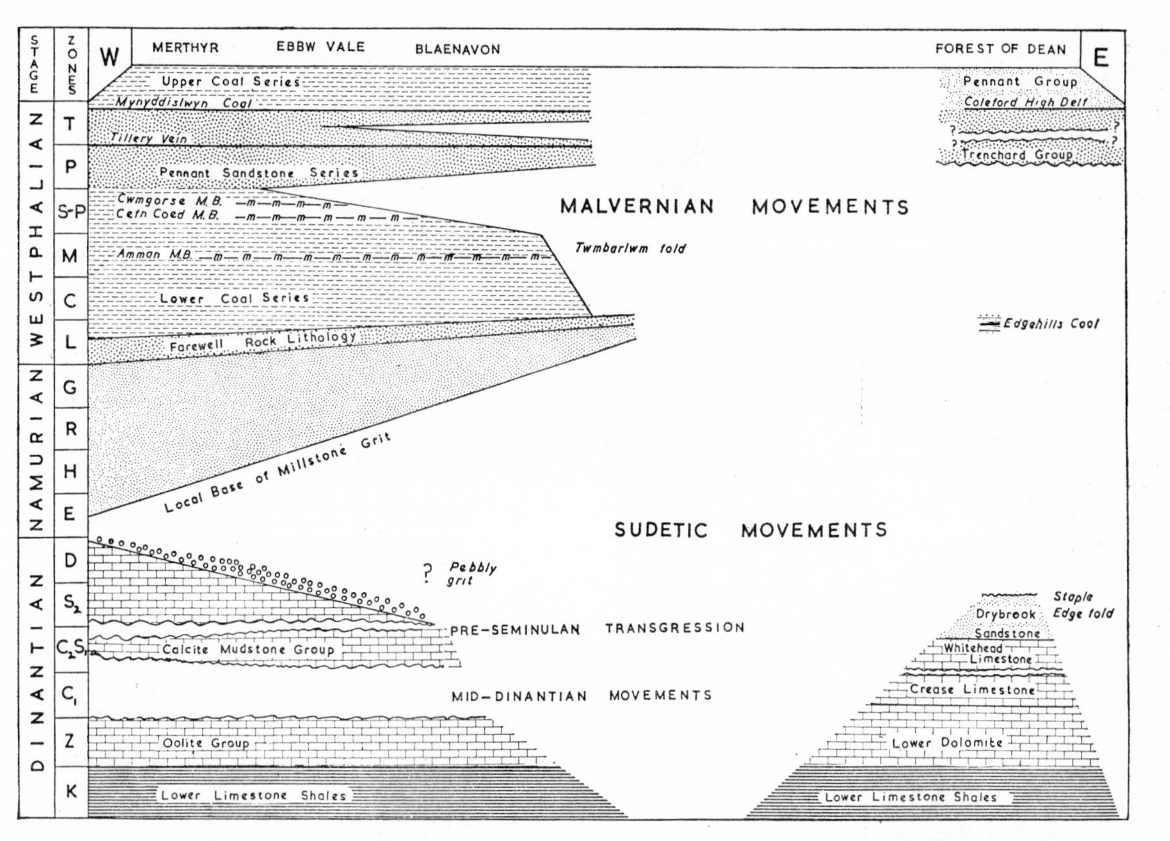

Fig. 9.17 Diagram to illustrate the stratigraphical and structural relationships within the Carboniferous rocks of the eastern portion of the S. Wales Coalfield and the Forest of Dean. (From H. J. SULLIVAN, 1964, *Palaeontology*, **7**, 351.)

However, three-quarters of the Coal Measures outcrops are concealed by Mesozoic strata so that isolated outcrops are frequent. In addition, there are separate basins, for example Avonmouth and Nailsea, to the west of the main outcrops (Fig. 9.18). The succession, which is unconformable on older rocks

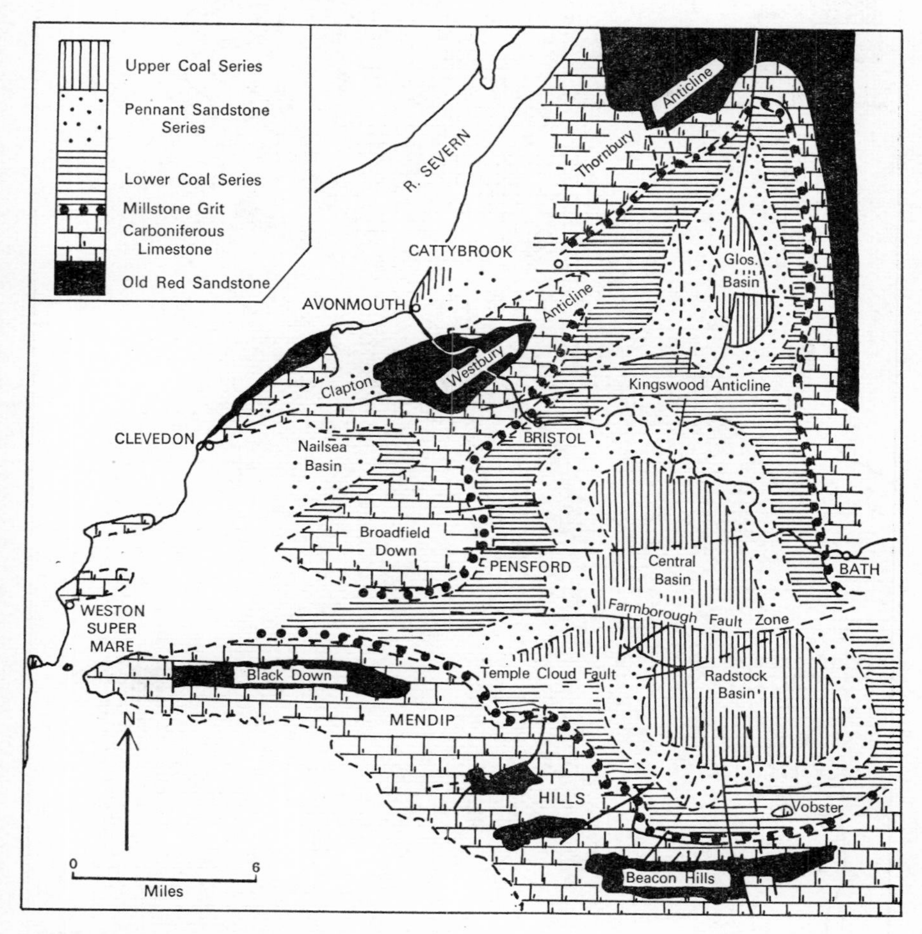

Fig. 9.18 The sub-Mesozoic coalfields of south-west England. (From L. R. MOORE, 1954, in *Coalfields of Great Britain*, ed. A. E. TRUEMAN. Edward Arnold, London.)

through the absence of much of the Namurian strata, is divisible into a Lower Coal Series, Pennant Series and Upper Coal Series. Seams have been worked in both the Lower and Upper Series. Uplift along a north—south axis, the Lower Severn axis, took place in pre-Pennant times and erosion resulted in

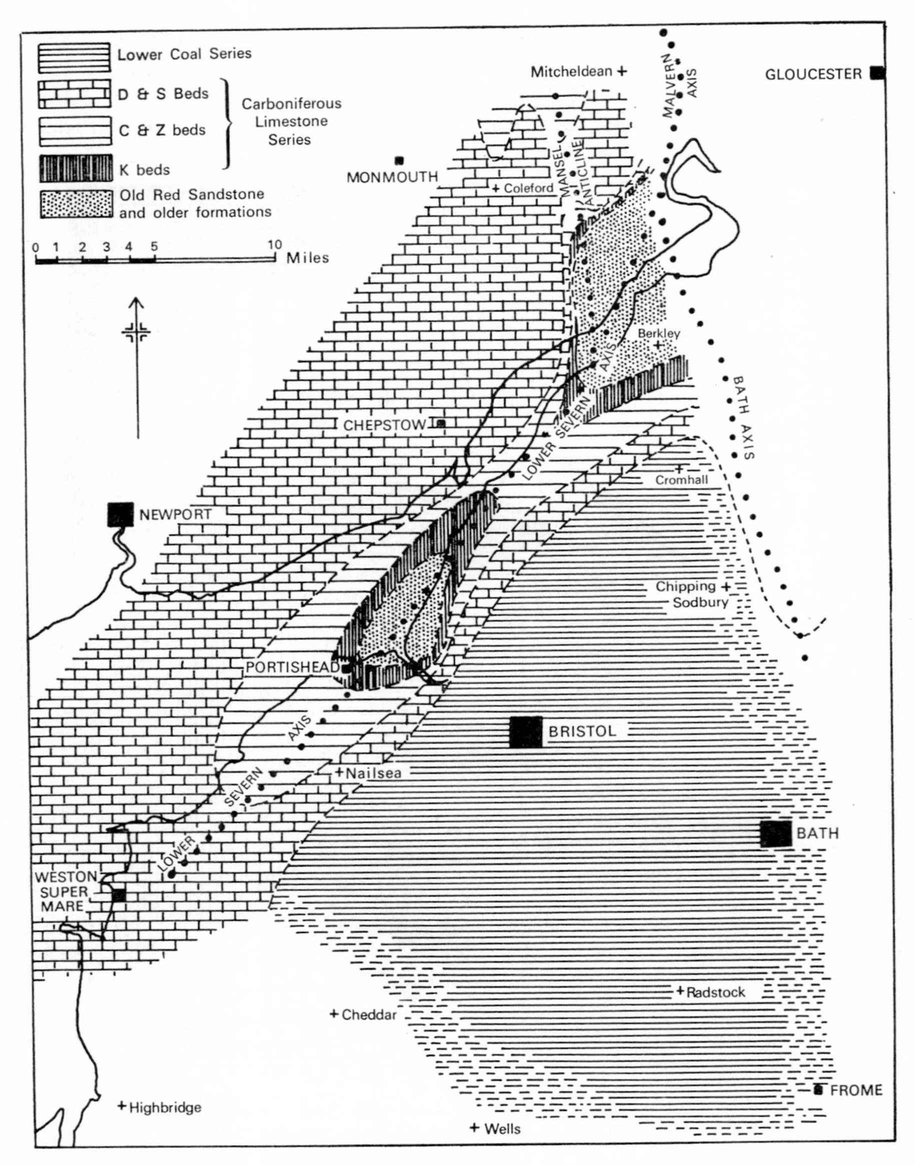

Fig. 9.19 The probable outcrops of the surface on which the Pennant Series was deposited. (From F. B. A. WELCH and R. CROOKALL, 1948, *British Regional Geology. Bristol and Gloucester District*, 2nd ed. Crown Copyright Geological Survey Diagram. Reproduced by permission of the Controller of H.M. Stationery Office.)

the Pennant Series being deposited on beds as old as Old Red Sandstone (Fig. 9.19).

Close to the Mendips, in the Radstock Basin, beds are steep and contorted and low-angle thrusts or 'slides' are present. Coal is now worked only in the Upper Series but structural complexities are well known from former workings in the Lower Series, for example in the Vobster area (Fig. 9.20). Little can now be deduced of the detailed structures from the limited outcrops available. The Lower Series coal seams are worked in the Kingswood Anticline but Upper Series seams are worked to the north in the Gloucester Basin, also known as the Coalpit Heath Basin. Upper Series seams are worked at Avonmouth where the measures occur partly under the R. Severn, but the Severn Tunnel is largely cut through Pennant Series sandstones.

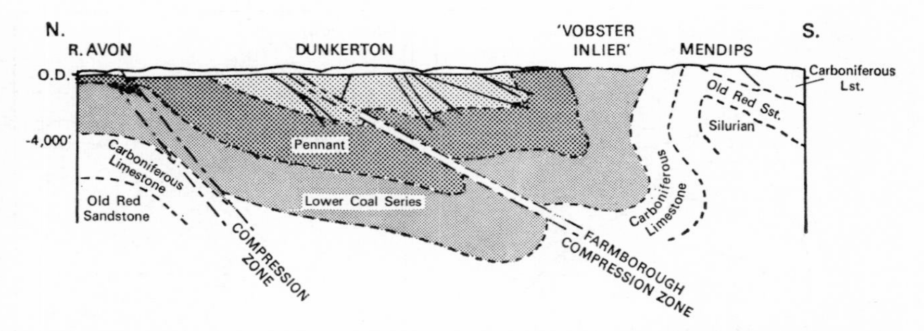

Fig. 9.20 A section through the Bristol-Somerset Coalfields. (Simplified from F. B. A. WELCH and R. CROOKALL, 1948, *British Regional Geology, Bristol and Gloucester District*, 2nd edn. Crown Copyright Geological Survey Diagram.)

Red beds are well developed in the Gloucester-Somerset Coalfields, especially in the Coalpit Heath Basin where they attain a thickness of 1,000 feet. Red bed conditions did not everywhere commence at the same time (later in the south) but they belong in part to the *Tenuis* and *Phillipsi* Zones. Since fauna is scarce it is uncertain how much of the succession is referable to the Stephanian. It is, of course, the type area for the development of the Radstockian flora.

KENT

The Kent Coalfield is completely concealed by younger strata. The Westphalian Coal Measures, resting unconformably on the Viséan (for the Namurian is absent here), form part of a platform of Palaeozoic rocks which formed a ridge from London to Kent and into Belgium. (During much of Mesozoic time it formed an important feature of palaeogeography.) The cover of Mesozoic and Tertiary rocks is 1,000 feet thick or more. The form of the coalfield is known with moderate precision, however, from the few mines and from boreholes (Fig. 9.21). It is a syncline plunging east-south-east. Westward the plunge terminates the Coal Measures abruptly and it is possible that they are terminated against a north—south fault. There is considerable stratigraphical similarity to the Somerset and South Wales Coalfields. The Lower Shale Division, about 700 feet of strata corresponding to the *Communis* to *Similis-Pulchra* Zones, shows that lowest Westphalian (of *Lenisulcata* Zone) is absent. It is followed by the Sandstone Division, closely analogous to the Pennant Measures of other coalfields, some 2,000 feet of strata belonging to the *Phillipsi* and *Tenuis* Zones. There is evidence of a pre-Pennant non-sequence. Coal seams occur in the Sandstone Division, for example the Kent No. 1 seam (the highest), as well as seams in the Shale Division. The latter are semi-anthracites.

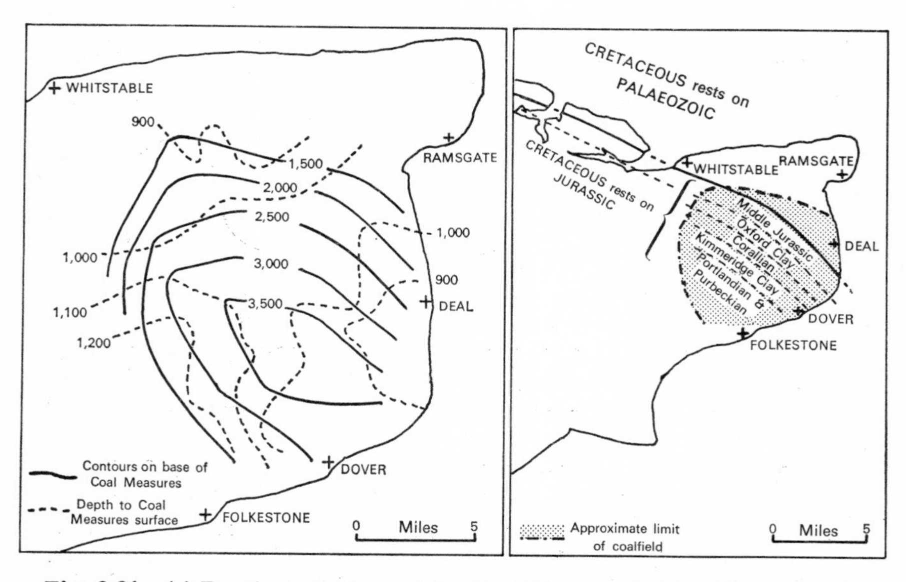

Fig. 9.21 (*a*) Depths to the base of the Coal Measures in Kent illustrating the form of the Kent Coalfield Basin. (*b*) The nature of the cover rocks of the Kent Coalfield.

Although the Kent Coalfield has not been traced far beneath the Channel, a corresponding basin occurs in the Pas de Calais, similar in a number of characters. There, too, the Ammanian is attenuated and rests unconformably on pre-Namurian rocks.

All these coalfields south of the Wales-Brabant Island have features in common. Although the thickness of Coal Measures exceeds 8,000 feet in South Wales, the maximum in Britain, and the Ammanian is approximately 3,000 feet, the *Lenisulcata* Zone is poorly developed. However, the Ammanian (Lower Coal Series) includes important seams. The Pennant Measures, despite their clearly limnic mode of occurrence demonstrated in South Wales, are everywhere represented, often with a break at the base because the abrupt changes in geography which introduced the Pennant were due to uplift. The higher Pennant Measures (Upper Coal Series) where still preserved (and locally as much as 1,000 feet remain) contain grey measures with coals followed by red beds, the latter extending into the Stephanian.

Coalfields on the northern flank of the Wales-Brabant Island

In the coalfields bordering the Pennines, described below, the Coal Measures attain a great thickness. Maximum downwarping of the crust to accommodate over 5,000 feet of strata took place in Lancashire. (The area of maximum downwarping was to become subsequently the site of maximum uplift, a common geological phenomenon.) The Coal Measures thin southwards towards the Wales-Brabant Island overlapping the Millstone Grit and Carboniferous Limestone. Occurring in horst inliers the Coal Measures rest on Lower Palaeozoic rocks, on Tremadocian in the Warwickshire Coalfield and on Wenlockian and Ludlovian (Downtonian) in the South Staffordshire Coalfield. As well as this unconformable relationship to earlier strata, these coalfields of the west Midlands and Welsh Borders have other characters consequent on their proximity to the landmass to the south. Thick coals, already described, occur in south Staffordshire and Warwickshire with seam splitting northwards away from the stable area. In the south there is also great development of fireclays which are worked opencast and mined.

SOUTH STAFFORDSHIRE AND WARWICKSHIRE COALFIELDS

Although these coalfields differ greatly in structure, for the Warwickshire Coalfield is a synclinal structure plunging southwards revealing the under-

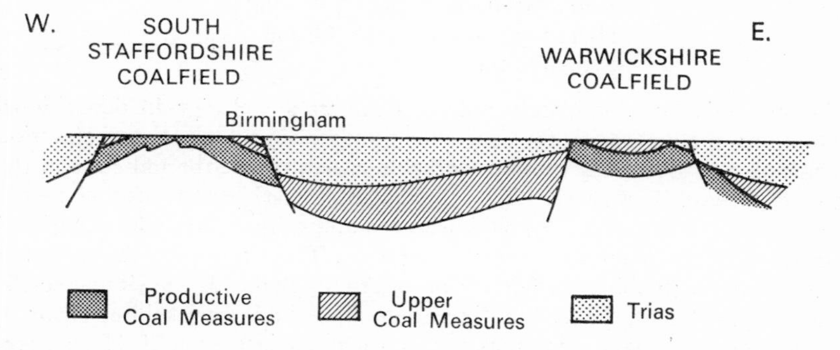

Fig. 9.22 Section through the Midlands Coalfields and intervening area to show the horst structure of the coalfields.

lying Cambrian rocks to the north-west and north-east while the South Staffordshire Coalfield is essentially anticlinal (Silurian rocks occur as inliers) (Fig. 9.22), both fields have a closely similar succession. In both cases the proximity of the high ground to the south is evident from the succession. Of course, Coal Measures occur between the two coalfields at considerable depth but they are unexplored and our reconstruction of the palaeogeography is dependent on the outcrops and workings of the coalfields.

The presence of lowest Westphalian is doubtful in both coalfields, and the chief coals were not formed until late *Communis* Zone times and later. The productive measures are not thick, though they thicken northwards away from the Wales-Brabant Island. In the south the Thick Coal is well developed in both coalfields, in south Staffordshire south of Dudley and in Warwickshire from Coventry southwards. In the former it is up to 30 feet thick and is still being worked opencast. Ironstones were formerly economically important, especially in the lower part of the sequence, and fireclays are still important in south Staffordshire near Stourbridge (where they are mined and worked opencast) and in Warwickshire. However, fireclay is less important than formerly and much has now to be brought into the area from the Coalbrookdale Coalfield to supply pottery works. The South Staffordshire Coalfield includes some igneous intrusions (of later date). The development of dominantly red measures of Upper Coal Measures (or Morganian) age is notable. They are undoubtedly partly Stephanian and are believed by some authorities to belong partly to the Permian, though faunal evidence is slight. These dominantly red beds attain a thickness of 3,000 feet in south Staffordshire where they occur on the flanks of the anticlinal structure and 5,000 feet in Warwickshire where they occupy the centre of the syncline in the southern part of the field. The succession in both areas is closely similar and is as follows:

Clent Group
Enville or Conglomerate Group
Keele Group
Halesowen or Newcastle Group
Etruria Group

The succession of dominantly red beds indicates a change in depositional conditions, gradual at first for no marine transgressions took place as elevation occurred, and grey beds without workable coals precede the red clays of the Etruria Group. The latter, often referred to as marl (erroneously since the lime content is less than 2%), comprise red and purple clays with occasional conglomeratic beds known locally as 'espleys'. These represent the spread of coarse material; pebbles of fairly local origin (in Staffordshire similar to the rocks of the Lickey Hills, but Cambrian Shale in Warwickshire) occur in coarse greenish brown sandy matrix. They suggest further uplift of the Mercian Highlands (the former Wales-Brabant Island), which also accounts for the lack of marine horizons and the termination of coal-forming conditions. The clays are of great economic importance and are very widely dug to make engineering bricks.

Further uplift took place locally since the next group, the Halesowen Group (also known as the Newcastle Group and, in Coalbrookdale, the Coalport Group), follows unconformably and was apparently not laid down at all between Shrewsbury and Lilleshall. Nevertheless, it reflects a return to grey cyclic sediments of typical Coal Measures type though no workable coals of good quality are found. Sandstones are dominant and freshwater lacustrine *Spirorbis* Limestones occur. The distribution of the Halesowen Group suggests that the Wales-Brabant uplands (Mercian Highlands) may not have been continuous at this time. Halesowen Beds, resting unconformably either on productive Coal Measures or on pre-Westphalian strata, make the search for concealed coalfields complex. The presence of coals in synclines depends upon the amount of pre-Halesowen erosion. It is known from geophysical work and a number of boreholes that pre-Mesozoic basins occur in southern England. Test boreholes have proved the presence of Westphalian beneath the Mesozoic cover (for example near Stratford), while in other bores (for example at Burford, Oxfordshire) the Halesowen Group was proved but whether it was underlain by productive Coal Measures or not was not discovered. It seems likely that south of the Wales-Brabant Island generally there may have been a fairly long pre-Halesowen period of erosion. To return to the northern flanks of the Wales-Brabant highlands, the red and purple clays with occasional sandstones, known as the Keele Group, follow the Halesowen Group—conformably except where the latter was not deposited over the Haughmond Hill area. Belonging to the *Tenuis* Zone in part, this group probably is partly of Stephanian age.

The ensuing red beds may, as Wills believes, be referable to the Lower Permian (Rothliegende) though they may be late Carboniferous. The Enville Group, including conglomerates, breccias and marls, indicates renewed uplift of the Mercian Highlands as spreads of conglomerate fans were deposited at the foot of the high ground. However, the general environment was similar to that in which the Keele Beds were deposited, widespread lakes occurred and at times dried up. Footprints of amphibians are known from both groups. Surprisingly no teeth have been preserved, but bones found half a century ago at Kenilworth in the south of the Warwickshire Coalfield support the view that the Enville Group is Permian. The pebbles of the conglomerate fans are Carboniferous limestone and chert chiefly, with Silurian limestones and Precambrian volcanics.

As many as five horizons of conglomerate have been recognized in Warwickshire, inferring repeated uplifting of the Mercian Highlands causing renewal of supplies of detritus. Northwards conglomerates pass into sandstones and marls, the Enville Marls, the latter thickening away from the Highlands.

The Enville Group is succeeded, generally unconformably where the relationship can be established, by the Clent Group. This name is given to stratified breccias which have many local names but which were formed on the flanks or at the foot of the Mercian Highlands, and have a rather similar distribution to the Enville Conglomerates. Of uncertain age through lack of fossil evidence, they have been regarded as from late Upper Carboniferous to Permian. Away from the Highlands the Clent breccias pass laterally into marls indistinguishable from the Enville Marls, and the palaeogeography was generally similar. The breccia fragments while varying from one locality to another, include much Precambrian volcanics and they are quite angular with a lack of sorting. Despite the renewed periods of uplift of the Mercian Highlands (indicated by the several conglomerate and breccia fans), and the subsidence of the area to the north to accommodate several thousand feet of red beds, there was a gradual peneplanation of the Midlands which we shall see had been accomplished by Upper Triassic times.

In the Welsh Borders Coal Measures flank the Triassic rocks of the Cheshire Basin to the west and south in the counties of Flint, Denbigh and Shropshire. The productive Coal Measures of the Coalbrookdale-Forest of Wyre Coalfield lay close to the Wales-Brabant Island and include large reserves of fireclay while marine bands and mussel bands are uncommon. The Halesowen Group (known as the Coed-yr-Allt Group here) rests with unconformity on older strata, the plane of unconformity originally being designated (in error) the Symon Fault. Sedimentation spread south of Shrewsbury in Morganian times and the Halesowen age beds rest on beds as old as Longmyndian.

Coalfields bordering the Pennines

Separated by uplift and erosion along the Pennine axis and by east—west uplift, the coalfields of North Staffordshire, Lancashire, Yorkshire-Nottinghamshire-Derbyshire, Cumberland and Northumberland have much in common. The Coal Measures of the southern Pennine flanks attain a thickness of 5,000 feet—thinning westwards in the Lancashire Coalfield and eastwards in the Yorkshire Coalfield, indicating that maximum deposition occurred near the site of the subsequent Pennine uplift. The *Lenisulcata* Zone is well developed in these fields but contains few important coals, the chief workable seams range from the *Communis* Zone to the *Similis-Pulchra* Zone, both inclusive. The plunging synclinal structures of the North Staffordshire and Lancashire Coalfields, though the latter is greatly complicated by faulting, have not been explored far beneath the Triassic cover. On the other hand, the concealed part of the Yorkshire-Nottinghamshire Coalfield exceeds in area the exposed coalfield. In addition to its importance as a coalfield—it includes the greatest known reserves of any of the British coalfields—it is in the Carboniferous rocks of Nottinghamshire that structures occur giving rise to Britain's chief oilfield, in the Eakring area. The many workable seams of the Lower and Middle Coal Measures, though local names have been given to individual seams, have been broadly correlated within each field and considerable progress has been made in correlation from one coalfield to another.

By Morganian (*Phillipsi-Tenuis* Zones) times coal-forming conditions had virtually ceased and in north Staffordshire the Etruria Marl and Newcastle (= Halesowen) Groups are typically developed. The towns of the Potteries have grown up on the outcrop of the former. Red measures occur, though not extensively, in Lancashire and are known as the Ardwick Group. East of the

Pennines red Coal Measures are only found at depth. They occur in Nottinghamshire less than 200 feet above the Top Marine Band but pass northwards into grey beds so that in Yorkshire red beds occur about 1,000 feet above the Top Marine Band. The red beds include 'espleys' like those found in the Etruria Group of the south Midlands.

The coalfields which flank the Pennines in the north, the Cumberland and the Durham-Northumberland Coalfields, have several similar characters. The succession in both cases is no more than 2,000 feet and the lower part, the Gannister Group of Northumberland and Durham, includes few coal seams. In Durham coking and gas coals of lower rank than the steam coals of Northumberland are found, although the Durham part of the field is partly overlain by Permian and Triassic strata whereas the Northumberland area is not concealed. It is interesting to speculate on the possible thickness of the former Mesozoic cover in Northumberland. In the coalfields on both sides of the Pennines the chief seams lie between mid-*Communis* and the top of the *Similis-Pulchra* Zone. After this time divergences occur: the upper part of the succession in Durham and Northumberland comprises grey measures with only very minor development of red beds whereas in Cumberland the Whitehaven Sandstone Group, some 600 to 700 feet of sandstones with thin *Spirorbis* Limestones is partly red or purple and its deposition was preceded by uplift and erosion. No seam can be correlated across the northern Pennines despite several intervening minor inliers of Coal Measures, such as Midgeholme. On the other hand, the correlation of seams within the Northumberland-Durham field itself is possibly the most highly developed in Britain as a result of work on mussel bands and *Lingula* bands.

Scotland

Four important coalfields of Westphalian Coal Measures occur in the Midland Valley of Scotland separated by post-Coal Measures folding. They are the Ayrshire, Central (or Lanarkshire), Midlothian and Fife Coalfields. Of course, important coals occur in the Limestone Coal Group and Millstone Grit, both of Namurian age. The Millstone Grit facies persisted in Scotland and the *Lenisulcata* Zone cannot be recognized since coal-forming conditions did not generally return until about the beginning of the *Communis* Zone. The Productive Measures continue up to about the end of the *Similis-Pulchra* Zone to be followed by red beds. The productive coal measures and the red coal measures both show a pattern of variations in thickness comparable to that of the other groups of the Carboniferous: they thicken generally northeastwards, attaining a maximum in Fife. The productive measures are about 700 feet thick in north Ayrshire but more than three times this thickness in Fife. The red measures show this pattern too, although two factors complicate matters; the amount preserved and the diachronous onset of red bed environmental conditions. To take an example of the latter from Ayrshire: in central Ayrshire (Mauchline) red beds occur 30 feet above the Ell Coal but in south Ayrshire (Cumnock) they occur 420 feet above the Ell Coal.

Outliers of Coal Measures which indicate the former extent of deposition are to be found in Kintyre (the Machrihanish Coalfield) and south of the Southern Uplands Fault, (the Sanquhar Coalfield). In the latter, Coal Measures overstep thin Namurian on to Lower Palaeozoic strata. Further south in Dumfries occurs the Canobie Coalfield providing a 'link' with the Cumberland Coalfield, and by Westphalian times there was little evidence of the existence of a positive element in the region of the Southern Uplands.

Ireland

Westphalian age Coal Measures are not of great extent in Ireland, the greater part having been removed by erosion. Small coalfields occur in the north at Coalisland just to the west of Lough Neagh and near Lough Allen. In the south the largest coalfield is the Leinster or Castlecomer field, and south-westwards from it there are a number of coalfields which lie to the north of the Armorican front occupying a structural position analogous to that of south Wales. These include anthracite seams. The Coal Measures of Ireland which remain belong to the Lower and Middle Coal Measures (= Ammanian) and are economically of some importance. They are geologically important since they extend our palaeogeographical knowledge and, as with the remnants of later formations found in Ireland, their palaeogeographical importance is great in relationship to their actual area. The area of Coal Measures in the south-west, south of the Shannon Estuary, is much less than formerly supposed and is restricted to the Crotaloe and Kanturk Coalfields. Concealed fields postulated on structural evidence may lie beneath the Antrim Plateau and geophysical work supports this view.

Carboniferous fossils

Corals and brachiopods are profuse in Carboniferous rocks of calcareous facies and the zonal sequence of the Dinantian has been given. Although brachiopods with looped brachial skeletons become commoner, the brachiopod fauna is still dominantly protremetous with Productids common as well as spire-bearing forms (Spirifers), these two groups continuing on to the end of the Palaeozoic. Only changes in depositional environment explain their absence from much of the Upper Carboniferous for in Russia they are used to zone the Moscovian of Upper Carboniferous age. Corals are important, not only those which give their names to faunal zones, becoming more diverse as Carboniferous time progressed. Thus the Viséan is marked by the incoming of clissiophyllid corals (prior to this simple Zaphrentids and multi-septate corals such as the Caninids and *Palaeosmilia* were present). The typical neritic strata possess a very varied fauna including many other forms of life. Though trilobites had been greatly reduced in importance and only Phacopidae, Cyphaspidae and Proetidae remain, the latter persisting into the Upper Carboniferous; echinoderms and molluscs were common.

Attached echinoderms are important, especially crinoids—though it is of interest that they were not as common nor so diversified as in the Mississippian of North America—and Blastoids attained their acme. The latter are commonly found in reef-knolls which have a neritic fauna distinct from adjacent bedded neritic limestones. Reef faunas include gastropods and lamellibranchs, brachiopods such as *Pugnax* and the coral *Amplexus*. Echinoids are not uncommon.

Goniatite cephalopods, important in zoning the Culm facies are yet more important in the typical Namurian Millstone Grit and occur in Coal Measures marine bands. The zonal sequence has been given in two tables.

The Upper Carboniferous where of marine facies, for example where limestone cyclic sedimentation persisted into the Namurian, is not radically different from the Lower Carboniferous in general terms of fauna. However, the development of widespread estuarine conditions in Westphalian times resulted in the preservation of non-marine and terrestrial organisms, including many plants. Non-marine lamellibranchs which made their appearance in Upper Devonian times become of major significance (the zonal sequence has been given). The fern-like Pteridosperms are very common (while true ferns were rare); also *Lepidodendron*, *Sigillaria* and *Calamites*, established in Lower Carboniferous times, became highly important in the Upper Carboniferous. Among vertebrates, fish continued without great change from the Devonian, though Ostracoderms were extinct by Carboniferous times. Amphibians, which originated in the late Devonian, became the most important vertebrates although the first primitive reptiles evolved during Carboniferous times. Thus the period covers the main transition of vertebrate life from water to the land.

10 The Variscan Orogeny

Great Britain: general structural picture and its relation to Europe

The Variscan Orogeny includes earth movements which took place during Devonian and Carboniferous times but which culminated in important mountain building late in Carboniferous times, in Britain after the deposition of the Coal Measures. There is also little evidence that any intense deformation of the Upper Palaeozoic rocks of Culm facies took place until Upper Carboniferous times. In Britain and in Ireland the intensity of deformation is greater in the south than the north, and it is yet further south that the effects were greatest. They can be seen in the Palaeozoic horsts of France, Germany and northern Spain. In Britain there is a difference in tectonic style and intensity of the structures found in south-west England and in Pembrokeshire from those further north. Similarly in Ireland, a line from Dingle Bay to Dungarvon separates regions of different structural pattern. The Variscan Orogeny was in magnitude, and perhaps duration, no less than the Caledonian Orogeny, but since its loci lay to the south of Britain the chief effects here were epeirogenic, giving rise to faulting and folding with relatively low dips. Only in south-west England (and Pembroke) and south-west Ireland were thrusts and overfolding produced accompanied by metamorphism. These areas lay in the region of 'Armorican' Fold Structures, with an east—west to north-west—south-east trend developed also in Brittany (Armorica), from which the synonymous term Armorican Orogeny is derived. Another synonym commonly used, the Hercynian Orogeny, is named from the Harz Mountains. Originally these terms were used to refer to different arcs, hence directional trends, of the late-Carboniferous Orogeny but have come to be interchangeable and the term Variscan is now the most widely accepted.

Among the most important structures produced to the north of the fold belts were the gentler folds of the Carboniferous strata which, on subsequent erosion, separated the Coal Measures outcrops into the present coalfields.

Igneous intrusion associated with the Variscan Orogeny was widespread in Britain, as was mineralization.

A glance at the geological map of England shows the conspicuous overlap which marks the unconformity of the Permo-Triassic rocks on older strata. Important structures can be seen to be post-Westphalian but pre-Permian. However, there is abundant evidence from the geological record of still earlier earth movements. The earliest, the Bretonic Phase of the Variscan Orogeny, produced great changes in palaeogeography in Brittany, but there is little evidence of effects further south in the geosyncline and none from the British area where there was a transition from Devonian to Carboniferous sedimentation. The mid-Avonian unconformity, conspicuous in Pembrokeshire, and the change in facies of the South-West Province (Bristol area and south Wales) indicate movements in pre-C_2 times, the Nassanic Phase, which at least affected the shelf-sea areas. A much more important earth movement, termed the Sudetic Phase, from effects formerly thought to be present in the Sudetic Mountains of Germany, occurred preceding the Namurian. It may have caused tilting of Britain about an east—west axis; certainly it led to the spread of rocks of deltaic facies southwards in late Viséan times. However, the principal effect was to produce widespread unconformity of the Namurian strata and their great attenuation in the Bristol area where Coal Measures, gently folded, rest on more strongly folded earlier Carboniferous and older strata. This effect can be clearly seen in the Forest of Dean. The whole complex history of the Lower Carboniferous inundation of Britain (discussed in the previous chapter) and the further elevation of St. George's Land and the Midland Barrier to form the Mercian Highlands are facets of the Variscan Orogeny. The cyclic nature of much of the Viséan, Namurian and Westphalian strata imply crustal instability preceding the main orogenic movements. The onset of red-bed conditions of deposition in late Westphalian-Stephanian times reflects geographical changes brought about by the orogeny.

Orogenic phases

Many phases and sub-phases of the Variscan Orogeny have now been recognized and while further research has demonstrated the purely localized effects of some, or has eliminated them, the more intensive the study the more phases which will be recognized. Gignoux drew attention to this more than a decade ago. In Britain it is clear that phases of earth movement can be demonstrated, at least in some areas, as follows:

- pre-C_2S_1 Zone (Nassanic Phase)
- pre-Namurian (Sudetic Phase)
- pre-Morganian (Malvernian Phase)
- pre-*Tenuis* Zone (i.e. preceding deposition of Halesowen Beds)
- pre-Keele Beds deposition (approximately pre-Stephanian*) (? Asturic Phase)
- pre-Clent Beds deposition.

* The Keele Beds may be diachronous.

The pre-Permian unconformity of north-east England may be due to the Asturic Phase or to the (later) Saalic Phase, the hiatus being of considerable duration—a point discussed in the next chapter.

The Variscan Fold Belt in Britain

Intense folding of Alpine proportions is known from northern Spain, from the Cantabrian Mountains and the Leonides. In the Palaeozoic massifs of central and north-western Europe overfolding and thrusting has occurred (the

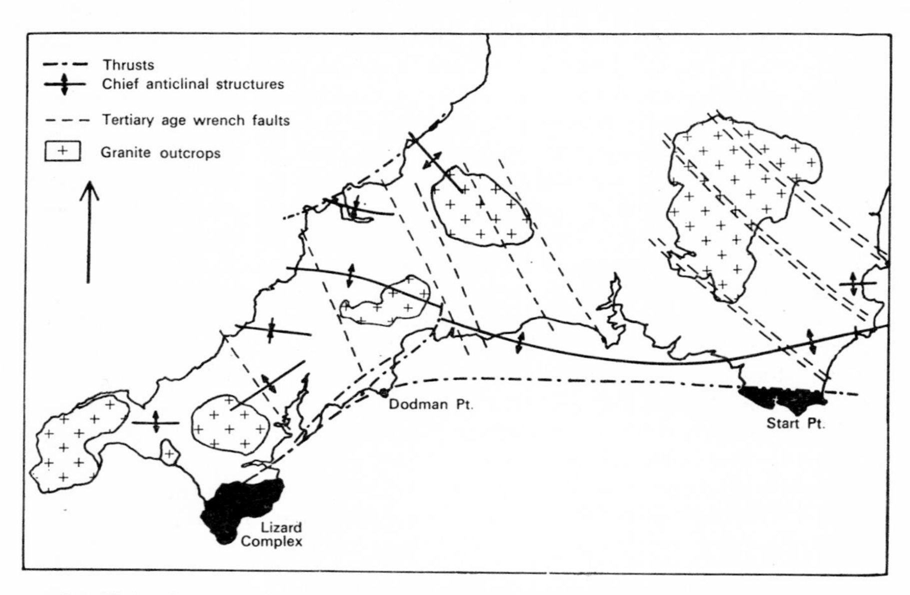

Fig. 10.1 Structural features of south-west England. (Based on G. HOSKING, 1962, in *Some Aspects of the Variscan Fold Belt*, ed. K. COE. Manchester Univ. Press, Manchester. Tertiary wrench faults after W. R. DEARMAN, 1963, *Proc. Geol. Ass.*, **74,** 265.)

folding can be seen to be polyphase), and low grade metamorphism has taken place. The grade of metamorphism increases generally southwards. Lying within this structural zone, in the south-west peninsula of England and in south-west Ireland complex folding and thrusting of nearly east—west trend are present. Interpretation of the major structures in Cornwall is dependent upon the correlation of the rocks of south-western Cornwall with the Middle Devonian, and in this area many problems remain unsolved. The boundary between the rocks of the Lizard Peninsula and the Devonian rocks to the north has been described as a northerly thrust, as has the junction between the Start-Bolt area schists and the Devonian as well as the Dodman Phyllite-Devonian junction. (The junctions have also been interpreted as a wrench

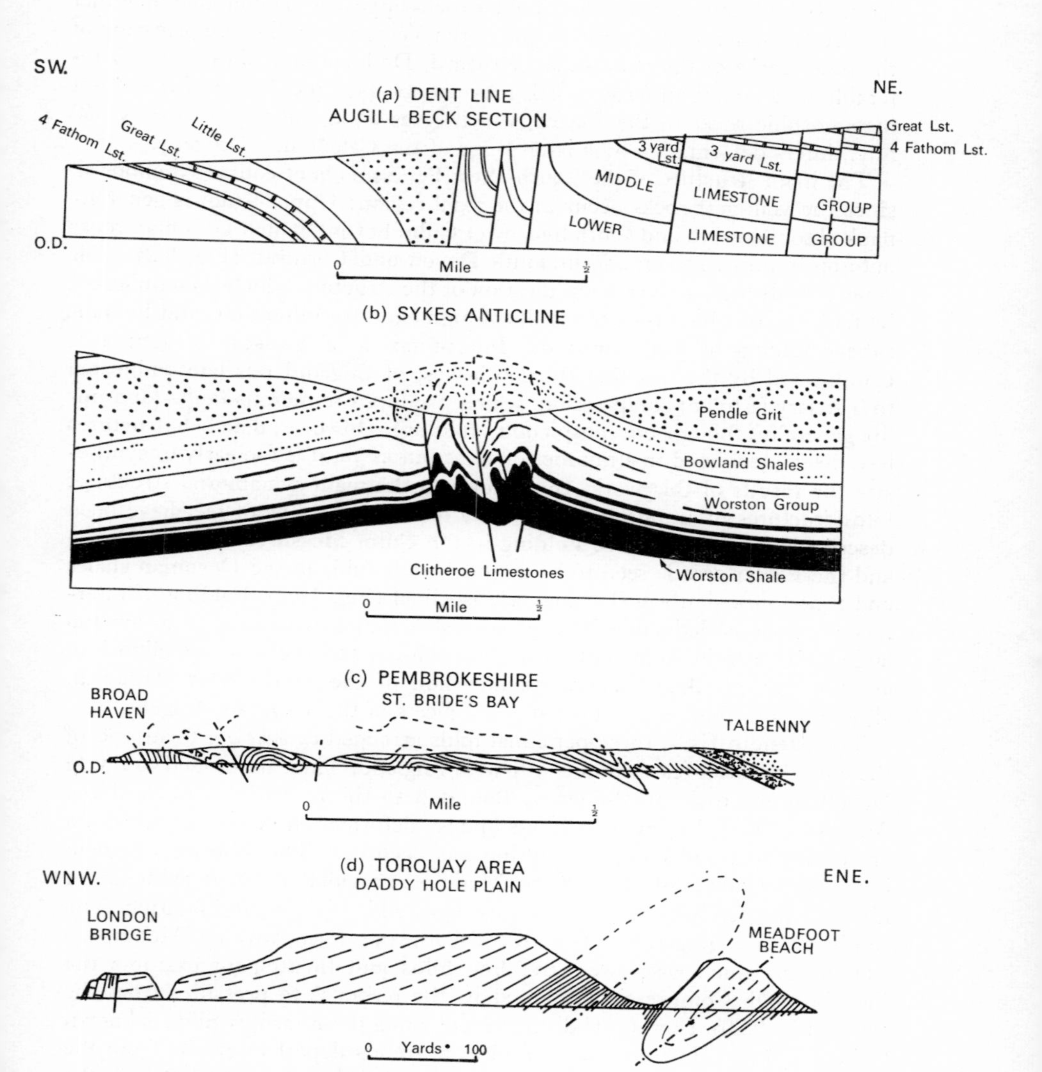

Fig. 10.2 Sections illustrating styles of Variscan folding: (*a*) the Dent Line, northern England; (*b*) the Sykes Anticline, mid-Pennines; (*c*) Pembroke, S. Wales; and (*d*) Torquay area, S. Devon. (*b* based on F. MOSELEY, 1954, *Proc. Yorks. geol. Soc.*, **33**, 287, and *c* on T. B. H. JENKINS, 1962, *Q. Jl geol. Soc. Lond.*, **118**, 65.)

fault.) In addition, thrust slices of older rocks appear to be bounded on either side by Devonian rocks, for example in the Veryan area. The deformation of the older rocks of the peninsulas of Lizard, Dodman and Start is partly referable to pre-Carboniferous earth movements. Isotopic dating of the oldest metamorphic rocks of the Lizard, which gives an age of approximately 360 m.y., infers deformation by a later phase of the Caledonian Orogeny.

The main structure of the South-West Peninsula is of course synclinorial, since the youngest rocks occur in the axial region: Upper Culm is generally flanked to the north and south by Lower Culm beyond which Devonian rocks outcrop in north Devon and in south Devon and Cornwall (Fig. 8.2). This is necessarily a gross over-simplification of the structure which is complicated by faulting, in which reversed faults of high hade are important, and by subsidiary folding of less amplitude. Interpretation of Variscan structures is complicated by the fact that the south-west of England was later subjected to Tertiary (Alpine) compression, which produced numerous faults (and uplift generally, though no conspicuous folding). However, detailed structures have been elucidated in a number of areas but as yet it is too early to synthesize the results of this work. In addition to the main synclinorial structure, fold structures of amplitude from a few feet to some tens of feet have been described from many areas. Folding in the Culm Measures of north Devon and the Exeter area is seen to be angular while folds in the Devonian shales and limestones south of Torquay are isoclinal (Fig. 10.2). Folding of intermediate scale is deducible from geological mapping where the folds are too large to be seen in single outcrops. This folding and the main synclinorium are seen to pre-date the granite intrusion of the South-West Batholith, although both folding and intrusion are facets of the Variscan Orogeny.

The Mendip Hills, three periclinal folds arranged *en echelon*, form one of the chief Variscan features of south-west England. They are eroded to reveal Devonian age rocks in the cores. Bounded to the north by thrusts of the Amorican Front, beyond which less intense deformation occurred, the strata are overturned and complex faulting has occurred. The Vobster Coalfield immediately north of the eastern Mendips provided detailed evidence of highly contorted beds with coal seams repeated by folding and faulting. The southward-plunging Lower Severn and Bath Axes form, with the Mendips, a triangle in which east—west trending folds and thrusts, for example the Farmborough Compression Zone, show the response to pressures from the south (Fig. 9.20). Horizontal displacement along the Radstock Slide is known to be 1,500 feet. Further north, fold axial trends depart markedly from the east—west direction and are even north—south (Malvernian trend) as in the Forest of Dean. In the latter folding is parallel to post-Viséan folding but the axial planes do not coincide.

In south Wales the principal structure of Variscan age is the main syncline of the coalfield. It is more complex than the approximately concentric outcrops infer; synclines occur *en echelon*. In the eastern part the Pontypridd Anticline separates synclines to the north and south and is responsible for bringing the seams up to a workable depth over a great area. The most conspicuous structure within the coalfield is the Maesteg Anticline running north-north-east from Swansea and over which the Pennant Measures have been removed by erosion. The 'Upper Coal Series' remains only in synclinal structures and between trough faults (Fig. 9.16). The Neath Disturbance, extending north-eastwards from Neath, is structurally complex; the Dinas wrench fault is associated with locally intense folding. Faulting, much of it almost north—south to north-west—south-east, immediately followed minor folding with a Caledonian trend—though of Variscan age. These faults, in some cases, have a lateral displacement though the chief wrench faults are somewhat later in age and have a north-east—south-west trend. Although the Caledonian structures, which are such a conspicuous feature in mid-Wales, are largely obliterated in south Wales by the east—west Variscan structures it is clear that there are important Variscan structures of Caledonian trend, such as the Neath Disturbance, and these may reflect the controlling influence of Caledonian structures in the underlying pre-Devonian rocks. In the east the coalfield 'basin' is bounded by plunging westwards away from the Usk Anticline, a structure initiated in pre-Carboniferous times and rejuvenated. To the west, in Pembrokeshire, the coalfield synclinal structure lies within the Variscan Fold Belt and has been subjected to greater compression. The Coal Measures outcrop narrows to little over a mile in width from north to south as a result of thrust faulting parallel to the plunging east—west minor folding. In Gower and south Pembrokeshire plunging folds, with minor strike faulting, chiefly cause repetition of outcrops of Upper Palaeozoic strata but Precambrian rocks outcrop to the south of—and thrust against—the coalfield in Pembroke.

The South-West Batholith

In the Tintagel area, where deformation is locally great, the strata appear to have been compressed against the Bodmin Granite. The production of cleavage has given rise to slates and phyllites while pillow lavas have been converted to green schists. An attempt to elucidate the structures of the South-West Peninsula by reference to the emplacement of granite intrusions and mineralization has been made. Granite is exposed in a number of areas, giving rise to high moorland scenery such as that of Dartmoor and Bodmin Moor. Surrounding each outcrop of granite is a metamorphic aureole, while in the granite peripheral areas and in the surrounding country rocks mineral veins are often present. The latter are of two phases, hypothermal and mesothermal, usually of different and characteristic trend. The earlier lodes are often parallel to dykes and include the economically important tin ore, cassiterite (worked since pre-Roman times), and wolframite. The later, mesothermal, lodes approximately at right angles to the earlier veins carry such minerals as galena and sphalerite with a little uranium ore as well. Gangue minerals occur in both phases of veins, of course. The granite outcrops are now believed all to belong to one quite large batholith, being cupolas on its surface connected by narrower ridges of granite still covered by sedimentary rocks (Fig. 10.3). The isotopic age determination certainly confirms that all the granites of the south-west were emplaced at approximately the same time, about 280 m.y. ago. Mineralization was undoubtedly related to the intrusion and is of the same date. On the revised time scale the Carboniferous-Permian boundary is placed at 280 m.y. The shape of the batholith is arcuate, extending westwards from Dartmoor to Bodmin Moor then south-westwards to include the St. Austell, Carn Menellis and Land's End Granites of the mainland. Beyond, it extends to include the Scilly Isles and recent investigation of the sea floor further to the south-west has revealed tor-like structures which

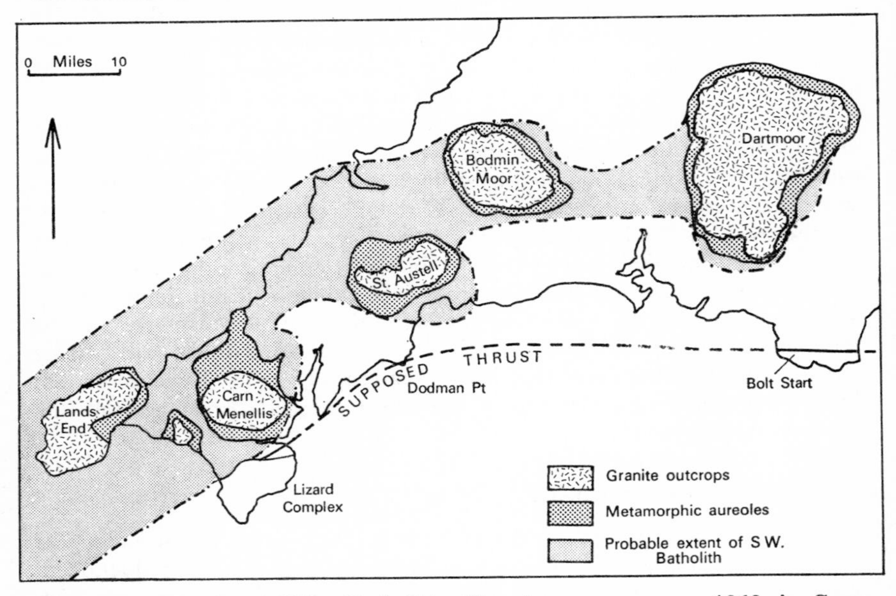

Fig. 10.3 The South-West Batholith. (Based on G. HOSKING, 1962, in *Some Aspects of the Variscan Fold Belt*, ed. K. COE, Manchester Univ. Press, Manchester.)

are probably part of the same batholith. It is postulated that the tors were subaerially weathered before the post-Pleistocene rise in sea level. It is probably more than coincidental that the arcuate shape of the batholith is roughly parallel to the postulated thrust running north of the Lizard and Start rocks. No granite has been found south of this line.

Structures of Malvernian trend

North of the South Wales Coalfield, between the Malvern line to the east and the Caledonian Towy Anticline to the west, lies a triangle, peculiarly protected from the effects of the Variscan Orogeny, in which nearly flat-lying or gently folded Devonian beds outcrop forming the table-like Black Mountains and the undulating country of Herefordshire. Bounding this area and separating it from the low-lying Vale of Evesham, floored by Triassic and Liassic strata, are the north—south trending Malvern Hills and Abberly Hills. Rocks from Precambrian to Westphalian age are exposed in complex fold structures of Variscan age. Many interpretations of the Malvern structure have been made, but the most recent shows it to be essentially a monocline affected by both steep and very flat strike faults. The nature of the eastern boundary is uncertain but undoubtedly the basin to the east continued to subside or downwarp during Mesozoic times for a very great thickness of Triassic is present.

The trend of the Malvern Hills, almost at right angles to the characteristic Variscan direction, is also characteristic of many of the structures of the Pennine area and to the structures in the Forest of Dean, Usk and at Tortworth. The latter indicates the initiation of the Lower Severn Axis as early as pre-Silurian times. The Malvernoid trend of these Variscan structures undoubtedly reflects rejuvenation of older structural lines. Clearly, however, the compressional forces giving rise to the Malvern structure—thrust and overturned to the west—cannot be of identical age to the forces which caused northwards thrusting and east—west trending folds in the Variscan Fold Belt. It is curious that east—west stress conditions apparently obtained over the whole of England north of the Variscan Front (there being only a few anomalous structures). Undoubtedly, the main structures of the Variscan Fold Belt were produced by a different phase—with a different stress field—from the structures north of the fold belt.

The pre-Permian structures and deformation of the Variscan foredeep have been dated relative to the deposition of the Culm succession and are post-Westphalian, possibly post-Stephanian, though it is uncertain whether they are attributable to the Asturic or Saalic Phase of the orogeny. The Malvernian structures are referred to the Malvernian Phase, slightly earlier, which was responsible for considerable changes in palaeogeography although not for the widespread production of major structures. There is evidence that deformational phases were of short duration and sometimes of comparatively local effect.

The effects of the Variscan uplift on the nature and distribution of Upper Carboniferous and early Permian strata in the Midlands have been discussed in the chapter on the Carboniferous. The fold structures produced, for example the anticlines in the southern part of the South Staffordshire Coalfield, the southward-plunging synclines of the Warwickshire and North Staffordshire Coalfields, etc., are of course of Variscan age. It must be remembered that the coalfield 'boundary faults' are post-Triassic and consideration of these is deferred. Between the coalfields and obscured by the Triassic cover similar structures, less well known, are present.

Structure of the Pennines

The Pennines are not a simple structural unit (Fig. 10.4). The response to Variscan stress depended on the thickness of Carboniferous rocks deposited and the nature of the basement. The southern Pennines, the Derbyshire 'Dome', is a Malvernoid anticline bounded on either side by the essentially synclinal coalfields. In the south, subsidiary folds are developed on the flanks, for example the anticlines of Ashover and Critch, and further east (now buried below younger rocks), the Eakring anticline. The latter, however, is of Char-

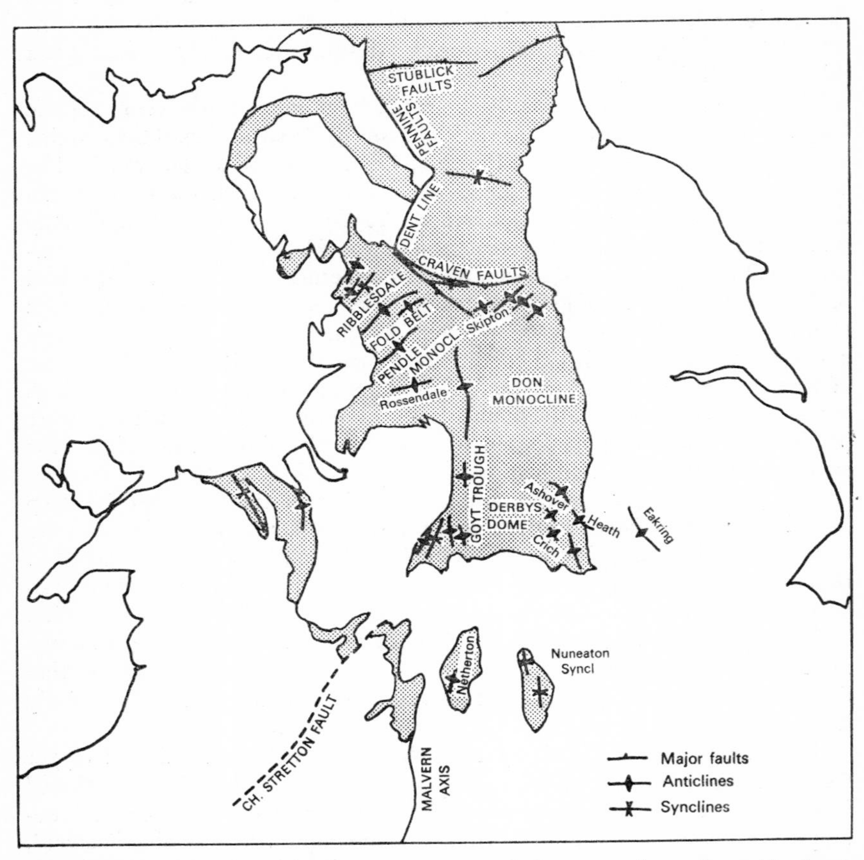

Fig. 10.4 The main Variscan age structural features of northern England. (Carboniferous outcrops shaded.)

noid not Malvernoid trend. On the west side of the main axis is the north—south Goyt Trough and in the Potteries (because of their economic importance better known than the intervening anticlines) are the plunging synclines of the North Staffordshire Coalfield and its subsidiary synclines of Shaffalong and Cheadle. The northern Pennines, north of the Craven Fault System, have a basement of Precambrian and Lower Palaeozoic strata covered by only 2,000 feet or less of Carboniferous rocks. These have responded to Variscan stress without significant folding as a result of the rigid basement. Only in the west where the Dent Line, a complex structure which has the effect of a reversed fault, bounds the Pennine Fault Block is there folding which produces steep dips. Beds are locally vertical in this area. As in the southern Pennines, structures clearly indicate that the stress field must have been near to an east—west compression. In the mid-Pennines where thick Carboniferous deposition occurred (and the basement is still unknown) folding took place in what is called the Ribblesdale Fold Belt. Anticlines such as the Slaidburn, Clitheroe and Skipton with an amplitude of several thousand feet have a Caledonian trend, north-east—south-west, anomalous in direction for the Pennines as a whole. Faulting is the more conspicuous feature of the coalfields. Primarily north-west—south-east, a second set nearly at right angles is also well developed. These are usually normal faults and the pattern is consistent with east—west compression. However, much attention has been paid to anticlinal structures since the initial discovery of oil at Hardstoft on the Heath-Brimington Anticline and later discoveries in the Eakring Anticline. To the west of the Pennine Fault Block, as well as those to the south, there are fold structures which include less well known folds near Lancaster.

While there is a considerable downthrow to the south as a result of the Craven Faults, and a famous fault-line scarp is present, it is largely attributable to Miocene movement. The differential effects of the Variscan Orogeny to the north and to the south of the fault system infer that the Craven Faults acted as wrench faults and the fault block was displaced westwards. (The main horizontal displacement was along the North Craven Fault.) The validity of this hypothesis was demonstrated by Wager over 30 years ago by a study of the direction of the joint system of the Carboniferous Limestone. The Pennine Fault Block is bounded to the west by the Dent Line which further north is replaced by the Pennine Faults which form the western boundary of the Alston Block. Again the main throw is clearly post-Mesozoic but Variscan structures, small thrusts such as the Brownber Thrust carrying Skiddaw Slates over Lower Carboniferous beds, are obvious. They have a north—south strike.

Mineralization and igneous intrusions

Heavy mineralization of the strata of the Alston Block took place in pre-Permian times. Lead, barytes and fluorspar have all been commercially important. The mineralization displayed a concentric pattern, especially in the distribution of the gangue minerals, showing that it was unrelated to the Great Whin Sill although this intrusion was itself pre-Permian in age. Not only have pebbles of it been found in Permian breccia (see p. 263), but its age by isotopic dating is 280 m.y. The Great Whin Sill is the largest hypabyssal intrusion in Britain. It is a quartz-dolerite sill up to 120 feet thick which extends from the Farne Islands southwards making a high north-facing escarpment across Northumberland along which Hadrian's Wall is built for part of its length (Fig. 10.5). It outcrops in the western Pennine scarp and in the deeply eroded valley of the Upper Tees. Intruded into the Middle Limestone Group over much of its length it changes horizon, sometimes abruptly in 'steps' (Fig. 10.6). A major intrusion postulated beneath the Alston Block to explain the pattern of mineralization was sought. Geophysical investigation and a subsequent deep boring proved the presence of a granite stock, the Weardale Granite. Surprisingly, the basal Carboniferous conglomerate rested on its eroded upper surface and isotopic age determination confirmed that it was emplaced before the Carboniferous rocks were deposited, some 150 m.y. before the mineralization of the Carboniferous rocks. It is a Caledonian granite and the origin of the mineralization remains one of the enigmas of geology.

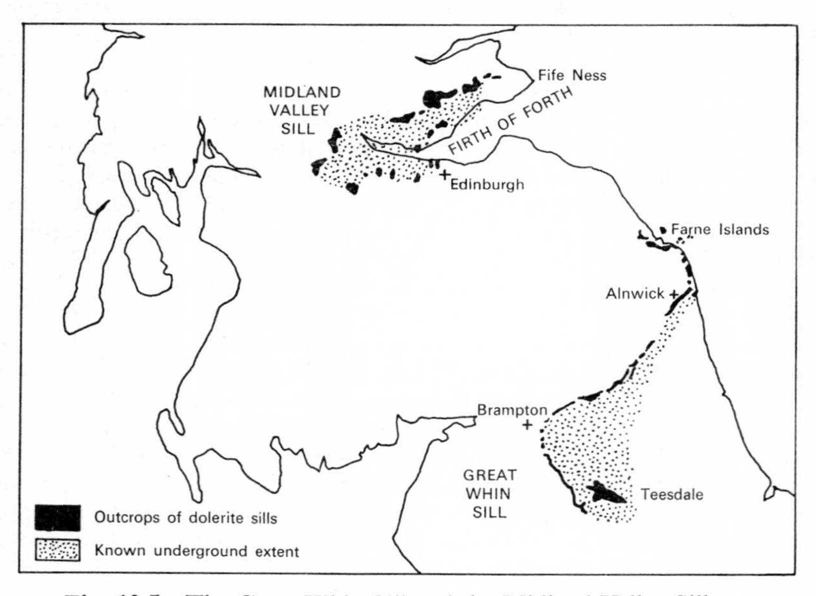

Fig. 10.5 The Great Whin Sill and the Midland Valley Sill.

A geophysical survey of the Askrigg Block also suggests the presence of another igneous body at depth, but this has not been confirmed.

It is hardly surprising that in Scotland the main effect of the Variscan Orogeny was the resuscitation of Caledonian structures and that north-east—south-west trends predominate. During Carboniferous times sedimentation was closely related to the Southern Uplands Fault and parallel faults. The Midland Valley Trough, the Northumbrian Trough and those of Northern Ireland, for example the Slieve Beagh Trough, had a Caledonoid trend although the area of sedimentation in Scotland (and Ireland) was not confined by the faults bounding the Midland Valley. Among the most conspicuous structures resulting from the Variscan Orogeny are the synclinal basins of the coalfields. Those of the Central Coalfield and Midlothian have

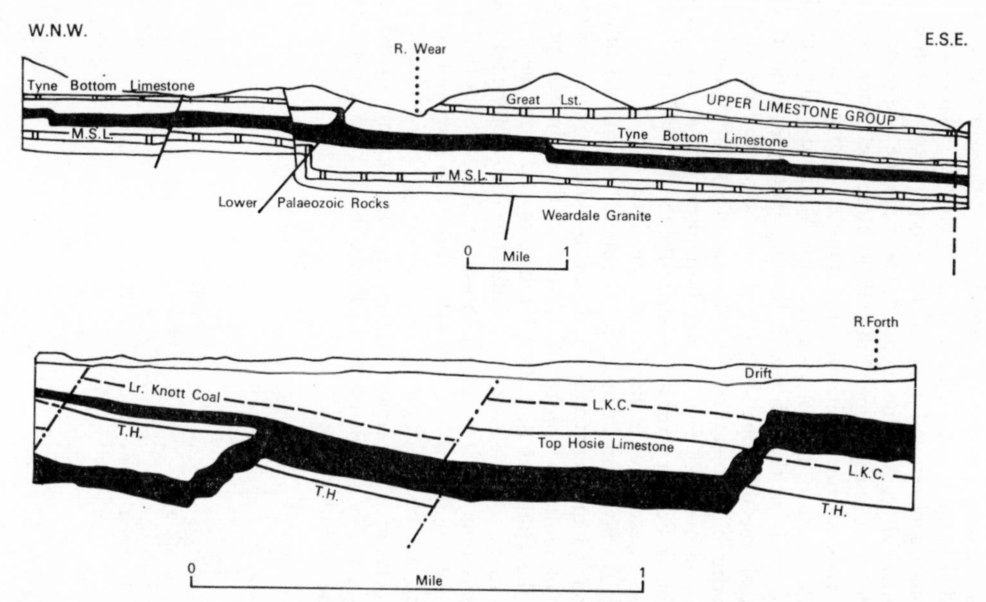

Fig. 10.6 Sections showing the changing horizon of Variscan sills: (*a*) the Great Whin Sill of Northern England, M.S.L. = Melmerby Scar Limestone and (*b*) the Midland Valley Sill. (*a* based on Crown Copyright Geological Survey Map, by permission of the Director, Institute of Geological Sciences; *b* simplified from E. H. FRANCIS, 1965, in *The Geology of Scotland*, Fig. 10.16, ed. G. Y. CRAIG. Oliver & Boyd, Edinburgh and London.)

a near-Caledonoid trend. The Pentland anticlinal fold, bounded on the east by a reversed fault which separates it from the steep limb of the asymmetrical Midlothian syncline, has a north-east—south-west trend. Faulting in the coalfields is chiefly east—west and the majority of faults are normal although many have sinuous outcrops. North-west—south-east faults which commonly occur and are parallel to dykes are a later phenomenon of Tertiary date.

Igneous intrusions associated with the Variscan Orogeny are widespread in the Scottish Midland Valley. Alkaline plugs, dykes and some sills occur. The latter are composite, such as the Lugar Sill and the Braefoot Outer Sill, and the range of rock types is from teschenite to peridotite. Quartz-dolerite sills appear to be of wide extent in the eastern part of the Midland Valley, making conspicuous scarps such as that on which Stirling Castle is built. It now seems certain that all the outcrops are part of one large sill, the Midland Valley Quartz-dolerite Sill. It measures 40 miles across (Fig. 10.5) and is seen to attain a thickness of over 500 feet in some boreholes. It changes horizon, often abruptly at faults, but for large areas is parallel to the bedding. Associated with it are east—west dykes of similar composition.

Post-Carboniferous movement of both the Highland Boundary Fault and the Southern Uplands Fault occurred. The latter, of course, is not a simple fault but a series of faults *en echelon*. Evidence of downthrown in pre-Permian times of the Highland Boundary Fault is to the north near Loch Lomond while it is to the south in Arran and Bute. The throw of the Southern Uplands Fault similarly is not consistently to the north or the south. Thus, although synclinal folding took place within the Midland Valley and faulting of Caledonian trend was renewed, the graben was not greatly revived. Succeeding Permian rocks, of which small outcrops remain, were deposited over Highlands, Midland Valley and Southern Uplands. Their remnant outcrops are described in the next chapter.

11 The Permian System

The geographical setting

The building of the Variscan mountain chains profoundly affected the geography of Europe, both in the distribution of land and sea and climatically. North-western Europe became part of the Hercynian Continent and strata of continental facies were laid down over wide areas in the succeeding Permian and Triassic Periods—as they had been also in late Carboniferous times. Frequently considered together as the Permo-Trias, there are however significant palaeontological differences, discussed later. The differences are such that the Permian is referred to the Palaeozoic Era and the Triassic to the Mesozoic. Unfortunately, rocks of continental facies, especially red beds, are sparsely fossiliferous so that in Britain the Permian-Triassic boundary cannot be readily drawn. Two facies of Permian rocks occur in north-east England, marine beds (chiefly limestone) and continental red beds. Since the transition from one facies to the other takes place both laterally and vertically in the succession, it can be safely assumed that the facies boundary is not a time-plane. In such circumstances the selection of a lithological boundary between the Permian and Triassic is a far from satisfactory choice but no alternative is open. The base of the Permian, on the other hand, is readily defined in northern England by the conspicuous unconformity at its base. Preceded by Variscan folding of the Saalic or Asturic Phase, the Permian rests with conspicuous overstep on beds of Carboniferous age. Only in the Midlands is difficulty encountered where the Permian age of the Enville and Clent Groups is in some doubt. However, even in other areas the age of the lowest Permian beds present is a problem, and the length of time represented by the pre-Permian unconformity must be related to the complete Permian succession.

The Permian System, the only one considered in this book not originally defined in western Europe, was first established by Murchison in his visit to

Perm, Russia, in 1841. The divisions of the Russian sequence (which includes stages lower than in Murchison's original definition) are given in Table 11.1, as well as the widely used terms based on the German Permian succession.

TABLE 11.1

NW. Europe	*Russian platform*
ZECHSTEIN { UPPER, MIDDLE, LOWER	TATARIAN
	KAZANIAN
UPPER ROTHLIEGENDE	KUNGURIAN
LOWER ROTHLIEGENDE = AUTUNIAN	ARTINSKIAN
	SAKMARIAN

The Permian was laid down in Britain in tectonic basins or cuvettes which resulted from the Variscan Orogeny, and Permian rocks are preserved in north-east England, north-west England, Scotland, the Midlands and south-west England.

North-east England

Permian rocks extend from the Durham coast near the mouth of the Tyne (an outlier occurs to the north of it) to the western suburbs of Nottingham. The succession in County Durham, though it cannot be widely applied, is as follows:

Marls with rock salt and anhydrite
Upper Magnesian Limestone and evaporites
Middle Magnesian Limestone
Lower Magnesian Limestone
Marl Slate
Yellow Sands

The lowest beds, the Yellow Sands, are not everywhere present, filling hollows in the pre-Permian land surface. Between Hetton and Moorsley they thin from 115 feet to a mere 5 feet thick in a distance of only 5 miles. In County Durham, where present (and they may be as much as 150 feet thick) they are yellow soft cross-bedded sands of aeolian origin. It has been suggested that they formed as dunes on the margins of the Zechstein Sea, discussed later. They are of considerable importance as a source of water in north-east England. During the sinking of collieries in the concealed part of the Durham Coalfield they constituted a hazard, frequently necessitating the freezing of the water which otherwise flowed into the workings at rates up to several thousands of gallons per minute. Yellow Sands occur at the base of the succession southwards through Yorkshire, but there seldom exceed 20 feet, and into Nottinghamshire where they pass into breccia with fragments of Carboniferous and Charnian rocks.

The Marl Slate, an inappropriately named bed since it is not a slate nor really a marl but a bituminous silty shale, sometimes calcareous, is a thin but laterally very persistent bed. Seldom exceeding 3 feet in thickness it can be traced for the whole length of the Permian outcrop, though in the south it is known as the Lower Permian Marl and comprises limy mudstone and near Nottingham a sub-littoral silty facies is found. It has been correlated with the Kupferschiefer of Germany on the similarity of their ganoid fish fauna. Apart

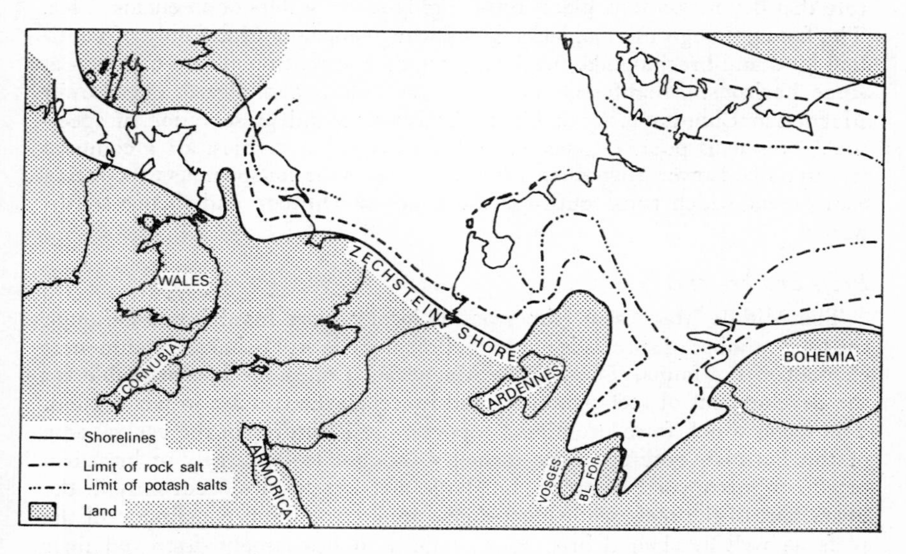

Fig. 11.1 The palaeogeography of the Zechstein Sea and the distribution of salt deposits. (Based on F. H. STEWART, 1963, *U.S. Geol. Surv. Prof. Papers* **440-Y,** 1.)

from fish and drifted plant material few other fossils occur in the Durham outcrops, but there is a normal fauna of marine benthos in the Nottinghamshire Lower Permian Marls. The sparseness of such a fauna in Durham, together with a high percentage of pyrite, infers deposition in reducing stagnant conditions inimical to life.

The Marl Slate is the lowest bed of a marine incursion of the Zechstein Sea in which the succeeding Magnesian Limestone was deposited. This sea occupied northern Germany and was bounded to the south by the Bohemian and Rhineland Massifs (Fig. 11.1). At its greatest extent it occupied not only north-east England but spread either across or round the northern Pennines (it may have spread via the Stainmore Syncline but there is no evidence of this). Thus thin representatives of the Magnesian Limestone are found in north-western England in the Vale of Eden and in Northern Ireland. The spread of marine conditions must have been rapid across a surface peneplaned in late Carboniferous and early Permian times for it is the Lower Magnesian Limestone which is of greatest southward extent. The area of marine deposition shrank as desiccation occurred so that while deposition of marine limestones continued in Durham, and 800 feet accumulated there, elsewhere limestone deposition ceased and continental facies became established, although later the sea spread westwards.

The Magnesian Limestone, a mixture of calcite and dolomite, is a commercially important source of magnesium. The lithology and fauna both indicate that deposition took place in the highly saline waters of an enclosed sea. The fauna, though marine, lacks corals while cephalopods are only of nautiloid kind and brachiopods, bivalvia, crinoids and gastropods are common at some horizons. Microfossils include foraminifera and algae. The Lower Magnesian Limestone is not highly fossiliferous and preservation of specimens present is poor. Geodes, crystal-filled solution cavities, are a common feature. The Lower Magnesian Limestone makes the conspicuous west-facing escarpment which runs south-eastwards across Durham and thence southwards.

Reefs and breccias

The Middle Magnesian Limestone is locally more fossiliferous although the total number of species is not high, brachiopods still being the commonest element of the fauna. The Middle Magnesian Limestone is remarkable for the development of reef-structures and breccias: the two are in fact related. The reefs, which form prominent topographical features in the central area of the Permian outcrop, were formerly thought to be largely of bryozoan origin. Bryozoans are common, but large mat-like species occur only in the lower part of the reefs while algal sheets are an important constituent of the reefs as well as shell debris. Recrystallization has largely destroyed finer structures so that the algae cannot be identified but they may resemble '*Collenia*' of the Carboniferous reefs. The reefs occurred near to the margin of the sea though they pass westwards into bedded limestones. East of the reefs breccias are well developed and are now exposed in the Durham coast. Some are undoubtedly accumulation of reef talus and slumped material since the outer edge of the reefs was probably steep. Reefs give steep dip readings, believed to be largely depositional dips (Fig. 11.2). Similar reefs are known from the southern margins of the Zechstein Sea in Thuringia.

Of the several kinds of breccias found in the Permian, cellular breccias are so called because of the solution of fragments leaving cavities in the matrix. The fragments are generally less dolomitic than the matrix. The remarkable angularity of the fragments in these breccias and their small size ($\frac{1}{4}$ to 1 inch across) inferred lack of transportation. It has been suggested that brecciation took place *in situ*, brought about by volumetric changes on the hydration of anhydrite to gypsum. Some massive breccias comprise large blocks, the blocks themselves being made of breccia. Later gash breccias are readily explained.

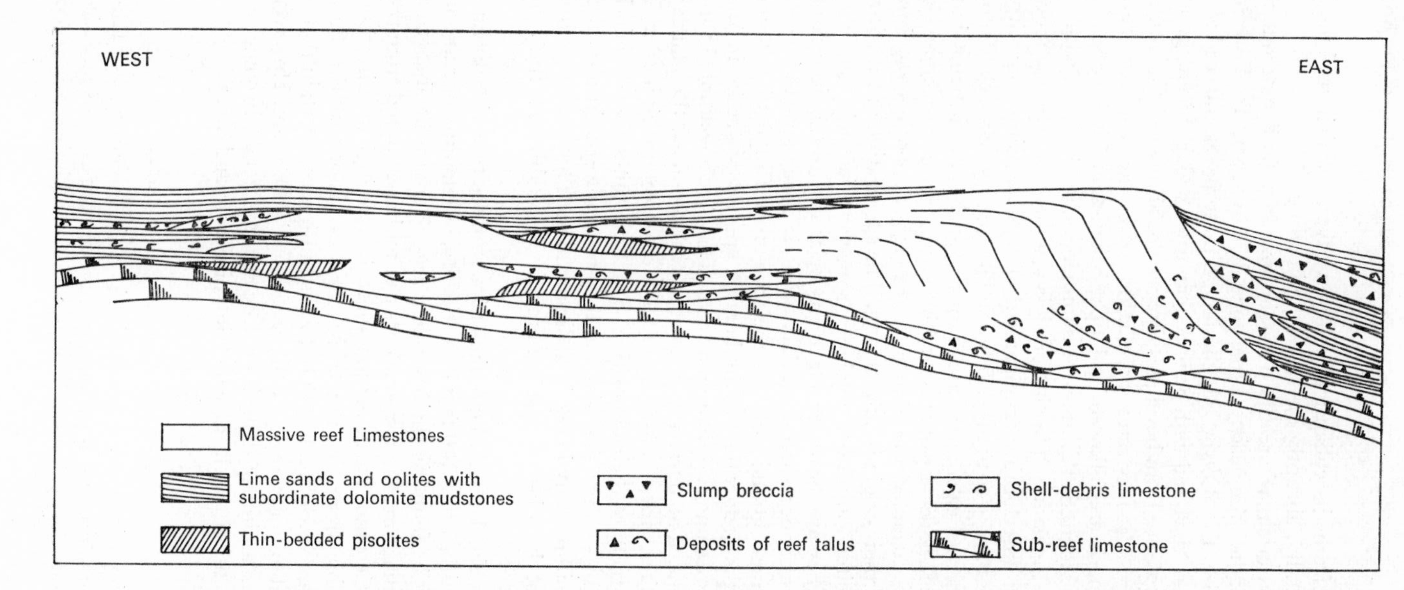

Fig. 11.2 Permian reefs showing the relationship of reefs and breccias. (From D. B. SMITH, 1958, *Bull. geol. Surv. Gt. Brit.*, **15**, 71. Crown Copyright, Geological Survey Diagram. Reproduced by permission of the Controller of H.M. Stationery Office.)

They resulted from the collapse of solution caverns which are breccia-filled and they, naturally, abruptly interrupt the bedding.

South of Catterick, where the Triassic boundary fault cuts out the Permian, the Middle Magnesian Limestone is unknown. In Yorkshire and Nottinghamshire the Lower Magnesian Limestone is succeeded by the Middle Permian Marl, red with beds of evaporites which thicken greatly eastwards as the evidence from boreholes shows. The absence of fossils other than scanty trace fossils frustrates correlation with the Durham succession, although the marls are followed by Upper Magnesian Limestone.

The Upper Magnesian Limestone commences, in Durham, with a 12-foot bed of flexible limestone. The flexibility is a somewhat rare phenomenon resulting from an interlocking arrangement of crystals. Limestones with many diverse concretionary structures follow. The commonest are 'cannon-ball' structures, spherical calcite structures up to 3 inches in diameter, but reticulate concretions known as pseudo-coralline structures are also common. These structures have attracted attention since the early days of geology but precise explanation of their origin is still lacking. The segregation of calcitic material, though 'deposited' around a nucleus—sometimes a shell fragment, is not understood. They are, however, clearly associated with the desiccation of the Zechstein Sea although their secondary origin is demonstrated by the passage of bedding planes through some of the structures. Progressive desiccation is indicated by the reduction in numbers of species of fossils in the Upper Magnesian Limestone and their small size. They became stunted through adverse conditions. The overlying yellow oolitic dolomites, the Hartlepool and Roker Dolomites, are unfossiliferous and chemical precipitation doubtless contributed largely to their deposition. The limestones are overlain by marls with very thick evaporite deposits and they pass upwards into similar beds of Triassic age without any break in sedimentation. Though the complete succession cannot be observed at outcrop it is established from numerous boreholes.

Southwards the Upper Magnesian Limestone is traceable as far as north Nottinghamshire. From 100 feet in Yorkshire, attenuation occurs southwards until it cannot be found in the Mansfield area. Here in the absence of Upper Magnesian Limestone the Upper Marls cannot be separated from the Middle Marls. Becoming sandy still further south in Nottinghamshire they become indistinguishable from the Lower Mottled Sandstone of the Bunter. The Upper Permian Marls are devoid of fossils but it seems certain that they were deposited contemporaneously with limestones in Durham. However, by late Permian times a similar depositional environment had become established over much of Britain with the exception of the upland areas.

Evaporites

Evaporite deposits occur at several horizons in the Permian of Durham and north-east Yorkshire and form an appreciable part of the succession. In the Fordon Borehole the $7\frac{1}{2}$ thousand feet of Permian includes 1,387 feet of evaporites. The mineralogy of these deposits is complex and more than 80 mineral species occur in marine evaporites, though not more than about 12 are important constituents. However, the mineralogical complexity is largely due to the secondary origin of many minerals by replacement. This is demonstrated both by the presence of crystal pseudomorphs and the absence or scarcity of minerals which would be expected to be present in quantity. Gypsum must initially have been more abundant than now, unless deposition took place in water above 34°C, which is regarded as unlikely. It has been very largely replaced by anhydrite and halite (rock salt) which are the commonest constituents of these evaporites. Gypsum in surface deposits often gives way to anhydrite at depth. However, examples of the replacement of anhydrite by gypsum occur as well. The dolomitization of the Magnesian Limestone may also be largely of secondary origin while, in turn, dolomite has sometimes been replaced by anhydrite. Clearly, the present mineralogy must be very different from that of the original deposits in both composition and texture. Much replacement was penecontemporaneous but some was consequent on burial which produced a rise in temperature and pressure.

The theoretical succession of deposition of minerals produced by the evaporation of sea-water is supported to a large extent by experimental work. The first salts to be deposited are carbonates, calcite and dolomite—although as stated much of the dolomite may be of secondary origin. Calcium sulphate next separates, at normal temperatures in the form of gypsum initially and later as anhydrite. Only if the temperature is high (see above) would all the sulphate be deposited as anhydrite. Halite separates with the gypsum, or anhydrite, followed by potash salts such as polyhalite. However, the detailed course of crystallization, which varies considerably with different temperatures, can only be adequately represented by ternary diagrams. A further complexity is introduced if the early-deposited salts are protected from reacting with the concentrating sea-water, for example by being covered with a thin layer of sediment, when an even more complex course of crystallization will ensue.

The evidence suggests that the composition of sea-water has not appreciably changed since Palaeozoic times. In an evaporating sea it might be expected that the minerals, which precipitate in order of increasing solubility, would be arranged concentrically so that the least soluble would appear near the margins of the sea while the most highly soluble would be found only near the centre of the basin. This in effect is observable in the evaporites of the Permian Zechstein Sea. The anhydrite and halite deposits are of greater extent while the potash salts are found further from the Permian shore and occur in north-east Yorkshire. It follows that in a borehole a cycle of desiccation should be found, the minerals occurring in the following order vertically:

Polyhalite
Halite
Anhydrite and/or Gypsum
Dolomite or Limestone

Essentially three such cycles have been found, with some evidence of a fourth, in the Upper and Middle Zechstein of Germany and north Yorkshire. The succession in a borehole near Whitby can be summarized as in Table 11.2.

TABLE 11.2

		Feet
Upper Permian Marl		600
Top Anhydrite		2–4
Salt Clay		7–12
Upper Evaporites	Upper halite Potash zone Lower halite Anhydrite Carbonate	200
Carnalitic Marl		30–60
Middle Evaporites	Upper halite Potash zone Lower halite Halite-anhydrite Anhydrite	400
Upper Magnesian Limestone and Marl		up to 108
Lower Evaporites	Upper halite-anhydrite Upper anhydrite Lower halite-anhydrite Lower anhydrite	1100
Lower Magnesian Limestone		365

It is not entirely clear how such vast thicknesses of evaporites could be precipitated. A 1,400-foot column of sea-water of normal composition would produce only a 1-foot bed of anhydrite. Obviously the evaporating water must be replenished. A theory put forward a century ago still is essentially accepted, that some form of 'bar' cut off lagoonal areas from oceanic circulation but occasional temporary connection occurred to permit replenishment. This theory is, to some extent, borne out by the minor cycles which comprise the zones of evaporites of Yorkshire. They are made up of repeated layers of carbonate, anhydrite and rock-salt, the layering demonstrating replenishment, perhaps seasonal. A count of the rhythmic layers of the Middle Evaporites of the Fordon succession infers that the cycle was deposited over a period of 25,000 years.

North-western England

West of the Pennines Permian rocks are found in the Vale of Eden with smaller but important outcrops on the north-west and south sides of the Lake

District. In the Vale of Eden, west of Appleby, the basal Permian rests unconformably on Lower Carboniferous beds and comprises a coarse breccia, the fragments of which are entirely Carboniferous Limestone. Included fragments up to a foot in breadth occur in a red sandy matrix locally full of derived Carboniferous fossils, chiefly crinoid ossicles which are conspicuous on weathered surfaces. In the south of the Vale of Eden this breccia, the Lower Brockram, is thick comprising much of the Permian succession. Northwards it thins and is of variable thickness filling hollows in the eroded Carboniferous surface, while the overlying Penrith Sandstone becomes thicker. The Permian was laid down on eroded Variscan structures and the palaeogeologic map shows the sub-Permian rocks (Fig. 11.3). Since, however, the deposition of the lowest Permian was probably diachronous this map does not show the outcrop geology of any particular moment of time. The Pennines were probably not elevated by Variscan structures, the Dent Line, the Inner Pennine Fault and the small thrusts such as the Brownber, discussed in the preceding chapter. The main uplift of the north Pennines is known to be of Tertiary age, since which time a great thickness of strata must have been removed by erosion, yet the Carboniferous rocks of the Pennines are younger (stratigraphically higher) than most of the rocks found beneath the Permian of north-west England. Minor outcrops of Coal Measures still occur in the Stainmore Syncline as well as immediately to the north and south of the Pennine Fault Block in the Midgeholme and Ingleborough Coalfields. It follows that the Pennines could not have been greatly uplifted and eroded as a result of the Variscan movements. The source of the fragments in the Lower Brockram is unknown and though it seems unlikely that it was derived from the Pennines this cannot be ruled out.

The Penrith Sandstone is largely coarse 'millet-seed' sand of aeolian origin. The cross-bedding of the dunes of which it was formed indicates transport by an easterly prevailing wind. The breccias constitute a piedmont scree deposit which passed laterally—and as erosion proceeded, vertically—into desert sand deposits. The Penrith Sandstone, soft in the Appleby district where the grains are cemented as well as stained by ferric oxide, is much more resistant to erosion further north through cementation by silica. There it forms the high ground of Penrith Beacon. Near the top of the Penrith Sandstone further beds of breccia occur but they are of lenticular form, not laterally persistent. Comprising several breccia horizons they have been collectively referred to in the past as the Upper Brockram. These higher breccias differ from the Lower Brockram: in addition to limestone fragments of Carboniferous age, Lower Palaeozoic rocks—chiefly volcanics—are common and Great Whin Sill pebbles have been found. These higher breccias may demonstrate the rejuvenation of the source of supply area and its further erosion to expose Lower Palaeozoic rocks at that time. It is possible, however, that these fragments were secondarily derived from the erosion of Carboniferous conglomerates. In the Kirkby Stephen area the Stenkrith Brockram occurs above the Hilton Series and forms a northward-thinning wedge. The overlying St. Bees Shales, which overlap the Permian eastwards,

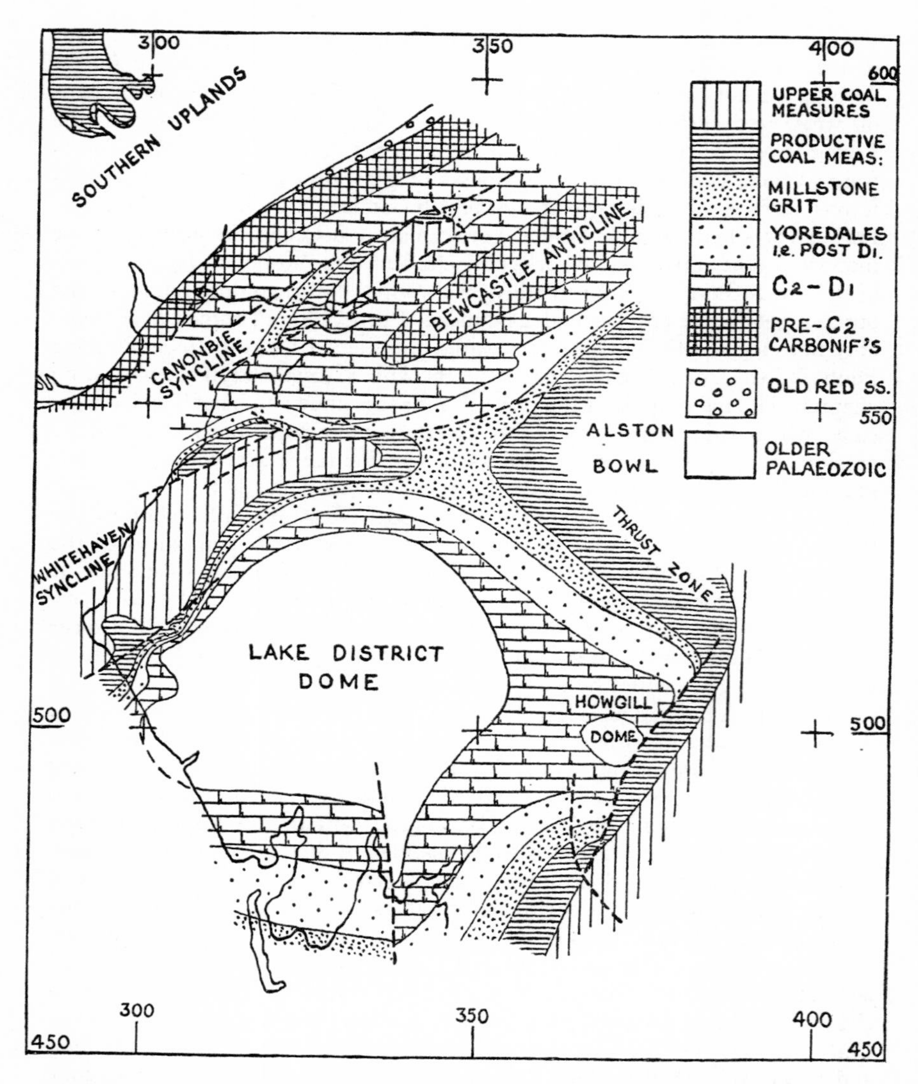

Fig. 11.3 Restoration of the geology of the sub-Permian floor of NW. England and SW. Scotland. (From F. W. SHOTTON, 1956, *Lpool Manchr geol. J.*, **1**, 450.)

are regarded as Permian or partly Permian and partly Triassic in age. The succession is:

St. Bees Shales
Hilton Series { Marls / Magnesian Limestone / Hilton Plant Beds }
Penrith Sandstone
Lower (or Penrith) Brockram

The Hilton Plant Beds, greyish-white shales and yellow sandstones, contain carbonaceous plant fossils. The evidence of the fossil flora suggests their equivalence to the Kupferschiefer of Germany for, although some species are of longer range, three are restricted to that formation. Similarly, the Marl Slate of the north-east of England contains fossil plants found in the Kupferschiefer and its equivalence to the Hilton Plant Beds had long ago been postulated. The fauna of the Magnesian Limestone of the Vale of Eden, a thin representative of those east of the Pennines (here only 20 feet thick), no longer sustains the belief that it is equivalent to the Lower Magnesian Limestone as formerly supposed. The limited fauna is characteristically of Upper Magnesian Limestone age. It seems improbable that the Hilton Series is really equivalent to the greater part of the Magnesian Limestone succession of Durham and that the Hilton Plant Beds-Marl Slate may be a strongly diachronous marginal facies of the advancing Zechstein Sea. In this case the base of the St. Bees Sandstone, of Bunter age, may also be diachronous. It cannot be ruled out, however, that the Hilton Plant Beds are equivalent to the Middle Permian Marls, since the Magnesian Limestone overlying them seems unquestionably to be equivalent to the Upper Magnesian Limestone.

The Magnesian Limestone of the Whitehaven-St. Bees area overlies Brockham and is 15 to 20 feet thick in coast exposures, but as much as 60 feet is known in borings. Overlain by beds which include two evaporite horizons it follows that, if these can be equated with the two evaporite beds designated B and C in the Vale of Eden (Kirkby-Thore), then the Magnesian Limestone of West Cumberland may be older than that of the Vale of Eden. The Permian of the Furness district is similar, although not everywhere have two evaporite beds been found above the Magnesian Limestone.

Ireland and the western limits of the Zechstein Sea

Permian outcrops in Ireland, though not extensive, are palaeogeographically important. The Permian is no more than 100 feet thick and the Magnesian Limestone, as much as 68 feet, is the most important member. As well as a small outcrop on the south shore of Belfast Lough, it has been proved in borings in the Grange District west of Lough Neagh. The faunal evidence of age is not entirely conclusive but strongly suggests an Upper Magnesian Limestone age and confirms, at least, a relatively high position in the Permian succession. The maximum westwards extension of the Zechstein Sea probably occurred in Upper Magnesian Limestone times but it is far from certain that the various outcrops of NW. England and Ireland were contemporaneous. The palaeogeographical map (Fig. 11.1) based on outcrops in the Vale of Eden, Furness and Whitehaven areas and Northern Ireland is therefore a composite representation.

The fragment of Permian of the Kingscourt Outlier in Ireland includes no Magnesian Limestone, further helping to define the position of the shoreline, but comprises breccia, shale with plant fossils and marls with evaporites. The configuration of the Zechstein Sea is supported by the palaeogeological map, the sea being bounded to the north by the Bewcastle Anticline and to the south by the Midlands Massif—the precise limits of the latter being unknown outside Nottinghamshire. The incursions of the Zechstein Sea in the western outcrops was brief. The location of the communication with the main Zechstein Sea seems unlikely to have been to the north of the Lake District in view of the Variscan structures and in the absence of Magnesian Limestone east of Whitehaven, which dies out. The Permian there comprises only breccias. The communication more probably occurred south of the Lake District, against which the Zechstein Sea shallowed and not against a North Pennine Massif for which there is virtually no evidence in Permian times (or in succeeding Mesozoic periods), and to the south of the Isle of Man where Permian sandstones outcrop.

The Desert Sands

As well as the Penrith Sandstone of north-west England and the Yellow Sands of Durham, both pre-Magnesian Limestone dune sands, there is widespread evidence of dune sands from northern Scotland to Devonshire. North of the Zechstein Sea red beds found in the south of Scotland are certainly, in part, considerably pre-Zechstein in age. The finding of plant fossils of possibly Westphalian D age (they could be as late as Autunian) in interbedded shales in the supposed Permian lavas of the Mauchline Basin, Ayrshire, infers that they may be diachronic. There is some evidence that they become progressively younger southwards from Ayrshire (Figs. 11.4, 11.5). In most cases dating of these outcrops of red beds, sandstones, shales and breccias, is not possible and the term New Red Sandstone is still a convenient if imprecise term. In Arran and Kintyre the successions are largely breccia; in the former a thick and perhaps continuous succession provides evidence of evolving geographical conditions. Piedmont breccia and aeolian sands are followed by water-laid deposits including cornstones, and in Triassic times playa deposits formed. The thickness of the Permian deposits, several thousand feet, in some of these isolated outcrops suggests deposition in tectonic basins resulting from the Variscan Orogeny. The furthest north known Permian outcrop is near Elgin on the Moray Firth where yellow aeolian sands have yielded vertebrate fossils which confirm their Upper Permian or earliest Triassic age. They are overlain by datable Middle Triassic rocks of similar depositional environment also with a vertebrate fauna. Further north Triassic rocks (found underlying lowest Jurassic) occur in the west, at Applecross and Gruinard Bay, and

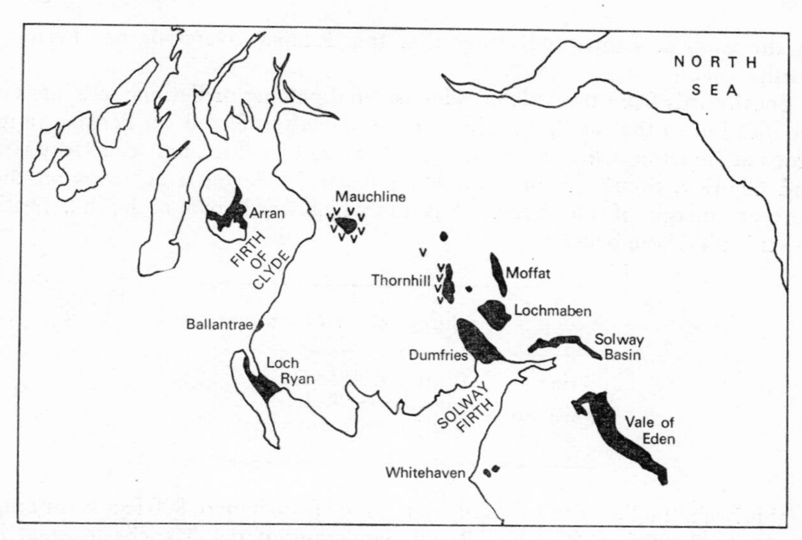

Fig. 11.4 'Permian' outcrops of north-western England and southern Scotland. Volcanic rocks indicated diagrammatically by v. (Based on W. MYKURA, 1965, *Scott. J. geol.*, **1**, 9.)

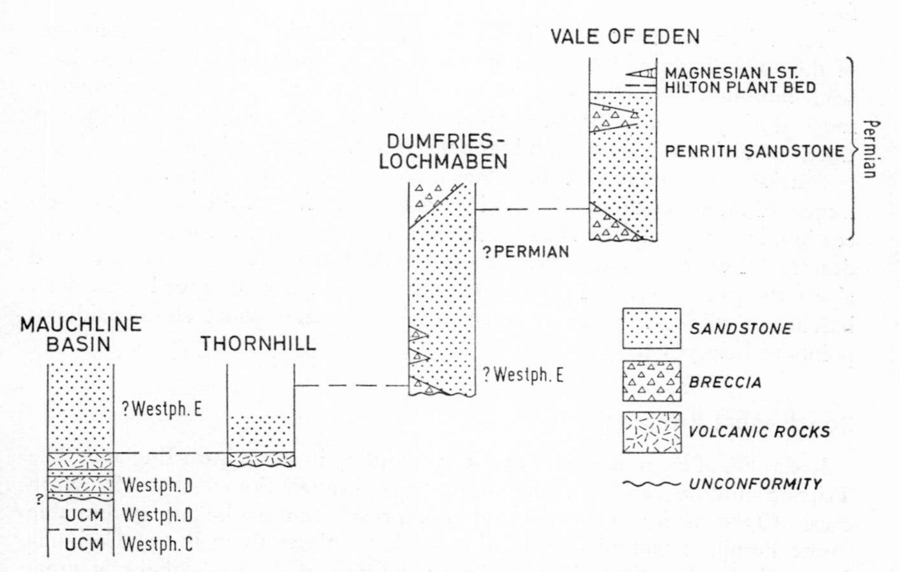

Fig. 11.5 Diachronism of red beds in the south of Scotland. (From W. MYKURA, 1965, *Scott. J. Geol.*, **1**, 9.)

in the east, at Brora, indicating that the Triassic oversteps the Permian northwards.

Southwards from the Vale of Eden isolated patches of Permian are known. In addition to that at Barrow-in-Furness already referred to, Permian outcrops at Ingleton, Clitheroe and, largely covered by drift, between Ormskirk and Chorley. South of the Lancashire Coalfield Permian outcrops on the northern margin of the Cheshire Basin in which it is known at depth beneath the thick Triassic beds.

Succession in Lancashire

Trias	Bunter Pebble Beds
Permian	Manchester Marls
	Collyhurst Sandstone

In this basin the Permian thickens southwards from a mere 800 feet at outcrop to almost 2,500 feet. The Collyhurst Sandstone of the Manchester area is demonstrably pre-Zechstein, but the greater part of the red sandstones of the west Midlands and that of the Vale of Clwyd cannot be so dated. The main dune sandstones of these areas are, however, older than the Bunter Pebble Beds. The Manchester Marls, which overlie the Collyhurst Sandstone, include fossiliferous calcareous bands which have yielded marine Permian fossils.

The thickness of the beds in the Cheshire Basin indicates a great amount of downwarping which continued into Triassic times. The basin may have been bounded by faults, such as the Red Rock Fault, which continued to move during Permian times. Similar fault-bounded troughs including very thick red beds are known from borehole records in Staffordshire.

Outside the confines of the Zechstein Sea, Permian deposits are very largely of dune sandstone, though breccias occur often at the base of the succession, and deposition was extensive in a sand sea comparable with modern deserts. Of course positive elements from which deposition was absent stood above the general level of the desert. The dunes appear to have been chiefly barchans resulting from a very constant wind direction which all the evidence points to being easterly.

South-west England

Red rocks of Permian and Triassic age outcrop in south-west England. The Triassic outcrop can be traced southwards through Somerset to the south coast of Devonshire so that the underlying red beds must be largely referable to the Permian. Isotopic dating of lower lavas places them in late Carboniferous times. However, in the absence of faunal evidence there is great difficulty in deciding the position of the Permian-Triassic boundary. The Budleigh Salterton Pebble Bed may be the equivalent of the Bunter Pebble Beds, but similarity of depositional environment is unreliable evidence of contemporaneity. The succession below the Pebble Bed, over 2,000 feet of strata tentatively ascribed to the Permian, is as follows:

Red marls with local sandstones
Red sandstones and breccias
Breccias and conglomerates
Watcombe Clay

Outcropping on either side of the mouth of the R. Exe, and as far south as Paignton Bay, the Permian Beds make beautiful cliff scenery. A long tongue, of breccias, extends westwards from Crediton along the axial region of the synclinorium. The greatest outcrop of Permo-Triassic is submarine, outcropping to the south of Cornwall (Fig. 12.2) and the basin of deposition was bounded to the west and north by the Cornubian Massif. Metamorphic rocks on which Eddystone Lighthouse is built project through the Permo-Triassic strata. Naturally, the latter are not known in detail but they may largely be Triassic Keuper Marl. The succession of the Permian, which rests unconformably on beds of the Devonian and Culm Series, the latter stained red beneath the unconformity, does not everywhere commence with the Watcombe Clays which are only locally developed in the Torquay area. Breccias are thick, formed as piedmont deposits, and they include fragments of Devonian limestone and Culm Series rocks with, in successively higher beds, a greater amount of igneous fragments. None of these can certainly be ascribed to the South-West Batholith and it was still in Permian times roofed by sediments although pebbles from the metamorphic aureole occur. In the Crediton Valley the Permian has been divided into four lithological groups: the Cadbury Beds, Bow Beds, Crediton Beds and St. Cyres Beds. In the first of these only Culm pebbles are known but in the Bow Beds and Crediton Beds pebbles of lavas, some tourmalinized, and tourmalinized sedimentary rocks occur as well as Culm pebbles. These igneous pebbles do not match the contemporaneous igneous rocks. A great variety of pebbles is found in the St. Cyres Beds, including those mentioned as well as cherts, microgranite and sanidine felspar pebbles. There is clear evidence that the material has been transported from areas to the west of the present outcrops.

Igneous rocks occur at the base of the breccias, for example at Dunchideock, or low in the succession at a number of places in the Exeter district. Known as the Exeter Volcanic Series, two groups are recognized and a number of rock types have been named from characteristic localities. Chiefly they are trachytic basalts and were extruded as lava flows. Somewhat unusual mineral assemblages have led to the suggestion that admixture of magmas took place before their extrusion. Although only remnants remain of more widespread flows, vulcanicity was not extensive in south-west England. It is possible that these volcanic rocks were approximately contemporaneous with those of the west of Scotland (though the latter may be somewhat older) and with extensive vulcanicity in Germany. Away from the Uplands of the

South-West Peninsula the breccias pass into sandstone which are succeeded by marls, the latter exhibiting a profusion of sedimentary structures such as flame-structures. In turn these dip eastwards under the Budleigh Salterton Pebble Bed, discussed in the next chapter.

Permian fossils

Because of the depositional environment of the greater part of the Permian succession in Britain, fossils are rare except in the deposits of the Zechstein Sea. The fauna of the Zechstein Sea, marine benthos including reef-building organisms, is not extensive, the commonest fossils being brachiopods. The small number of genera present, as well as the absence of coelenterates and the rarity of cephalopods, reflects the somewhat adverse conditions which persisted. The fauna has a number of characters which give it a Palaeozoic aspect, for example the brachiopods are chiefly of genera common in the Carboniferous. Trilobites and Blastoids, so characteristically Palaeozoic, make their last appearances in the Permian while outside Britain the last straight cephalopods occur in the Permian. The flora, similarly linked to the Palaeozoic, comprises many genera which persist from the Carboniferous in addition to some new forms.

It is perhaps not surprising that the vertebrate fauna, as a whole, is not so distinctly Palaeozoic. Palaeoniscid fish are found and amphibians became more numerous and successful so that some, such as *Eryops*, were able to compete with the reptiles. The latter had become numerous and diversified and include the mammal-like forms.

V The Mesozoic Era

12 The Triassic System

Rocks of the Triassic System outcrop extensively in the Midlands of England and extend north-westwards and north-eastwards on either side of the Pennines as well as south-westwards. Minor outcrops occur in the far north of Scotland and in Northern Ireland, proving the former extent of the Triassic area of deposition. However, south-eastwards under the cover of newer Mesozoic and Tertiary rocks the Triassic thins and was probably not deposited in south-east England, except for the late Triassic deposits laid down by the Rhaetic transgression. The Triassic rocks are almost entirely of continental facies with the exception of the Rhaetic beds which are now generally included in the Triassic. The British Triassic deposits include dune sands, but most lithological types indicate deposition in water, conglomerates and sands transported and deposited by running water and marls deposited in standing sheets of water.

Divisions of the Trias

The Triassic is so called from its (former) threefold division in Germany where a middle division of marine beds occurs. Not present in Britain, here there is a twofold division into Bunter and Keuper which are recognizable lithological or facies divisions but not strictly equivalent to strata similarly designated in Germany. The extensive marine successions of Triassic rocks which occur from the Alps, via the Himalayas to Timor and elsewhere, have been divided into the Palaeotriassic, the Mesotriassic and the Neotriassic which are further divided into stages (Table 12.1). Naturally, these divisions of the geosynclinal Triassic are only approximate equivalents of the divisions of the German succession.

TABLE 12.1

Germany	*Marine, Geosynclinal*	
RHAETIC KEUPER	NEOTRIASSIC	RHAETIAN NORIAN KARNIAN
MUSCHELKALK	MESOTRIASSIC	LADINIAN ANISIAN
BUNTER	PALAEOTRIASSIC	OLENEKIAN INDUAN

The absence of beds of Muschelkalk facies in Britain does not mean that beds of Middle Triassic age are absent, merely that deposition of continental facies continued throughout the Triassic Period (except for evidence of a minor marine incursion). Many problems of dating and correlation occur as a result of the lack of fossils in much of the British succession and the difficulty of relating terrestrial faunas to marine faunas. The rarity of isotopically datable rocks in the Triassic System of Britain and north-western Europe presents a further problem. In Britain igneous activity was virtually unknown and where igneous rocks are found cutting Triassic strata they are almost invariably much later, usually of Tertiary age.

The Triassic environment

Something of the difficulty of defining the Permo-Triassic boundary in Britain has already been discussed. Changes in geography which occurred in early Permian times, the production of upland areas followed by their erosion and the spread of similar conditions of depositional environment over wide areas, gave rise to the deposition of red beds largely of aeolian origin and to the deposition of evaporites. Initially in Triassic times the geography was similar to that of late Permian times and upland areas formed the source of material for piedmont deposits, while sands were deposited on the flat plains more remote from the hills. The replacement of Permian deposition by the deposition of beds regarded as Bunter (Lower Triassic) in age did not take place simultaneously over the whole of Britain. The lowest Bunter is locally indistinguishable from the Permian or passes laterally into beds identified as Permian, for example in Nottinghamshire. A further complexity is the variable succession of the Lower Triassic, discussed below. The drawing of the Permo-Triassic boundary is fraught with difficulties and such a boundary based on lithological criteria, as it must be, is probably not a time-plane.

The Triassic rocks vary greatly in thickness, basins in Cheshire and north Shropshire, Staffordshire and east of the Malverns subsiding to accommodate several thousands of feet of beds. Two important features of Triassic geography were the hills of Charnwood and the Mendips. These were inselbergs rising above the general level of the Triassic landscape but gradually buried as the deposits accumulated. Recent erosion has exhumed these hills and locally the old land surface can be studied in detail. At the present day the hills, much eroded down, again rise above flat Trias-floored lowlands. The gradual erosion and burial of the upland areas during Triassic times led to a progressive change in the type of sedimentation. Early aeolian sands and piedmont delta deposits gave way later to water-lain sands and to marls deposited in standing water. In general deposits of progressively finer grain were laid down as the Triassic Period continued.

The Lower Trias (Bunter)

A threefold division of the Lower Triassic (or Bunter) into the Lower Mottled Sandstone, the Pebble Beds and the Upper Mottled Sandstone is recognizable, for example in Shropshire and Worcestershire and in Cheshire. The Lower Mottled Sandstone (= Dune Sandstone Group), well developed around Bridgnorth and seen in the cliff on which the town stands as well as in the escarpment of Kinver Edge, is aeolian dune sand formed in the 'New Red Desert' referred to in the previous chapter. The sandstone, iron-stained and soft, is invariably cross-bedded. A high percentage of the grains are rounded and this evidence of its aeolian origin is confirmed by the occurrence of wind-faceted pebbles known as dreikanter. In the area west of the South Staffordshire Coalfield a breccia, the High Habberly Breccia, occurs at the top of the dune sandstones and underlies the Bunter Pebble Beds. Generally, however, where breccias are present, for example on the east of the South Staffordshire Coalfield, the Lower Mottled Sandstone is absent. It appears that these breccias, the Barr Beacon Breccia and the Quartzite Breccia—so called because of the predominance of Cambrian quartzite fragments (unlike the Clent Breccia with its majority of Uriconian Volcanics pebbles)—were derived from an uplifted ridge which limited the spread of Lower Mottled Sandstone eastwards. Was this area a rocky desert swept bare by the winds which carried sand to the desert destined to become the Lower Mottled Sandstone? The similarity of the Lower Mottled Sandstone to the Penrith Sandstone, in colour, cross-bedding and aeolian origin, suggests contemporary formation, and its passage into the Permian Marls in Nottinghamshire further implies that the lower part of the British Bunter is of Permian age.

In Warwickshire the Westphalian Coal Measures which outcrop in the north are overlain by successively younger beds to the south, the Enville Beds succeeded by breccias the chief of which is the Kenilworth Breccia. It appears that the succession continued without any major interruption from Upper Carboniferous into what must undoubtedly be Permian, although fossiliferous confirmation is lacking. The major unconformity occurs at the base of the Keuper division of the Triassic for it rests unconformably on beds ranging in age from Precambrian to Permian.

The Pebble Beds

The Bunter Pebble Beds characteristically occur in the Midlands, reaching a thickness of 600 feet in Shropshire. They die out to the north-west and north-east. The name Pebble Beds is a descriptive term and the formation comprises sandstone with pebbles. Only locally, for example on Cannock Chase, are the pebbles sufficiently numerous for the beds to be described as a conglomerate. Elsewhere the sandstone may contain few or no pebbles, the name being inappropriate but the formation still being recognizable as a more yellow or buff-coloured sandstone than the underlying Lower Mottled Sandstone. The pebbles are well rounded and make a marked contrast with the angularity of the fragments either in the Clent or later Permian breccias. The Pebble Beds reflect a major change in geography of the Midlands area since the pebbles have been water transported by torrents. The lack of sorting, pebbles range in size from $\frac{1}{4}$ to 9 inches or more, infers deposition as delta fans. The sandy matrix is not of aeolian origin and wind-faceted pebbles have only rarely been reported so that deposition, while still terrestrial, was of an entirely different character: the New Red Desert had virtually ceased to exist. Detailed studies of the pebble content of the Pebble Beds has shown a great variety of lithologies to be present, although quartzite and vein quartz make up 80 to 90% of the total (by weight, for the majority of the larger pebbles are quartzite, and perhaps 80% numerically). A very few pebbles are fossiliferous so that they can be dated and, in some cases, their probable area of origin can be deduced. Some quartzite pebbles are of Cambrian age and others of Llandovery (Silurian) age; their source may be the south Midlands. Ordovician quartzite pebbles with the brachiopod *Orthis budleighensis* may be presumed to have been transported from the south-west of England, as may Devonian pebbles and tourmalinized rocks, but the source of the majority of the quartzite pebbles, liver-coloured or grey and unlike any rocks now exposed in the Midlands, is unknown. Chert pebbles, Carboniferous Limestone, schists and a wide variety of igneous rock pebbles have been described. Some are highly weathered and decomposed and possibly less tough rocks have been completely destroyed during transport. The pebble suite of rocks varies locally and in east Warwickshire (Polesworth) is distinct. Around Bridgnorth it includes much Carboniferous Limestone. These variations in pebble content infer transport by different rivers (Fig. 12.1). In pre-Pebble Bed times uplift of the Mercian Highlands and erosion took place so that the Pebble Beds rest unconformably upon rocks ranging in age from Westphalian Coal Measures to Lower Mottled Sandstone. The Midlands may have been a cuvette, or intermontane basin, filling up gradually so that higher beds overlap near the margins. Not everywhere present, the Lower Mottled Sandstone is not found in the east Midlands, being absent from Nottinghamshire. It follows that the Pebble Beds of that area, seen in Nottingham Castle Rock and much of Sherwood Forest, may not be precisely contemporaneous with those of Shropshire and Staffordshire.

The Pebble Beds are followed by the Upper Mottled Sandstone ≡ Moulding Sand. Although known as the Upper Mottled, the sands are not everywhere mottled but are frequently very bright brick red. The term moulding sand refers to their widespread use and the foundry industry of the Birmingham area owes its location to this source of valuable sand. East of the Birmingham Fault, a fault with a large easterly downthrow (imparted in post-Triassic times), which forms the eastern boundary of the South Staffordshire Coalfield, the Moulding Sand is unknown. It has been suggested that this fault formerly had a downthrow to the north-west and that it bounded the Midland

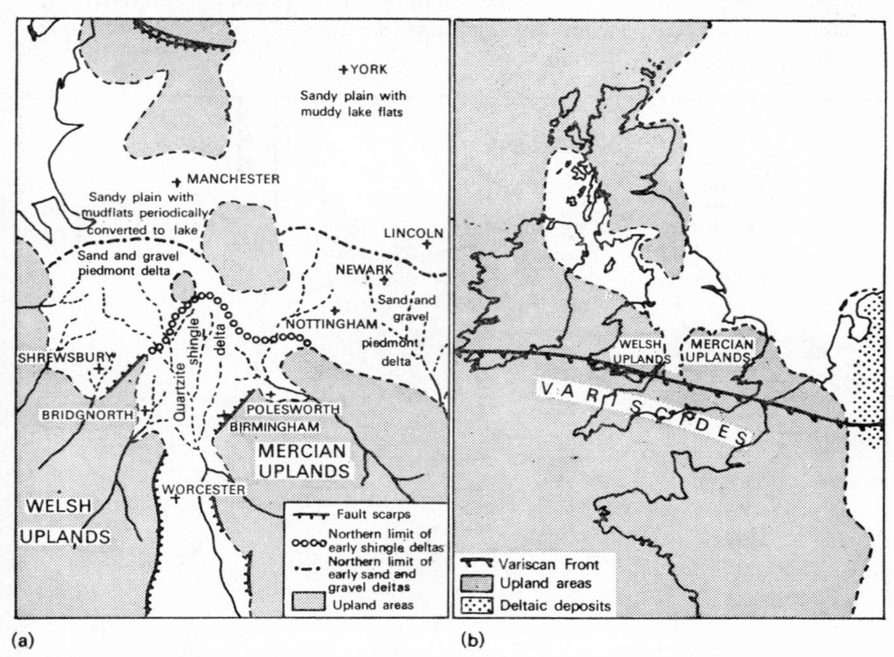

Fig. 12.1 Bunter Palaeogeography: (*a*) the geography of the Midlands in Pebble Bed times; and (*b*) The general geography of NW. Europe in Lower Triassic times. (*a* from L. J. WILLS, 1956, *Concealed Coalfields*, Fig. 17. Blackie, London.)

Cuvette separating it from the Mercian Highlands (as a fault-line scarp) in Bunter times. The Upper Mottled Sandstone is water-laid, although it includes some rounded grains perhaps carried by winds and deposited in water. It is without dreikanter, or any pebbles, and the type of cross-bedding, channeling and other sedimentary structures leave no doubt about its deposition in water. Nevertheless, the sandstones above and below the Pebble Beds are usually red in colour and northwards—where the Pebble Beds are absent or cannot be recognized—the Bunter cannot readily be subdivided, for example in the Garstang area of north Lancashire.

The Bunter is largely unfossiliferous, except for the very rare fossiliferous derived pebbles found in the Pebble Beds, the aeolian sands and piedmont screes being deposited in environments where life would not be abundant and the chance of subsequent preservation as fossil slight. However, thin lenticular beds of marls occur, especially in the Upper Mottled Sandstone, which include specimens of the small crustacean *Euestheria*. The Bunter Sandstones and Pebble Beds are a vitally important source of water and are tapped by a multitude of wells in the west Midlands. The site of the brewing industry can largely be related to the distribution of Bunter beds (although the source of water, in some cases, is from other sandstones, either of Upper Carboniferous or Keuper age).

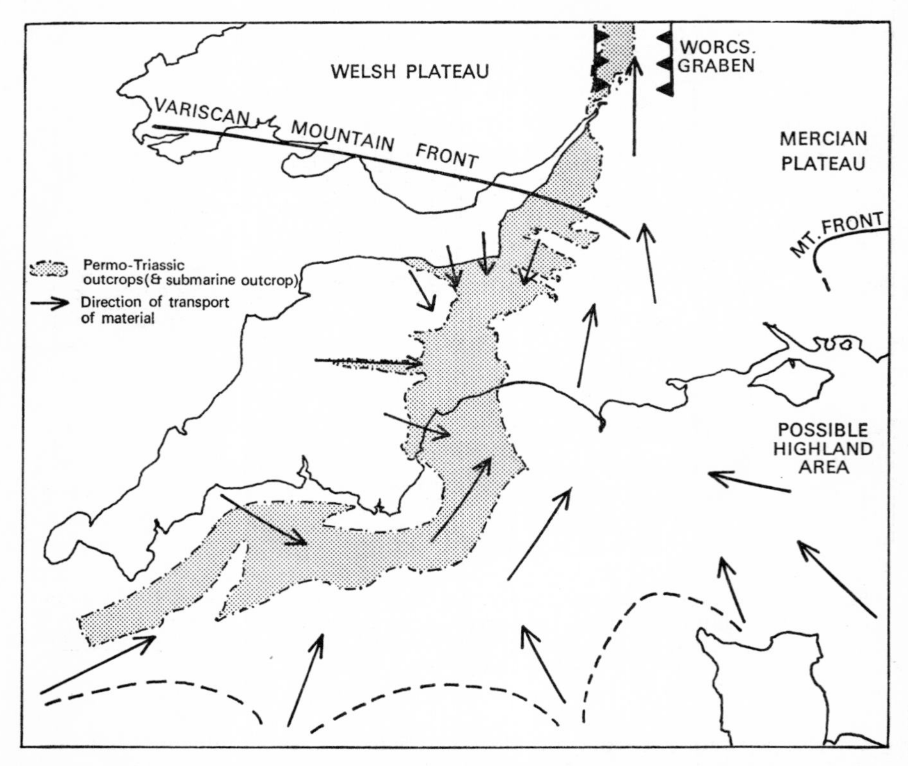

Fig. 12.2 The outcrops and known submarine extent of the Permo-Triassic rocks in south-west England and a reconstruction of the geography of the South-west Cuvette. (Based in part on F. J. FITCH, J. A. MILLER and D. B. THOMPSON, 1966, *Palaeogeogr., Palaeoclim., Palaeoecol.*, **2,** 525.)

The Upper Trias (Keuper)

The Upper Division of the Triassic, the Keuper, generally divisible into the Keuper Sandstone below and the Keuper Marl above, follows the Bunter often unconformably. Although this unconformity cannot be established at outcrops, south of Birmingham it is proved in boreholes and it is marked by breccia. Further west the Keuper-Bunter relationship may be conformable but in the east Midlands the Keuper overlaps the Upper Mottled Sandstone and rests on Pebble Beds. The Lower Keuper is overlapped by the Upper, in general, and this can be proved in Worcestershire northwards towards Birmingham as well as in the better known examples against the flanks of the Charnwood and Mendip Hills. The Keuper as a whole is very thick and in graben reaches 4,000 feet in Cheshire, although the Lower Keuper, somewhat variable in thickness, does not attain more than 400 feet.

The Keuper Sandstone is locally conglomeratic but chiefly comprises fine grained cross-bedded water-laid sands. The sands were perhaps deposited in a temporary lake and the conglomeratic beds may have been deposited as delta fans or as true deltas built out into the lake. This lake, less saline than during the deposition of the Moulding Sand, gave rise to less bright red sands than the Moulding Sand. The presence of pellet-rocks and lenses of marl infers deposition as sand banks, not dunes, as does the nature of the cross-bedding. The climate, although still arid, was less so than throughout most of the period and the Keuper Sandstone may be equivalent to the German Muschelkalk. Plants found in the Lower Keuper known as the Bromsgrove Group occur in the beds which underlie the Muschelkalk of Germany but the fish fauna is similar to that of the Muschelkalk and the overlying Lettenkohle. Approximate correlation with the Muschelkalk seems undoubted, but the Lower Keuper of England seems likely also to be partly equivalent to the Upper Bunter of Germany. The Keuper Sandstone has yielded a considerable fauna, the lacustrine conditions supporting a variety of life. Fish and amphibians are present as well as terrestrial forms including scorpions and reptiles. The highest part of the Keuper Sandstone, flaggy with marl or shale partings, is known as the Waterstones. This is a supposedly descriptive name and nothing to do with the water-bearing properties. Keuper Sandstones are a source of underground water, locally important, but less than the Bunter. The presence of the Brachiopod *Lingula* (some in the position of growth) in marl within the lower part of the Waterstones at Eakring, Notts., clearly indicates a temporary marine incursion, the only one preceding the Rhaetic for which there is conclusive evidence. The Waterstones extend from Cheshire to Nottinghamshire and form a passage group to the overlying Keuper Marl. Found southwards as far as Gloucestershire they cannot be recognized as a separate formation beyond. The Keuper Sandstone is there no more than 200 feet thick and less than half that thickness in Somerset, while eastwards the Keuper oversteps the Bunter on to Coal Measures and again Waterstones cannot be identified in the boreholes from which this evidence is derived.

The Keuper Marl is seldom true marl since the percentage of lime is too low and it is only slightly calcareous. There is considerable divergence of opinion on how much lime occurs in marl, and the term has been used to describe a great diversity of sediments although a calcareous argillaceous deposit is implied by the term. The beds of the Keuper Marl are red- and chocolate-coloured mudstones with some beds of sandstone and shale and show a cyclic pattern of sedimentation.

Three scales of rhythms have been postulated and megacyclothems have been recognized as well as cyclothems; the latter, five in number, range from 40 to 290 feet. These have been related to the eight formations of the Keuper recognized in Nottinghamshire (Table 12.2). Each cyclothem commences at a relatively sharp boundary and the lower part comprises laminated units but the higher part is more massive and structureless. In some formations smaller scale cyclothems (*c.* 20 feet) can also be recognized.

TABLE 12.2

	Formations	*Cyclothems*
RHAETIC		
KEUPER	Parva	5th
	Trent	4th
	Edwalton Harlequin	3rd
	Carlton Radcliffe	2nd
	Waterstones Woodthorpe	1st
BUNTER		

The Keuper Marl comprises clay minerals with a high percentage of fine silica of desert origin but accumulated in shallow water. There is frequent evidence of desiccation in the form of mudcracks, curling and overturning of clay laminae, evaporite deposits and salt pseudomorphs. The presence of ripple marks, cross-bedding and slump structures confirms their deposition in water. 'Skerries', sandstones—usually calcareous—occur in the lower part of the Keuper Marl and often give rise to higher ground. The Keuper Marl generally forms low ground, forming the Midland and Cheshire Plain and flooring the valleys of the Rivers Trent and Lower Severn (though largely covered by drift).

Of the sandstones which occur within the Keuper Marl Group, the best developed occurs about 120 to 150 feet below the top of the Keuper in Warwickshire, Worcestershire and Gloucestershire. Known as the Arden Sandstone it is some 40 feet thick. It is fossiliferous with plants, vertebrates and molluscs and may have been laid down in conditions similar to the sandstones of the Lower Keuper.

Evaporites

The Keuper Marl is remarkable for its uniformity over wide areas. Its precise depositional environment is still in doubt and perhaps no analogous conditions exist at present although it has been suggested that a similar environment is to be found in parts of Western Australia, and more recently the semi-arid basins of central Asia have been cited as a similar example of alternating lacustrine and subaerial deposits. Certain sedimentary structures, scoured oscillation ripple marks and air-heave structures, have been quoted as evidence of deposition in an intertidal zone of a lake or sea. The presence of standing water subject to evaporation is clearly confirmed by the presence of salt deposits which are extensively worked (almost entirely by pumping brine) around Northwich in Cheshire and near Droitwich in Worcestershire. In Cheshire the salt totals 800 feet. Salt deposits also occur in northern Lancashire and in Northern Ireland. The extent of salt beds can be partly deduced from the trail of subsidence of the ground which follows their extraction. Water from the Keuper Sandstone where it is overlain by Keuper Marl is hard and sometimes saline accounting for the situation of spas, for example Leamington and Droitwich. Two widespread gypsum horizons occur in the Keuper, the Newark Gypsum and the Chellaston Gypsum, approximately 60 and 140 feet below the top of the Keuper Marl. The higher of these is widespread but may not be exactly correlatable with the gypsum which also occurs in Somerset about 60 feet below the top of the red marls, since here Celestine (strontium sulphate, which often occurs with gypsum and salt deposits) is present but does not occur in the Newark Gypsum. Gypsum is common in the Keuper Marl as stringers. It is postulated that the main gypsum horizons represent deposition in lakes or 'Dead Sea' conditions subject to intense desiccation.

In the Mendips area the Keuper Marl has a 'marginal' facies, dolomitic conglomerate occurs and its distribution can be closely related to the Variscan anticlinal folds. Similarly, the Rhaetic Beds (and the Lias) have littoral facies in the vicinity of the Mendips.

The highest beds of the Keuper Marl are green clays and marl, the Tea Green Marl, best seen in the coastal exposures of Somerset but occurring wherever uppermost Triassic rocks are preserved from Yorkshire to Somerset. Affected by reducing conditions they may reflect an abrupt change in depositional environment, perhaps the onset of a wetter climate, not necessarily exactly contemporaneous over the whole country, but the green coloration may be secondary in origin.

The Rhaetic

The Tea Green Marls are followed by the Rhaetic Beds, now generally included in the Triassic System although marking a widespread marine incursion, the beginning of a long period of dominantly marine deposition over much of Britain. The Rhaetic Beds, not exceeding 100 feet thick and usually only about half that thickness, are important as a datum of reference at the top of the Triassic and can be presumed to be approximately contemporaneous throughout the outcrop, for the Rhaetic sea spread rapidly over the now-peneplaned landscape and established uniform conditions. The Tea Green Marls pass upwards into Grey Marls in Somerset and Glamorgan but these are of variable thickness. Above this the Westbury Beds (= Lower Rhaetic) follows with a slight unconformity which northwards is more marked. Despite this, the Rhaetic essentially forms a passage group between Triassic and Jurassic with affinities for both systems. The flat plains of late Keuper times with its gypsiferous lakes were invaded by the sea which remained shallow and with little lateral variation in facies. The extensive downwarping of sedimentary basins, a character of Jurassic times, had not yet commenced and the Rhaetic passes unchanged over the Market Weighton Axis of Yorkshire (discussed in the next chapter). As well as lithologically transitional, the Rhaetic is faunally an interesting link between the Triassic and Jurassic Systems. Its fauna can be described under three categories: species which survived from the earlier Keuper such as *Euestheria* and *Rhaetavicula contorta* (the latter occurring in the Grey Marls [or Sully Beds]), typically Rhaetic species—some restricted to the Rhaetic Beds generally and some to a particular horizon—such as *Ostrea bristowi*, *Mytilus crowcombeia* and *Ceratodus latissimus*, and thirdly forerunners of the Jurassic fauna such as *Liostrea liassica*. The invertebrate fauna comprises chiefly lamellibranchs, and their small size together with the absence of other animals common in marine benthos probably reflect adverse environmental conditions despite their widespread occurrence from northern Britain to the Swabian Jura of southern Germany. Vertebrate fossil fragments are abundant in the bone beds, chiefly *Ceratodus* teeth and other fish fragments but including bones of aquatic saurian reptiles. However, the Rhaetic is perhaps most famed for its record of the earliest British mammal, of which teeth only have been found, in Rhaetic deposits filling fissures in the Mendips limestones near Frome. The teeth are of a small rodent, *Hypsiprymnopsis rhaeticus*.

The succession of the Rhaetic in Somerset and Glamorgan is as follows:

UPPER RHAETIC	Watchet Beds White Lias or Langport Beds Cotham Beds
LOWER RHAETIC	Westbury Beds or *Contorta* Shales with Bone Beds
KEUPER	Grey Marls [or Sully Beds]

The Westbury Beds are black shales which follow the Grey Marls, where present—for they are impersistent—or the Tea Green Marls. At the base is a non-sequence and though no angular discordance can be seen as a rule, fragments of the underlying rocks occur in the basal Westbury Beds and at higher levels in Glamorgan. Some bands are very fossiliferous including two thin pyritiferous limestones with *Chlamys*. A conglomeratic sandy limestone, the

Bone Bed—usually several thin fossiliferous beds totalling about 8 feet—occurs about 20 feet above the base of the Rhaetic at Blue Anchor, Somerset, but north of the Mendips is a basal conglomerate resting on the Tea Green Marls. It includes many fragments of bone, scales and teeth and perhaps represents a condensed fauna—currents removing sedimentary grains constantly condensing the fossil remains into thin beds by this 'winnowing' process. Littoral deposits occur in the Bristol-Mendips area and the Bone Bed is locally absent, overlapped by higher Westbury Beds. In Glamorgan, similarly, the Lower Rhaetic is dominantly sandy, with a succession of sandstone, shale, sandstone, in the Bridgend area this being a littoral facies while further east around Penarth it comprises black shales.

The Upper Rhaetic comprises lighter coloured clays and limestones and, while the detailed divisions of the Rhaetic recognized in Somerset cannot be identified everywhere, the lower black shale division and an upper light-coloured calcareous division can be found throughout the main outcrop from Devonshire to Yorkshire. The Cotham Beds follow the Westbury Beds with non-sequence and are grey-green silty clays with thin limestones. A hard calcite-mudstone only 6 to 8 inches thick is known as the 'Landscape Marble' from dendritic markings resembling trees and bushes—especially on polished surfaces—now known to be algal growths. Laid down in shallow water, some beds have a marine shelly fauna like that of the Westbury Beds but others (for example the *Naiadites* Bed) include plants such as liverworts and algae together with *Euestheria*. Minor structures produced by slumping have been described from the lower part of the Cotham Beds of both Yorkshire and Glamorganshire.

The White Lias, or Langport Beds, is lithologically distinct from the Cotham Beds below although comprising muddy limestones (calcilutites) with marl partings. Conglomeratic bands, sun-cracks and organism-bored or -burrowed surfaces all indicate deposition in very shallow water but with oscillations in sea level. An environment comparable with that of the Great Bahama Bank of the present day has been postulated. The White Lias shows a littoral facies in the vicinity of the Mendips and it thins towards the region of the Lower Severn Axis which must have formed a margin to the area of deposition although no littoral facies has been found in its proximity, surprisingly. Thickening southwards, the most complete and best exposed section of the White Lias can be seen in the south coast near the Devon-Dorset boundary. The fauna comprises, chiefly, lamellibranchs and trace fossils, the latter very common.

The Watchet Beds are calcitic mudstone, like the marly beds of the White Lias with *Ostrea*, occurring in Somerset and Glamorganshire—coincidentally approximately in the area where the Grey Marls are present at the base of the succession.

The outcrop of the Rhaetic is traceable over a wide area and it makes a very low scarp feature across the Midlands with the low Keuper plain to the north-west and the Jurassic clays flooring low vales to the south-east. It is poorly exposed through much of its length being quite concealed by drift in

Lincolnshire, but it is seen in stream sections in the west scarp of the Yorkshire Moors. Further north in the Tees Valley its outcrop is drift-covered but it is known in detail from numerous borehole records. The Rhaetic transgression spread far: Rhaetic beds are proven in Kent where dark shales with *Rhaetavicula contorta* are found and where, subsequently, Mesozoic sediments accumulated to great thickness. The Rhaetic is known from Mull and Arran and fragments occur in Skye. It is sandy in Mull, but dark shales with Lower Rhaetic lamellibranchs are found in Arran. Elsewhere in Scotland, in Raasay and near Brora, Sutherland, beds immediately below the Jurassic may be of Rhaetic age though not of characteristic facies and no palaeontological confirmation of age can be obtained. In Northern Ireland Triassic beds including the Rhaetic are preserved attaining a total thickness of almost 3,000 feet although good exposures are now few. The Rhaetic can be seen at a number of places on the periphery of the Tertiary Antrim Plateau Basalts (to which the Mesozoic rocks in north-eastern Ireland largely owe their protection from erosion) and a complete succession, though less than 50 feet thick, has been described with Bone Bed, Black *Contorta* Shales and thin Upper Rhaetic marls on the east and south of the plateau. South-west of Lough Neagh the *Contorta* shales are absent in boreholes and a major non-sequence must occur.

The palaeogeography of Rhaetic times poses a number of problems. The wide extent of the marine transgression is demonstrated from the outcrops occurring from southern Germany (the Swabian Jura) to Northern Ireland and the Western Isles of Scotland. Despite this vast spread it was laid down in shallow-water conditions which must imply a eustatic rise in sea level. Some crustal warping took place in Rhaetic times and thickness variations have been shown to occur, particularly in the east Midlands where borehole records show strong attenuation in the vicinity of Grantham but increase in thickness west and north-west of Lincoln. In addition to the development of littoral facies already described, in the Mendips area and in Glamorgan, sandy beds develop in the Lower Rhaetic of north Lincolnshire and reddish clays of Keuper type have been recorded in the Upper Rhaetic. In view of this and other evidence it seems unlikely that the marine transgression invaded from the east, but in the absence of evidence of marginal facies except those near archipelagoes the Rhaetic shorelines cannot be drawn with certainty (Fig. 12.3).

The present search for gas beneath the North Sea is adding to the data available on the Permian and Triassic rocks. Prospecting in Denmark and Holland has already shown the subsurface distribution of Triassic rocks, including the Rhaetic, but the relationship to Britain is still to be established. Much of the gas so far obtained from the North Sea field has come from sandstones of Bunter age but the oil struck north of Cromer (October 1966) is from the Magnesian Limestone of Permian age. Flowage of the salt deposits of Upper Permian and Lower Triassic age is known to have produced salt-dome structures which commonly give rise to oil traps, salt being capped by gypsum. Such structures are well known beneath the North German Plain and are now being investigated under the North Sea.

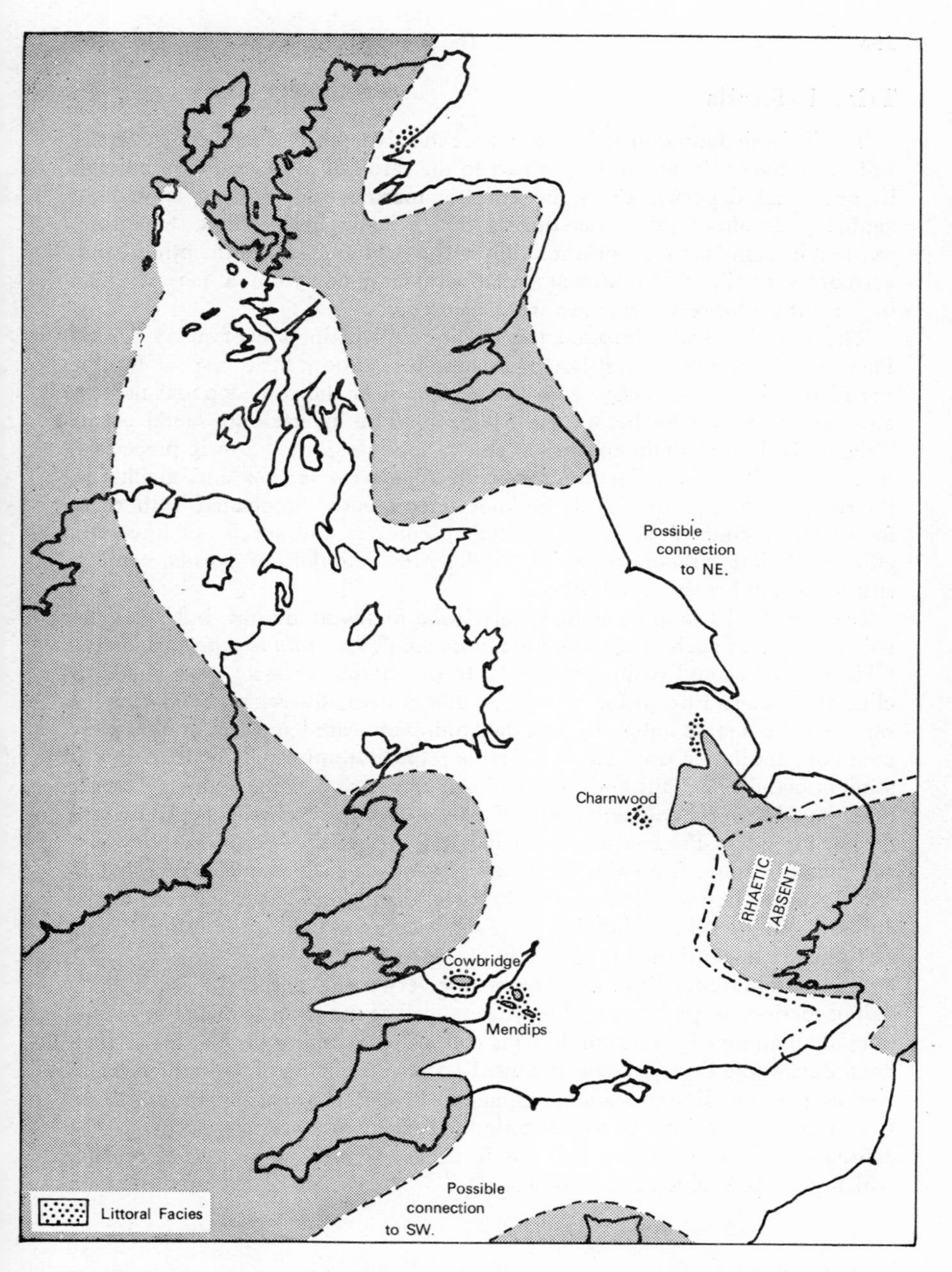

Fig. 12.3 The probable palaeogeography of Rhaetic times. The Rhaetic beds are absent through erosion in eastern England and the form of the London Massif can therefore be predicted only from present Rhaetic contours. The absence of littoral facies, except where shown, makes the position of shorelines of the Rhaetic sea speculative.

Triassic fossils

The Triassic faunas of Britain are not extensive, partly due to the generally unfavourable environment but in part to the rarity of preservation of animals in terrestrial deposits. Even the Rhaetic marine transgression is without cephalopods although it was during the Triassic Period that this group reached its acme with the proliferation of the Ammonites. On the other hand, vertebrate fossils are abundant (relatively) and are from a period when important evolutionary changes were occurring.

The Permian had witnessed the decline and extinction of many typical Palaeozoic invertebrates, although the mollusca show little respect for the Permian/Triassic boundary, with the exception of the cephalopods: no true ammonites are known before the Mesozoic. The invertebrate fauna of the Triassic beds in Britain comprises chiefly molluscs, often poorly preserved, together with the phyllopod crustacean *Euestheria minuta* and arachnids. Poorly preserved plant fossils are not uncommonly associated with these fossils and include *Equesitites* (horsetails), conifers and species of uncertain affinity. By late Triassic times a typically Mesozoic flora of cycads, conifers and ferns had become established.

Only in the Rhaetic beds are invertebrate fossils numerous, but again are chiefly molluscs such as *Rhaetavicula contorta*, *Protocardia rhaetica*, *Modiolus*, *Chlamys*, *Ostrea* and small gastropods. In the marine Triassic, of the geosyncline of southern Europe for example, corals flourished—scleractinians having replaced the Tetracoralla, the first Belemnites appeared and Ammonites were common, the latter used as zone fossils. *Nautilus* appeared in place of the straight orthocone Nautiloids, the majority of which become extinct. Triassic brachiopods are characteristically of Mesozoic types, chiefly telotrematous forms, for many Palaeozoic genera did not survive the Permian. However, the faunal change from the Palaeozoic to Mesozoic is scarcely revealed in some groups. Echinoderms of the Triassic form a link between those of the Jurassic and the Palaeoechinoids while some common Palaeozoic animals, such as Conularids, continued to abound.

The vertebrate fossils of the Triassic are very varied and include fish, bony fish occurring, amphibians and reptiles as well as the earliest British mammal (already mentioned). The amphibians continued to be important, as they had been during the Permian, and included Labyrinthodonts of several genera as well as possible dinosaur ancestors, such as *Cheirotherium* known only from its five-toed footprints. In the Rhaetic the first marine saurians occurred, the Ichthyosaurs—so common in Jurassic times, as well as terrestrial reptiles which perhaps included early dinosaurs.

13 The Jurassic System

Subdivisions of the Jurassic

The Jurassic System was the first system to be given a 'geographic' name, for it takes its name from the development of strata of that age in the Jura Mountains and was defined as early as 1829. Jurassic strata also outcrop in south-west Germany, partially encircling the Paris Basin, and in Britain in a broad expanse extending from the south coast of England to the Yorkshire coast. Minor outcrops occur in Scotland and Northern Ireland (Fig. 13.1). The system has been the subject of intense study and more has been written on it than on any other system except perhaps the Carboniferous. The thick succession of strata which comprise the Jurassic System was divided into a great number of formations and many of the names still used persist from the classical work of William Smith in the first half of the nineteenth century. Divisions based on lithology have been defined relative to the faunal zones, a zonal system being established over a century ago for the German Jurassic by diligent work of Oppel and Quenstedt. Eleven stages are internationally recognized (but in Britain it is convenient to continue with the usage of a twelfth, the Purbeckian [see Table 13.1]); these comprise a great number of zones and subzones. Through the abundance of ammonites during Jurassic times and their rapid and complex evolution at that time it has been possible to carry the zonal division of the Jurassic System further than in the case of any other system. For example, the Lower Jurassic (Lias) is divided into more than 20 zones, further divided into 49 subzones. Because of recent discussion on proposed changes in the definition of the boundaries between certain major divisions of the Jurassic, the full zonal list is given in Table 13.1.

A proposition to include the Aalenian (Lower Bajocian), generally regarded as part of the Middle Jurassic and in Britain comprising the Lower Inferior Oolite, in the Lower Jurassic would have several important consequences. While the Lias and the Lower Jurassic would no longer be synonymous, the

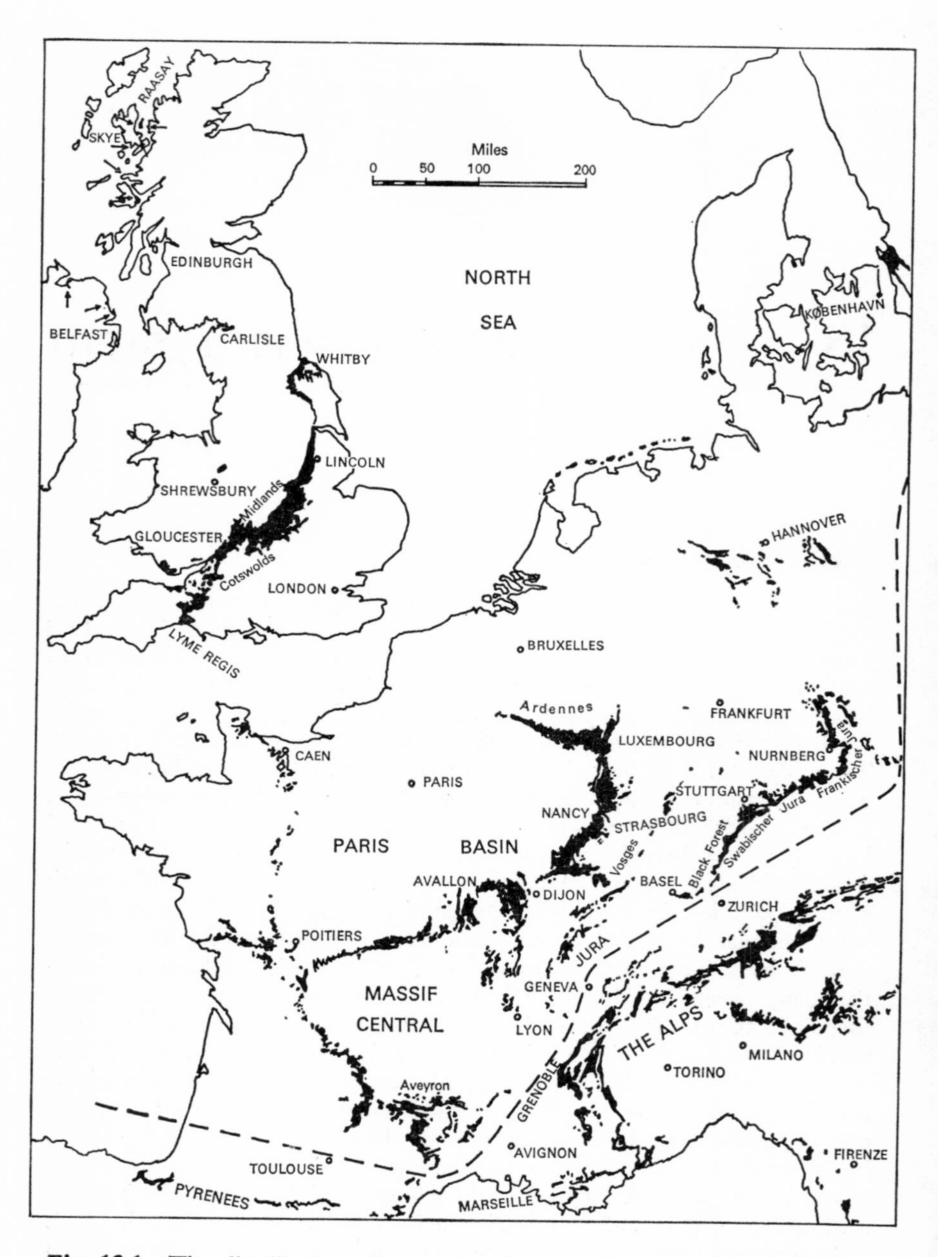

Fig. 13.1 The distribution of outcrops of Lower Jurassic rocks in western Europe. The broken line marks the northern margin of the Tethyan geosyncline. (From W. T. DEAN, D. T. DONOVAN and M. K. HOWARTH, 1961, *Bull. Br. Mus. nat. Hist.*, **4**, 437.)

difficulties of placing the boundary between Middle and Lower Jurassic, which resulted from diachronous sandy facies, would be overcome by the inclusion of all these zones in the Lower Jurassic. However, that problem is simplified by merely including the *Opalinum* Zone in the Upper Lias (as

TABLE 13.1

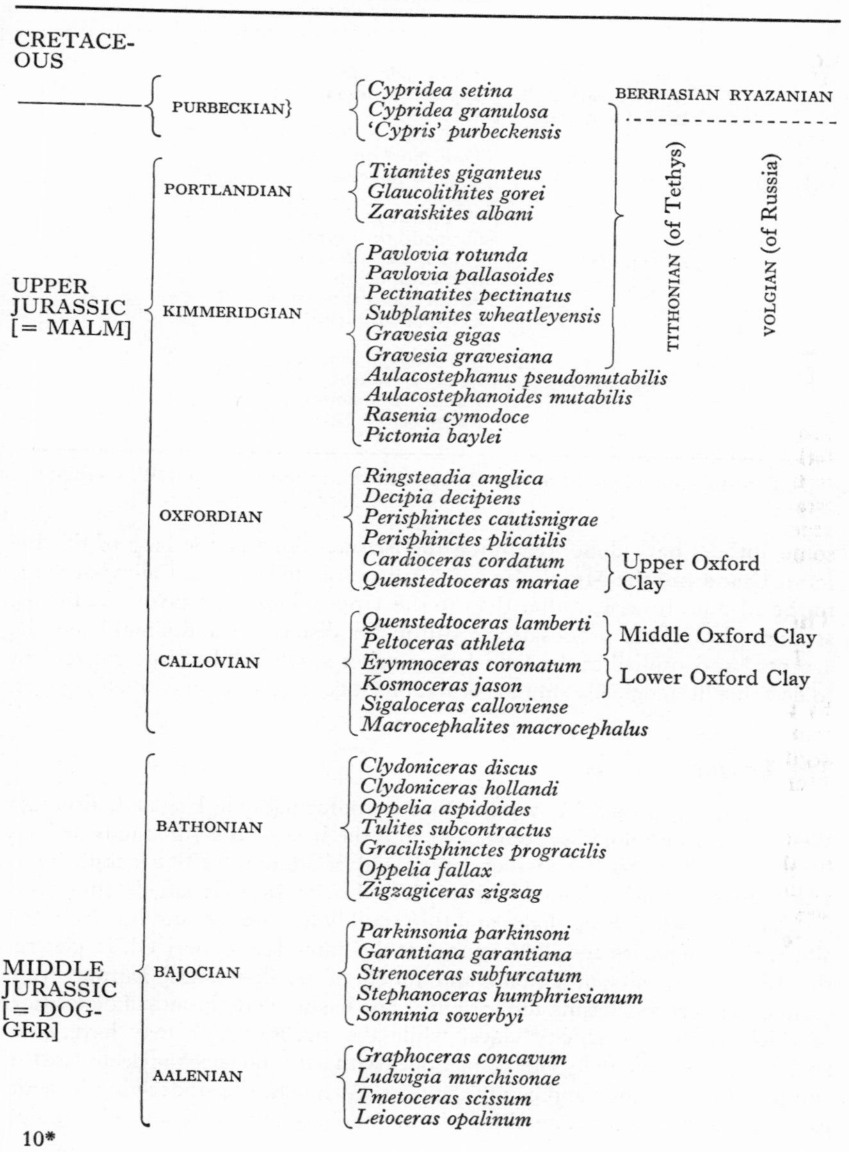

Series	Stage	Zones		
CRETACEOUS				
	PURBECKIAN	*Cypridea setina*	BERRIASIAN	RYAZANIAN
		Cypridea granulosa		
		'*Cypris*' *purbeckensis*		
UPPER JURASSIC [= MALM]	PORTLANDIAN	*Titanites giganteus*	TITHONIAN (of Tethys)	VOLGIAN (of Russia)
		Glaucolithites gorei		
		Zaraiskites albani		
	KIMMERIDGIAN	*Pavlovia rotunda*		
		Pavlovia pallasoides		
		Pectinatites pectinatus		
		Subplanites wheatleyensis		
		Gravesia gigas		
		Gravesia gravesiana		
		Aulacostephanus pseudomutabilis		
		Aulacostephanoides mutabilis		
		Rasenia cymodoce		
		Pictonia baylei		
	OXFORDIAN	*Ringsteadia anglica*		
		Decipia decipiens		
		Perisphinctes cautisnigrae		
		Perisphinctes plicatilis		
		Cardioceras cordatum	Upper Oxford Clay	
		Quenstedtoceras mariae		
	CALLOVIAN	*Quenstedtoceras lamberti*	Middle Oxford Clay	
		Peltoceras athleta		
		Erymnoceras coronatum	Lower Oxford Clay	
		Kosmoceras jason		
		Sigaloceras calloviense		
		Macrocephalites macrocephalus		
MIDDLE JURASSIC [= DOGGER]	BATHONIAN	*Clydoniceras discus*		
		Clydoniceras hollandi		
		Oppelia aspidoides		
		Tulites subcontractus		
		Gracilisphinctes progracilis		
		Oppelia fallax		
		Zigzagiceras zigzag		
	BAJOCIAN	*Parkinsonia parkinsoni*		
		Garantiana garantiana		
		Strenoceras subfurcatum		
		Stephanoceras humphriesianum		
		Sonninia sowerbyi		
	AALENIAN	*Graphoceras concavum*		
		Ludwigia murchisonae		
		Tmetoceras scissum		
		Leioceras opalinum		

TABLE 13.1—(*Contd.*)

Series	Stage	Zone	Substage
LOWER JURASSIC = LIAS	TOARCIAN = Upper Lias	*Pleydellia aalensis*	
		Dumortieria levesquei	YEOVILIAN
		Grammoceras thoursense	
		Haugia variabilis	
		Hildoceras bifrons	
		Harpoceras falcifer	WHITBIAN
		Dactylioceras tenuicostatum	
	UPPER PLIENSBACHIAN = Middle Lias (Domerian)	*Pleuroceras spinatum*	
		Amaltheus margaritatus	
	LOWER PLIENSBACHIAN	*Prodactylioceras davoei*	
		Tragophylloceras ibex	
		Uptonia jamesoni	
	SINEMURIAN	*Echioceras raricostatum*	
		Oxynoticeras oxynotum	
		Asteroceras obtusum	
		Caenisites turneri	
		Arnioceras semicostatum	
		Arietites bucklandi	
	HETTANGIAN	*Schlotheimia angulata*	
		Alsatites liasicus	
		Psiloceras planorbis	
		Pre-*planorbis* Beds	

[ ] Terms commonly used on the Continent, e.g., in the Jura and SW. Germany

some authors have done) regarding the *Scissum* Zone as the base of the Inferior Oolite and the Middle Jurassic. A proposal to refer the Callovian Stage to the Middle Jurassic rather than to the Upper finds little favour in British stratigraphy. These proposals are still under discussion and should they be accepted and applied to the British succession it will still be more convenient to describe lithologically similar formations together as in the following pages.

The Lower Jurassic

The Lower Jurassic follows the Rhaetic conformably in England, although the change in lithology is often sharp, and the base of the Jurassic is defined by the Pre-*Planorbis* Beds which are devoid of ammonites but include *Liostrea liassica* and other lamellibranchs. The Lower Jurassic largely comprises argillaceous rocks, clays, shales and thin muddy limestones, and it is from the alternations of shales and limestones that the name Lias is derived. In general the Liassic deposits of Britain and much of north-western Europe were related to a series of basins and 'swells'. The basins were the site of deposition of thick sequences, chiefly clays, while the swells, which may have been shallow shoals or actually emergent ridges with a littoral facies adjacent, are the site of thin calcareous and sandy deposits. Although the underlying Rhaetic is remarkably uniform laterally in both lithology and thickness, the basins began to subside differentially early in the Lower Jurassic. The areas of little subsidence which separate the basins have been called axes; three in number they are the Mendips, Moreton-in-the-Marsh and Market Weighton Axes. All affect the Lias and the most northerly was most persistent, affecting sedimentation throughout the Jurassic. Great lateral variation in thickness of the Lias occurs (Fig. 13.2) and the axes often separate strata of somewhat different facies, though in the Lower Jurassic all are marine. In the Middle and Upper Jurassic the axes separate more diverse facies acting as thresholds to the southwards spread of deltaic material. As well as the basins known from the main Jurassic outcrop, a basin of subsidence occurred south of the London-Belgian Ridge.

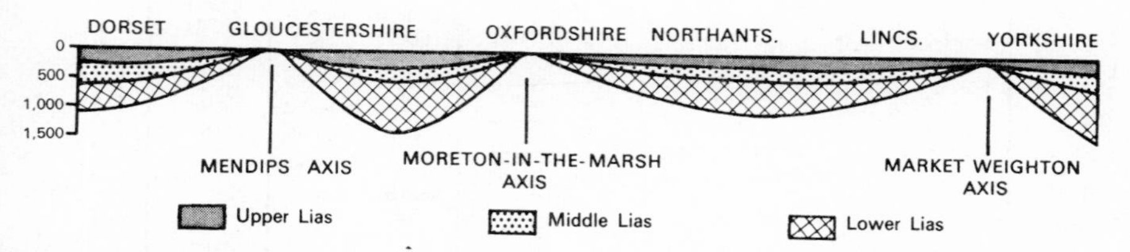

Fig. 13.2 Lateral variations in thickness of the divisions of the Lias along the main outcrop, showing basins of subsidence separated by relatively stable axes.

Mendips—Moreton-in-the-Marsh

In the area between the Mendips and the Moreton Axis Lias outcrops over the broad Vale of Evesham and southwards to Gloucester and is seen in the lower part of the Cotswolds escarpment. The most complete single section of Lias is visible in an Inferior Oolite capped outlier at Robin's Wood Hill on the southern outskirts of Gloucester. Brickpits such as the one here provide almost the sole opportunities of seeing the Lias in inland areas. The Lower Lias of this area is of shallow-water facies. A cyclic sequence of thin limestones and clays of Blue Lias type comprise the Hettangian followed by clays chiefly, although locally limestones may occur as high as the top of the Sinemurian. In the vicinity of the Mendips (and North Hill, another Variscan anticline), beds of littoral facies occur. The Mendips may have been an archipelago and certainly they separated the relatively shallow shelf facies of Gloucestershire from the deeper water facies of Somerset. Except near the Mendips, the Lower Lias chiefly comprises clay shales, there being little lateral variation in lithology. It seems unlikely that the Lower Severn Axis persisted from Rhaetic times. In south Wales the Lower Lias is generally similar with a littoral facies developing in the Cowbridge area.

The Middle Lias of Gloucestershire is up to 230 feet thick comprising marly silts with ironstones, formerly thought to be the Marlstone (which normally occurs at the top of the Middle Lias) but belonging to the lower zone (*Margaritatus*) of the Middle Lias. The upper zone (*Spinatum*) comprises clays succeeded by sandstone (Fig. 13.3). The Upper Lias succession is very variable due to the development of a diachronous sandy facies which can be related to the position of the Bath Axis (Fig. 13.4). The sandy formation forms an outcrop traceable from the Cotswolds to the Dorset Coast, although it is locally named Cotswold Sands, Midford Sands, Yeovil Sands and, in the south, Bridport Sands. Only about 10 feet thick near Cheltenham

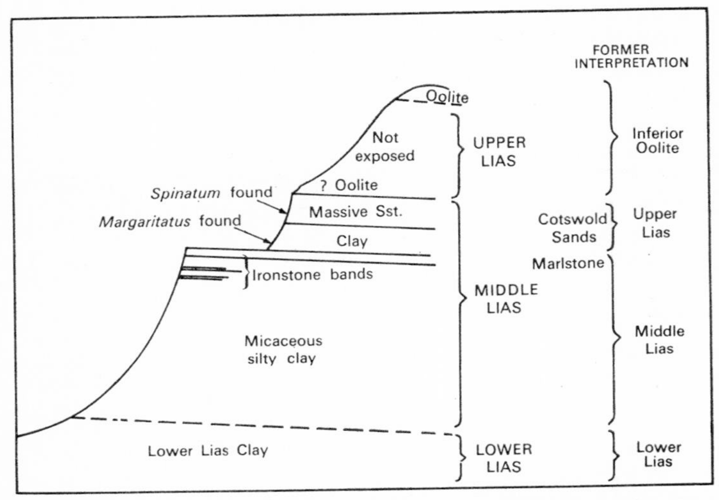

Fig. 13.3 The Lias succession exposed in Robin's Wood Brickpit, Gloucester

and resting on the strata of the *Bifrons* Zone, it thickens up to 150 to 180 feet in the south Cotswolds where it precedes the *Levesquei* Zone which here is an ironshot oolite with a remarkable condensed fauna. Known as the Cephalopod Bed it is also diachronous. In the Mendips the Liassic sequence is only very partially represented and near Radstock is:

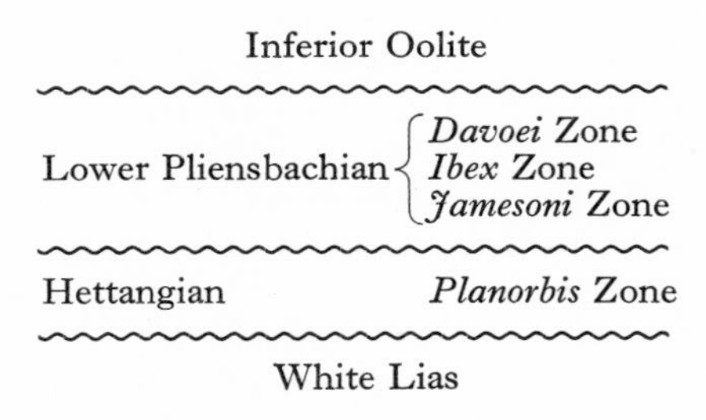
Inferior Oolite

Lower Pliensbachian { *Davoei* Zone, *Ibex* Zone, *Jamesoni* Zone

Hettangian — *Planorbis* Zone

White Lias

During the Sinemurian minor folding occurred and no beds of this age are found. Southwards towards the Mendips the Upper Lias oversteps the Middle Lias which is not seen south of Bath.

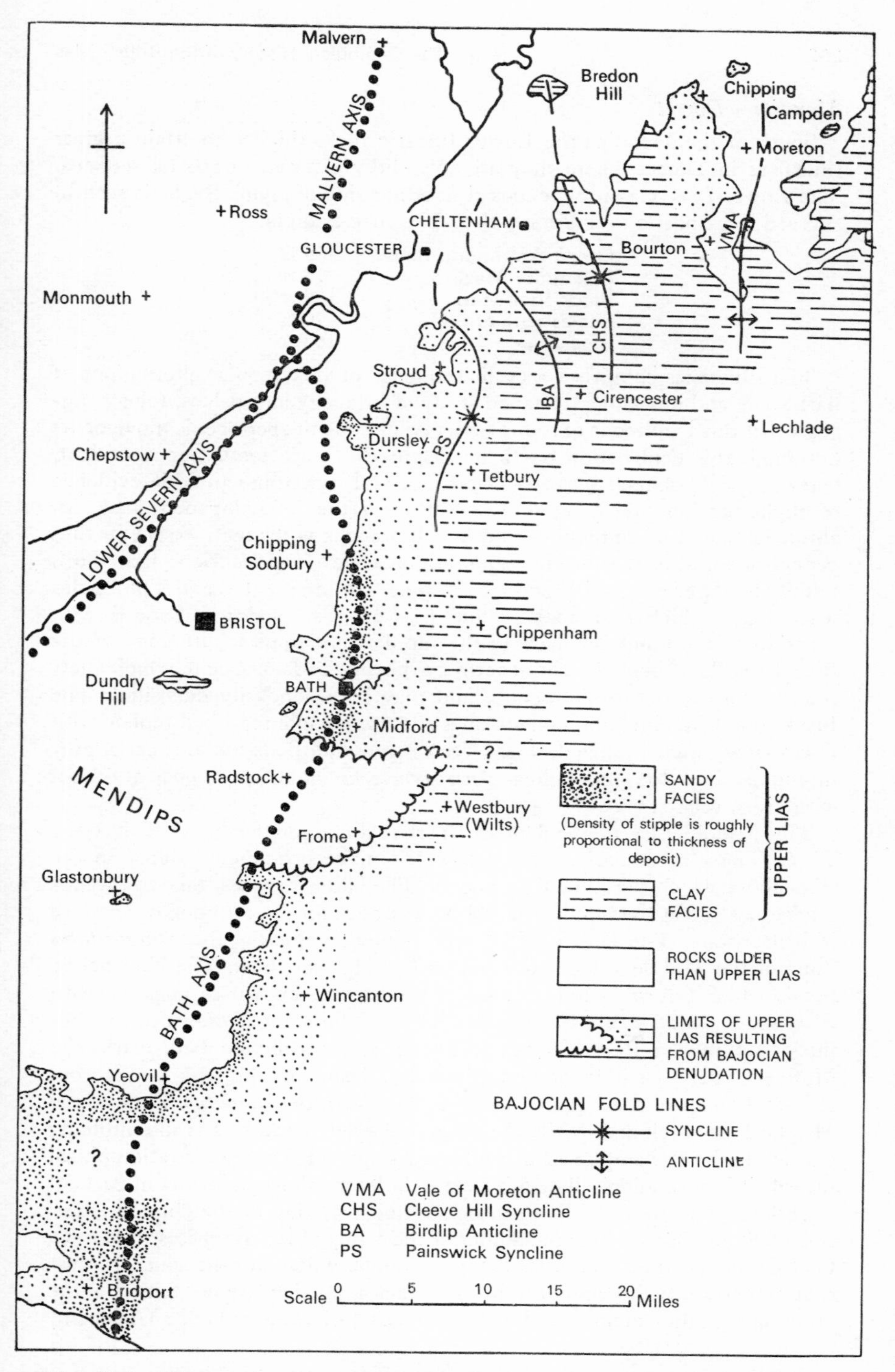

Fig. 13.4 The relation of the sandy facies of the Upper Lias to the fold axes. (From F. B. A. WELCH and R. CROOKALL, 1948, *British Regional Geology. Bristol and Gloucester District*, 2nd edn. Crown Copyright Geological Survey Diagram. Reproduced by permission of the Controller of H.M. Stationery Office.)

Mendips—Dorset

South of the Mendips the Lower Jurassic rocks thicken to attain a great thickness in Dorset where they are splendidly exposed in coastal sections. The Lower Lias, classically exposed to either side of Lyme Regis, is seen to succeed the Rhaetic near Pinhay Bay. The succession is:

Green Ammonite Beds	105 ft
Belemnite Marl	75 ft
Black Ven Marl	150 ft
Shales with 'beef'	70 ft
Blue Lias	105 ft

The Blue Lias comprises a cyclic sequence of very regular alternations of dark shale and limestone. Detailed study of lithology and palaeontology suggests that this banding is of sedimentary origin due to epeirogenic movements (although this explanation has been disputed). Trace fossils are abundant, vertical and U-shaped burrows occurring, and limestones provide evidence of slight erosion preceding the deposition of the following marl bed. An abundant fauna of ammonites and lamellibranchs is present, except in thin paper-shales, and at some horizons crinoids and large saurians have been found (Ichthyosaurs and Plesiosaurs), some within recent years. The Shales with 'beef' which succeed the Blue Lias are also shales and marls with occasional limestones but lacking the rapid alternations in lithology of the Blue Lias. The 'beef' is a local term for fibrous beds of calcite which show cone-in-cone structure commonly and may be secondarily deposited. The Black Ven Marls include cementstones and are notable for a bed replete with *Pentacrinus*, often pyritized. The Green Ammonite Beds too are marls, grey in colour. It is the ammonites, *Androgynoceras lataecosta*, which are filled with green calcite.

The Middle Lias of the Dorset Coast includes in the basal 35 feet the Three Tiers, calcareous sandstones which make prominent 'steps' in the cliffs above which the Eype Clay occurs. The latter includes some micaceous sandstones and at the top the Starfish Bed famous for the common occurrence of brittle-stars. The Down Cliff Sands follow completing the *Margaritatus* Zone of the Middle Lias, here from 340 to 410 feet thick. The Thorncliffe Sands which follow include masses of sandstones, known as doggers, with *Pleuroceras spinatum* and brachiopods but the Marlstone is only a few inches thick. A thin bed, 2½ to 4 feet and known as the Junction Bed, marks the Middle-Upper Lias junction and it includes, despite its thinness, fossils from at least 4 zones of the Upper Lias as well as from the *Spinatum* Zone of the Middle Lias. Included fossils are worn and eroded and this is undoubtedly one of the best examples of a condensed sequence from the stratigraphical record: it is chronologically equivalent to at least 350 feet of strata in parts of Yorkshire. The Junction Bed can be seen outcropping in the cliffs to either side of Eype, Dorset. The remainder of the Upper Lias comprises the Down Cliff Clays and the succeeding Bridport Sands. These are the southernmost representatives of the sandy diachronous facies, for they are not found on the other side of the channel, and are here in part of Inferior Oolite (Aalenian) age since *Opalinum* Zone fossils are found in them. Approximately 140 feet thick, they form spectacular yellow cliffs east of Bridport Harbour and, though they are soft and poorly cemented, they are reinforced by hard cemented bands. Cross-bedding and washout channels are conspicuous.

Moreton-in-the-Marsh—Market Weighton

Between the Moreton and Market Weighton Axes, in Northamptonshire and Lincolnshire, the Lias is thick—over 1,100 feet despite the absence in most areas of much of the Yeovilian. There is considerable development of cementstones in the Hettangian and Sinemurian extensively worked around Rugby. The commercially important Frodingham Ironstone, a chalybite-oolite (siderite with limonitic ooliths), occurs extensively (35 square miles) in Lincolnshire at the base of the Sinemurian. The Middle Lias, of which the lower zone is clay and the upper zone the ferruginous Marlstone, is also locally a source of iron ore around Banbury and Grantham. This, in fact, is the most widespread of the Jurassic ironstones and was formerly important in north Yorkshire. Oolitic, the ironstones are chamositic with some chalybite. 'Nests' of brachiopods occur in the ironstone, for example in the Banbury area, and include *Tetrarhynchia tetrahedra* and *Lobothyris punctata.* More resistant to erosion than underlying and overlying clays, the Marlstone forms a prominent escarpment, for example Edge Hill, and caps a number of outliers to the west of the escarpment. The Upper Lias attains a thickness of about 200 feet, but the area of maximum subsidence shifted southwards during Upper Liassic times so that successive zones thicken to the south but probably overlap earlier zones. Following the deposition of the Upper Lias slight regional uplift occurred, for the succeeding beds overstep different zones of the Upper Lias and are of different facies. These are the deltaic Northampton Sands, some ferruginous, some calcareous, with ironstones and are of Inferior Oolite age.

North of Market Weighton

North of Market Weighton the Jurassic rocks generally differ considerably from those of the south of England and the Midlands but the differences are slightest in the Lower Jurassic. The Lower Lias comprises shales with cementstones followed by shales and shales with ironstones. Ironstones in the Middle Lias were formerly economically important in Cleveland, north Yorkshire. The ironstone, bedded oolitic chamosite with some sideritic ooliths (the latter probably of secondary origin), was formerly one of the richest sources of bedded iron ore in England. The succession is:

Spinatum Zone	Shales		
	Main Seam	11 ft	28 ft
	Ferruginous shales		
	Pecten Seam	1 ft 9 in–6 ft	
Upper part of *Margaritatus* Zone	Two foot seam	2 ft	
	Shale		
	Avicula Seam	2 ft	

Here in Yorkshire the *Spinatum* Zone is only about 25 feet thick compared with the 100 feet plus of the *Margaritatus* Zone but this disparity is much more pronounced in Dorset. However, it is not only in thickness variation that there is evidence that Liassic sedimentation had become divided into

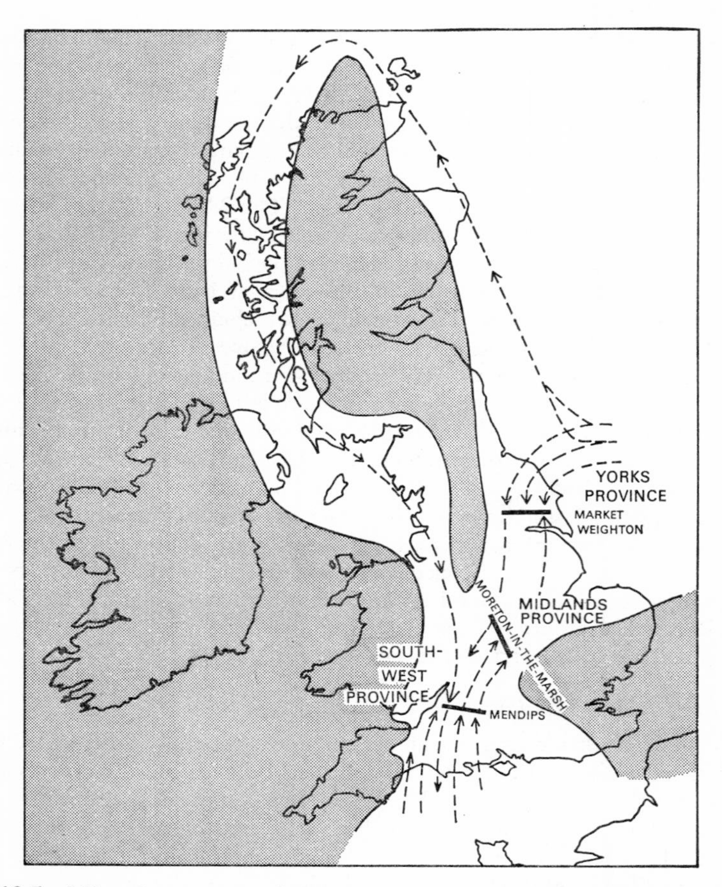

Fig. 13.5 Migration routes of Middle Lias (*Spinatum* Zone) brachiopods, showing the main faunal provinces. (After D. V. AGER, 1956, *Q. Jl geol. Soc. Lond.*, **112,** 157.)

provinces; there were lateral faunal variations and this has been demonstrated in the case of the brachiopods (Fig. 13.5). It is true that some species of brachiopod crossed the postulated barriers and that different faunal provinces may be an expression of different ecological conditions in different areas. It has been pointed out that there are many regional faunal differences in modern seas not separated by any physical barrier, but nevertheless the faunal provinces may have been separated in Jurassic times by actual barriers. Certainly the Middle Lias beds cannot now be found for a distance of 8 miles, and the Upper Lias for a distance of 12 miles, over the Market Weighton Axis. The distribution of brachiopod faunas has been worked out in great detail for the *Spinatum* Zone of the Middle Lias. It is interesting to

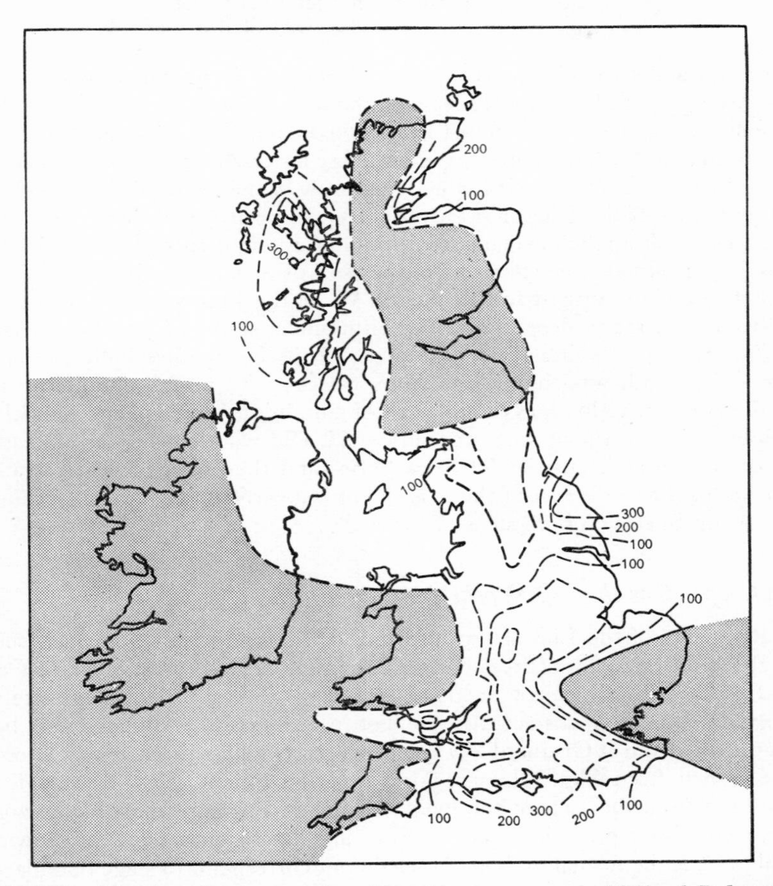

Fig. 13.6 Liassic palaeogeography. (Based on L. J. WILLS, 1951, *A Palaeogeographical Atlas*. Blackie, London; with isopachytes [in metres] adapted from P. BITTERLI, 1963, *Geologie Mijnb.*, **42,** 183.) N.B. Mochras Lias, see p. 297.

note that the presence of a Pennine barrier to the spread of faunas is postulated and is perhaps supported by Liassic isopachytes (Fig. 13.6), although lithological evidence in the form of littoral facies is lacking and there is scanty evidence of its uplift by the Variscan Orogeny.

The Upper Lias of Yorkshire presents a problem since to the east of the Peak Fault the succession is more complete and is:

UPPER LIAS (= TOARCIAN)	Yeovilian	Blea Wyke Beds
		Striatulum Shales
		Peak Shales
	Whitbyian	Alum Shales
		Jet Rock Series
		Grey Shales

West of the Peak Fault the Yeovilian is absent, the Dogger of Inferior Oolite age resting upon the Alum Shales, and the succession is 200 feet thinner. The absence of the Yeovilian infers a contemporaneous fault movement, perhaps the presence of a fault shoreline, the thicker succession to the east reflecting inter-Jurassic movement since the downthrow of the Alum Shales is 400 feet whereas that of the Dogger is only 200 feet. A wrench fault has been postulated as an alternative explanation, bringing different successions and facies into juxtaposition. The steepness of the fault plane and the absence of slump structures lends support to this theory. While the Lias was laid down generally in moderately deep water, the bituminous seams of jet, formed from drifted wood, as well as fossilized stranded marine reptiles indicate that at times the depth was slight. The Alum Shales, with a high aluminium and pyrite content, yield alum (double salt of aluminium sulphate and potassium sulphate) by treatment with sulphuric acid. Old workings are still a feature of the cliff tops. Both the Jet Rock Series and the Alum Shales are rather thicker to the south-east of the Peak Fault but correlatable horizons occur in the Alum Shales on both sides of the fault.

Outcrops other than the Yorkshire-Dorset belt

Remnants of the Lias are widespread, occurring in west Somerset, south Wales, south Cheshire (Prees Outlier), the Solway Basin west of Carlisle and in Scotland. In all except Scotland even the Lower Lias is incomplete. In Scotland Jurassic rocks occur in the east in Sutherland and in the west both on the mainland at Gruinard Bay and Morvern as well as in the Isles. Outcrops are known from Skye, Raasay, Mull and the Shiant Isles. Lithologically different from the Lias of the main outcrop in England, most of the zones have now been proved present in Skye and Raasay however. As shown in Table 13.2, the formation boundaries do not correspond to stage boundaries. The lowest beds, the Broadford Beds, resting unconformably on Trias or older rocks, include limestones—both oolitic and coralline—shales and sandstone. The silty Pabba Beds pass up into the Scalpa Sandstone (*Ibex* to *Tenuicostatum* Zones). The thin Upper Lias comprises shales and includes the formerly worked Raasay Ironstone. Except in Mull, the lowest Liassic is not represented and only there was sedimentation continuous from the Rhaetic. The two zones of the Middle Lias in the west of Scotland are approximately the same thickness, *c.* 140 feet, and the *Spinatum* Zone in

TABLE 13.2

	Stage	Formation	Thickness
(Upper Lias)	TOARCIAN	Raasay Ironstone	8 ft
		Portree Shales	70 ft
(Middle Lias)	DOMERIAN	Scalpa Sandstone	450 ft
(Lower Lias)	LOWER PLIENS-BACHIAN	Pabba Beds	600 ft
	SINEMURIAN	Broadford Beds	235 ft
	HETTANGIAN		

particular is very different in lithology and fauna from the occurrences in Yorkshire and Dorset.

In south-east England only the highest Jurassic rocks (Purbeckian) outcrop but earlier Jurassic beds are known to be extensive from many borehole records. The Lias attains a maximum of 1,250 feet, over 800 feet referable to the Lower Lias, near the centre of the basin of subsidence. Thickness variations are great (although lateral variations in lithology seem to be slight), in part due to the overlap of successively higher zones against the London-Belgian Ridge to the north. The Lias itself is overlapped northwards by the Inferior Oolite as the area of sedimentation further encroached the ridge. The succession can be summarized:

Upper Lias: shales, mudstones and shelly marl;
Middle Lias: limestones overlie shales and sandy shale;
Lower Lias: shales with thin limestones, locally a pebbly basal bed.

In Ireland Lias outcrops at a number of places in the north-east, for example on the shores of Belfast Lough and at Island Magee. Only Lower Lias remains but derived fossils of Middle and Upper Lias in overlying Cretaceous beds suggest their former presence. Whether Middle and Upper Jurassic deposits were also laid down but subsequently removed by erosion or whether a prolonged period of non-deposition followed the deposition of the Lias is not known.

A pointer to the greater extent of Liassic seas than formerly suspected is the recent discovery of over 600 feet of Upper Lias beds in a borehole at Mochras on the Welsh Coast, south of Harlech. It indicates the lack of confidence which can be placed in those areas of palaeogeography maps remote from outcrop information.

The Middle Jurassic

The lithological twofold division of the Middle Jurassic strata into the Great Oolite and the Inferior Oolite is only readily recognizable in certain areas because of the very great lateral variation of all but the highest formations of the Middle Jurassic. The chief break in sedimentation occurred after the deposition of the Middle Inferior Oolite when folding and erosion preceded the widespread Upper Inferior Oolite transgression. Zonal ammonites are often scarce since much of the Middle Jurassic is of limestone facies but brachiopods have proved invaluable stratigraphically. In the north, where much of the Middle Jurassic is of deltaic facies, ammonites are absent from rocks of that facies.

The Cotswolds

The Inferior Oolite is well exposed in the Cotswolds in sections now classic as a result of the work of William Smith over a century ago and by Richardson 50 years ago. Between the Mendips and the Moreton Axes it attains a maximum of about 350 feet in thickness east of Cheltenham. The succession seen at Leckhampton Hill (with the zones of strata missing indicated in parentheses) is given in Table 13.3.

TABLE 13.3

		Lithological Divisions
Upper Inferior Oolite	*Parkinsoni* *Garantiana*	*Clypeus* Grit Upper *Trigonia* Grit
	(*Subfurcatum*) (*Humphriesanum*)	
Middle Inferior Oolite	*Sowerbyi*	Notgrove Freestone *Buckmani* Grit Lower *Trigonia* Grit
	(*Convacum*)	
Lower Inferior Oolite	*Murchisonae*	Upper Freestone Oolite Marl Lower Freestone Pea Grit Lower Limestone
	Scissum	*Scissum* Beds

(Zones in parentheses absent in Leckhampton section)

The Limestones vary greatly in character although all were deposited in shallow marine conditions. All but the rubbly limestones are oolitic but some more conspicuously so. The 'grits' are so called from the abundance of shell fragments while cross-bedding is common, especially in the massive jointed freestones, confirming shallow conditions. Only the Oolite Marl is argillaceous, reflecting a brief interval of the incoming of terrigenous material and the deposition of calcitic mud in the usually clear sea. Numerous organism-bored surfaces, for example in the Notgrove Freestone, also confirm the shallowness of the sea.

Even at Leckhampton the succession is not complete, for the highest beds of the Lower Inferior Oolite, the Tilestones, Snowshill Clay and Harford Sands were only deposited in the Cleeve Hill Syncline. This was initiated at that time as a basin, higher beds overlap towards its margins, while elsewhere slight elevation resulted in erosion which preceded the deposition of the Lower *Trigonia* Grit, the lowest member of the Middle Inferior Oolite. It was, however, in post-Middle Inferior Oolite times after the deposition of the *Witchellia* Grit and *Bourguetia* Beds that the main folding took place. As a result of erosion, only in the Cleeve Hill Syncline—and less completely in the Painswick Syncline—are the higher beds preserved (Fig. 13.7). The Upper Inferior Oolite transgression which followed deposited the Upper *Trigonia* Grit and its equivalents (though somewhat earlier beds were laid down further south) on beds ranging in age from Middle Inferior Oolite to Lias, overstepping on to yet older beds in the eastern Mendips.

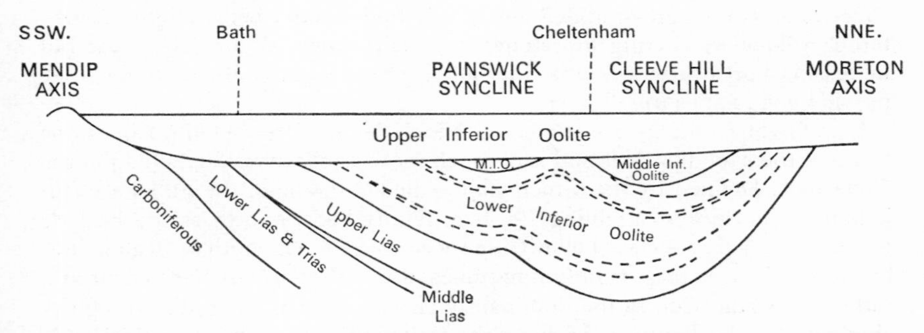

Fig. 13.7 Variations in thickness of the Inferior Oolite of the Mendips-Moreton-in-the-Marsh area. (Simplified from F. B. A. WELCH and R. CROOKALL, 1948, *British Regional Geology. Bristol and Gloucester District*, 2nd edn. Crown Copyright Geological Survey Diagram. Reproduced by permission of the Controller of H.M. Stationery Office.)

The lower beds of the Bathonian are very variable. The Fuller's Earth, divisible around Bath into Lower Fuller's Earth clays with oyster beds, a thin limestone known as Fuller's Earth Rock and the Upper Fuller's Earth Clay, is less than 150 feet thick at Bath. In Somerset it becomes much thicker while northwards in the Cotswolds it thins rapidly and is succeeded by the Great Oolite Limestone. The latter is locally divided into several members which include the famous Bath Stone, an oolite used in building. The Stonesfield Slate, a flaggy sandy limestone used in the past for roofing, outcrops at the base of the Great Oolite Limestone in the north Cotswolds. Succeeding the Great Oolite east of Bath the Bradford Clay, though only 10 feet thick, is an important horizon. The crinoid *Apiocrinus elegans* [*A. parkinsoni*] is abundant with many specimens in the position of growth. They apparently flourished on the limy mud of the sea floor, now limestone, and were overcome by the influx of mud which gave rise to the Bradford Clay. The latter possesses a distinctive brachiopod fauna of value in broader correlation although this facies is not found south of the Mendips. The Forest Marble, shelly cross-bedded limestones with clays, completes the Middle Jurassic succession. Although thinning towards the Moreton Axis, the Forest Marble thickens southwards passing over the line of the Mendips Axis and into Somerset where it makes a conspicuous escarpment.

Southern England

In Dorset and Somerset the diachronous sands which range upwards to include the *Opalinum* Zone have been discussed. The overlying Lower and Middle Inferior Oolite are thin and are represented by only a few feet of strata in the Dorset Coast and at their maximum, in the Sherborne area, do not exceed 30 feet. They are sandy, pebbly and crystalline limestones showing evidence of wave action. The Upper Inferior Oolite transgression, as in the Cotswolds, swept across eroded and gently folded older beds. Highly fossiliferous yellow-weathering limestones were laid down, thin in the coast but increasing northwards towards Sherborne where quarries for building stone provide good exposures.

The Great Oolite Series (Bathonian) includes no Great Oolite Limestone south of the Mendips Axis and is of much more argillaceous facies than in the Cotswolds. Defined by the 6-inch *Zigzag* Bed at the base, the Fuller's Earth is here at its thickest attaining 290 feet. Sandy beds, known as the Fuller's Earth Rock, separate clays below from those above as far south as Beaminster, but in the Dorset coast rubbly limestones of the *Wattonensis* Beds occur at a rather higher horizon in the dominantly clay succession. A readily recognizable horizon, the *Boueti* Bed follows the Fuller's Earth. Only a foot thick, it is crowded with the brachiopod *Goniorhynchia boueti* and must have been deposited very slowly. It is at the horizon of the Bradford Clay but the terrigenous material of the latter did not spread so far south. On the other hand, the succeeding Forest Marble formation in Dorset is more argillaceous than it is in the Cotswolds. These lateral variations in lithology, though known in detail, cannot yet be interpreted in terms of changing palaeogeography.

Moreton-in-the-Marsh—Market Weighton

North of the Moreton Axis the Middle Jurassic rocks are, in part, of deltaic facies. The Inferior Oolite north of this axis is sandy and ferruginous, and the beds immediately overlying the Upper Lias Clays are the Northamptonshire Sands and Ironstone. Consequent upon the development of the recent topography with valleys cut down into the Lias, while the competent sands and ironstone cap the hills, superficial structures have developed. They are the result of the differential loading and flowage and contortion of the clays. The commoner structures are illustrated (Fig. 13.8). The remainder of the Inferior Oolite comprises the Lower Estuarine Series, sands, silts and clays of deltaic facies which further north are in turn succeeded by a fissile sandy limestone, the Collyweston Slate. The latter forms a transition to the Lincolnshire Limestone which, appearing in Northamptonshire, thickens northwards to form the conspicuous west-facing escarpment on which Lincoln Cathedral stands. Accompanying the lithological changes, from the terrigenous material of the deltaic facies to the calcareous deposits overlying them, is a faunal change. The drifted plant fragments of the deltaic beds are followed by marine faunas, chiefly molluscan in the Collyweston Slate and more varied in the

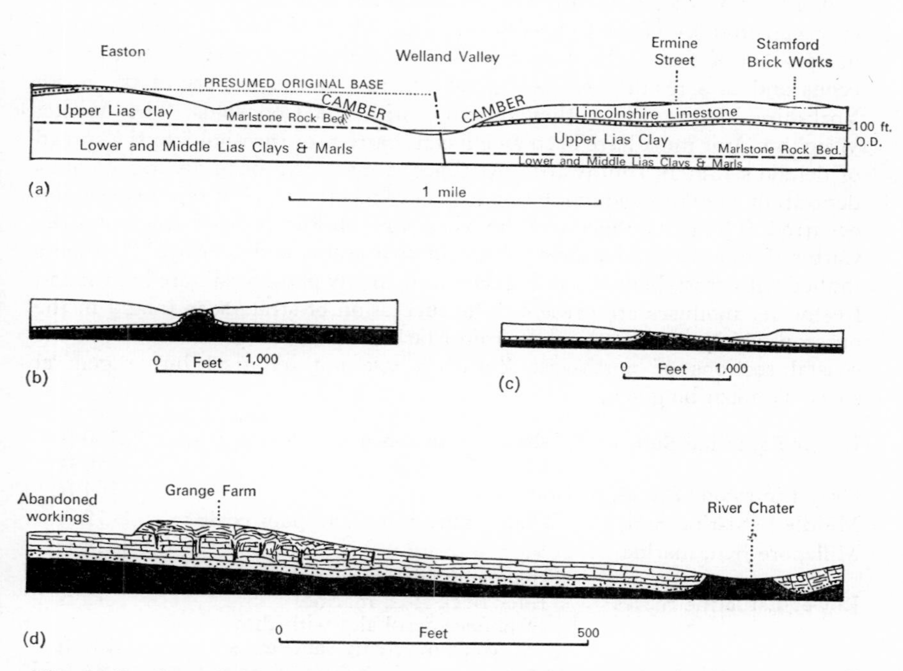

Fig. 13.8 Superficial structures in the Jurassic rocks of Northamptonshire and Rutland. (*a*) Camber structures in the Northampton Sands and Ironstone. (*b, c*) Valley bulges near Oundle and near Slipton; (*d*) Gulls in the Lincolnshire Limestone. The Lias clays shaded black in *b–d*. (Redrawn from S.E. HOLLINGWORTH, J. H. TAYLOR and G. A. KELLAWAY, 1944, *Q. Jl geol. Soc. Lond.*, **100**, 1.)

Lincolnshire Limestone. The latter is divisible into several members and the upper part is unconformable on the lower, overstepping it southwards in Northamptonshire and northwards on the flank of the Market Weighton Axis.

The Great Oolite Series, like the Inferior Oolite, in this basin, is partly of deltaic facies. After the clear-water deposition of the Lincolnshire Limestone a spread of argillaceous and sandy material occurred. The Upper Estuarine Series, as these beds are called from Northamptonshire to Market Weighton, contains marine fossils locally but around Lincoln includes the freshwater gastropod *Viviparus*. The Great Oolite Limestone follows, succeeded by the Blisworth Clay (as the Great Oolite Clay is here called) which passes southwards into the Forest Marble. There were, therefore, two occurrences of non-marine deposition in the Middle Jurassic in this basin (Fig. 13.9).

North of Market Weighton

North of Market Weighton the Middle Jurassic rocks are of very different character to those of the Cotswolds or of any other area to the south. Largely of non-marine facies, the deltaic sediments of this basin were largely arenaceous and as a result there is a great contrast in the scenery of the north Yorkshire Moors, where they outcrop, with the Cotswolds. The Market Weighton Axis must have been an efficient barrier, or threshold, to the spread of deltaic sediment southwards. At times it may have been an area of non-deposition, at others sedimentation may have taken place, but repeated erosion occurred. The deltaic facies of the Yorkshire Middle Jurassic resembles the Carboniferous Coal Measures with thin coal seams and fireclays. Washouts infilled with cross-bedded sands occur while many plant fossils are known and freshwater molluscs are present. The succession of strata seen inland in the outcrop which runs northwards from Market Weighton and that seen in the coastal sections of north-east Yorkshire are not identical but a general succession can be given.

Upper Estuarine Series	shales, mudstones and sandstones	160 ft
	grit	40 ft
Grey Limestone Series, marine		104 ft
Middle Estuarine Series	shales, sandstones and poor coals	72 ft
Millepore Bed, marine		14 ft
Lower Estuarine Series	shale and thick sandstone	110 ft
	Eller Beck Bed, marine	15 ft
	carbonaceous shales with thin coals overlying sandstones	160 ft
Dogger, marine sandstones and oolites		40 ft

Although this basin of sedimentation was generally separated from areas to the south during Middle Jurassic times, there were occasional marine incursions which show that it was probably a gulf open occasionally to the east. The Dogger rests, as a rule, unconformably on Lias and is well exposed in the coast. It yields a varied brachiopod fauna from the basal bed followed by marine lamellibranchs and ammonites of the lowest Inferior Oolite (*Opalinum* Zone). Changes in lithology and facies have been described but palaeogeographical interpretation is difficult. It is sandy passing into chamositic oolite while locally a thin bed with corals, the *Nerinea* Bed is present. Similarly the Eller Beck Bed, which marked a return to marine conditions after the deposition of over 100 feet of poorly fossiliferous deltaic sediments, has a ferruginous facies and a calcareous facies. At some horizons the deltaic beds yield abundant plant fossils.

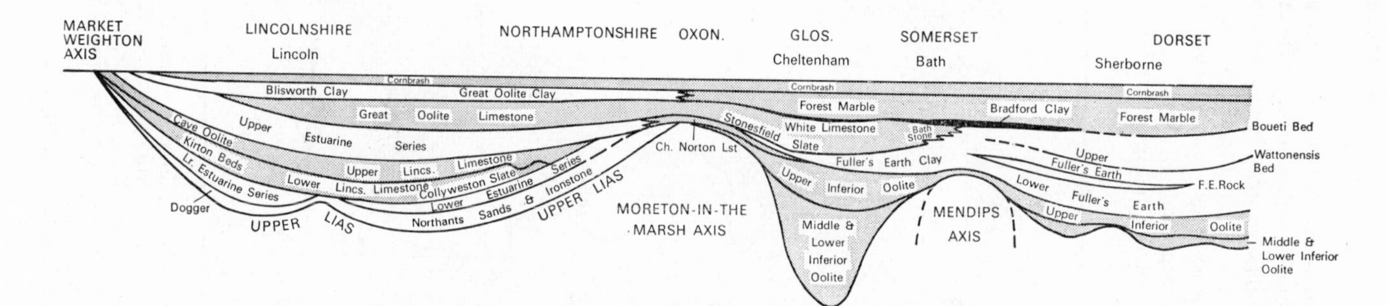

Fig. 13.9 Lateral variations in the Middle Jurassic rocks from Market Weighton to Dorset. The chief developments of limestone are shaded. (Thicknesses are not accurately to scale.)

The Millepore Bed is a bryozoan limestone represented in the inland escarpment by the Whitwell Oolite with a largely different fauna though including the same bryozoan *Haploecia straminea* as the typical Millepore Bed. The latter is tentatively equated with the Lincolnshire Limestone. The overlying Middle Estuarine Series contains many plants which include *Equesitites*, cycads and primitive angiosperms.

The Grey Limestone or Scarborough Limestone has an abundant marine

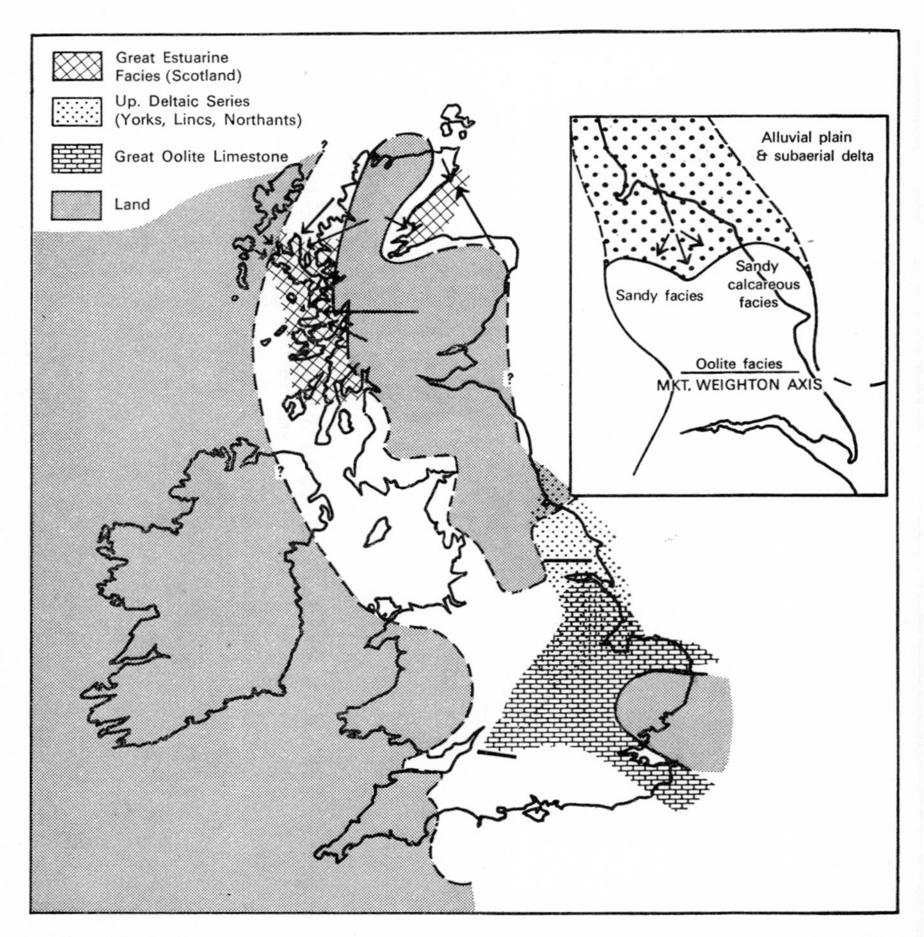

Fig. 13.10 Middle Jurassic palaeogeography. *Inset:* Palaeogeography of Yorkshire during the deposition of the Millepore Bed. (Based on L. J. WILLS, 1951, *A Palaeogeographical Atlas*, Plate XVI, Blackie, London; detail of Scotland based on J. D. HUDSON, 1962, *Trans. Edin. geol. Soc.*, **19,** 139; inset based on F. SMITHSON, 1942, *Q. Jl geol. Soc. Lond.*, **98,** 27, and on R. H. BATE, 1967, *Bull. Br. Mus. nat. Hist.*, **14,** 111.)

fauna which proves that it also belongs to the Bajocian (Inferior Oolite). The Upper Estuarine Series therefore represents the whole of the Bathonian as well as part of the Inferior Oolite. Its terrigenous material may have been of northerly origin.

Although a general succession has been given, lateral changes in facies have been observed. For example, in the Millepore Bed oolites pass northwards into sandy calcareous deposits and sandy beds. Yet further north marine beds grade into deltaic (Fig. 13.10).

Scotland

The Middle Jurassic rocks which in Scotland succeed the Lias are unlike those of either southern England or Yorkshire. Strata of Bajocian age are chiefly marine sandstones, the Bearraig Sandstone, with limestones in Skye and Raasay. Great lateral variations in thickness are known and the maximum development, 1,600 feet in Skye, is the greatest thickness of deposits of this age known in Europe. Sedimentation was cyclic: three cycles are recognized, the beginning of each marked by important changes in the ammonite fauna, and they approximately correlate with widespread marine incursions recorded in western Europe. The Bathonian rocks are of quite different facies comprising the Great Estuarine Series. While no major marine incursions comparable with those of the Yorkshire Middle Jurassic (Inferior Oolite age) penetrated the Scottish province, the series is not devoid of marine fossils. The conditions of deposition did not closely resemble those of the Estuarine Series of Yorkshire—the nomenclature clouds the issue. In Scotland (except in Sutherland) coals are absent, as are fireclays with rootlets, and washouts do not occur. The sediments were not laid down in the classically deltaic conditions of the Yorkshire 'estuarine' beds but more probably in lagoons. Ripple marks indicate shallow-water conditions of deposition but some beds are laterally very persistent. The fauna comprises molluscs and algae with, in some beds, *Euestheria*. The term estuarine is not inappropriate for the conditions in which the Great Estuarine Series was laid down if it is not narrowly defined to exclude brackish environments which became, at times, marine with a fauna of oysters. Lithologically the beds are chiefly sandstones although the 600-foot succession commences with an oil shale and shaly horizons occur at intervals. In the east of Scotland, in Sutherland, only the highest Bathonian strata are seen where non-marine beds include the Brora Coal—the only non-Carboniferous age coal worked in Britain. The palaeogeography of these northern areas (Fig. 13.10) has been largely deduced from a study of the detrital minerals.

The Upper Jurassic

In Middle Jurassic times there had been a considerable recession of seas with the accompanying spread of sediments of deltaic and estuarine facies in northern Britain. A widespread marine transgression followed, giving rise to the deposition of the Cornbrash Limestone which is separable into Upper and Lower Cornbrash on both lithological and faunal evidence. This junction within the Cornbrash is considered to be the base of the Callovian which in Britain is regarded as the lowest stage of the Upper Jurassic. The Lower Cornbrash is therefore of Bathonian age. The Upper Jurassic is divided into a number of stages related to the ammonite zones, the latter divisions based on faunal criteria originally being worked out in south-west Germany. The stages (Table 13.4) do not coincide with major lithological divisions which occur in Britain.

TABLE 13.4

Britain	*Western Europe*	*British Lithological Divisions*
PURBECKIAN (in part)	PORTLANDIAN	Lower and part of Middle Purbeck Beds
PORTLANDIAN		Portland Stone Portland Sand
KIMMERIDGIAN	KIMMERIDGIAN	Kimmeridge Clays and Shales
OXFORDIAN	SEQUANIAN	Corallian Limestones (Ampthill Clay = Upper part of Corallian)
	RAURACIAN	
	ARGOVIAN	Upper Oxford Clay
	OXFORDIAN	
CALLOVIAN	CALLOVIAN	Lower and Middle Oxford Clay Kellaways Beds Upper Cornbrash

The Purbeckian Stage is not recognized by some Continental geologists, although the predominantly non-marine facies found in Britain extends into western Europe. However, because of its facies and complete lack of ammonites it is difficult to relate it to the ammonite zones of either the Tithonian of the southern European geosynclinal deposits (of Tethys) or with the Russian Volgian. In Britain it is zoned by reference to the ostracod faunas. The inclusion of the upper half of the Purbeckian in the Cretaceous is discussed in the next chapter.

The Lower Cornbrash, marly rubbly limestone and massive limestone, includes a fauna similar to that of the underlying Forest Marble. The sandy limestone with hard flaggy bands of the Upper Cornbrash has a fauna resembling that of the overlying Kellaways Beds. Laterally persistent from Dorset to the Yorkshire Coast except over the Market Weighton Axis, with sandy beds of this age found as far north as Sutherland and Skye, the Cornbrash represents a great expansion of the epicontinental seas which spread northwards from the Tethyan geosyncline of southern Europe. Still, however, the London-Belgian Ridge was a major feature of Jurassic palaeogeography and was not inundated.

The Cornbrash is divided into four brachiopod zones:

	Ammonite Zones	*Brachiopod Zones*
UPPER CORNBRASH	*Macrocephalites macrocephalus*	*Microthyridina lagenalis* [*Ornithella lagenalis*] *Digonella siddingtonensis* [*Ornithella siddingtonensis*]
LOWER CORNBRASH	*Clydoniceras discus*	*Obovothyris magnabovata* [*Terebratula obovata*] *Cererithyris intermedia* [*Terebratula intermedia*]

The Kellaways Beds, clays with a fauna similar to that of the Upper Cornbrash and sands with *Gryphaea* and belemnites, are also laterally persistent and are found from Dorset to the Humber. They are succeeded by the Oxford Clay, incredibly uniform in character and thickness although it thins from a maximum of 550 feet in Dorset to 500 feet in the Vale of Oxford and thins further both northwards along the main outcrop and eastwards towards the East Anglian Massif. Deposited in conditions generally similar to those of the Lias it gives rise to similar scenery. A threefold lithological division has been made:

		Ammonite Zones	
LOWER OXFORDIAN	Upper Oxford Clay	*Cardioceras cordatum* *Quenstedtoceras mariae*	Shaly, with *Gryphaea dilata* and *Belemnites owenii*
CALLOVIAN	Middle Oxford Clay	*Quenstedtoceras lamberti* *Peltoceras athleta*	Clays with *B. hastatus*
	Lower Oxford Clay	*Erymnoceras coronatum* *Kosmoceras jason*	Shales and clays

Fossils are very common and well preserved, though crushed, and include mud-loving ammonites, with rostrum or lappets, and belemnites. Large vertebrates, including Plesiosaurs and Pliosaurs, are common. Although as Market Weighton is approached the Callovian and the Oxfordian are not present due to the overstep of Cretaceous beds, their former presence—and subsequent erosion—is inferred, since the Cretaceous includes derived Callovian ammonites. Furthermore, lateral change in facies can be related to the axis since north of Market Weighton the Callovian is sandy. The deposition of the Kellaways Rock was succeeded, probably after an interval of non-deposition, by the Hackness Rock. The Oxford Clay is nowhere more than 130 feet thick, usually considerably less, and the 'Oxford Clay environment' did not spread into this basin of subsidence until much later than south of the axis. The Oxford Clay is infrequently exposed.

Marked geographical changes took place terminating the Oxford Clay deposition which was followed by deposition of a sequence of strata commencing with arenaceous beds but dominantly of calcareous facies. These beds are known as the Corallian from the abundance of scleratinian corals and, locally, the development of reefs at some horizons. This calcareous facies of Oxfordian age is highly developed south of the Oxford Axis as far as the Dorset Coast where excellent exposures can be seen immediately to the south of Weymouth and between Osmington and Ringstead Bay.

The succession of about 200 feet of strata is:

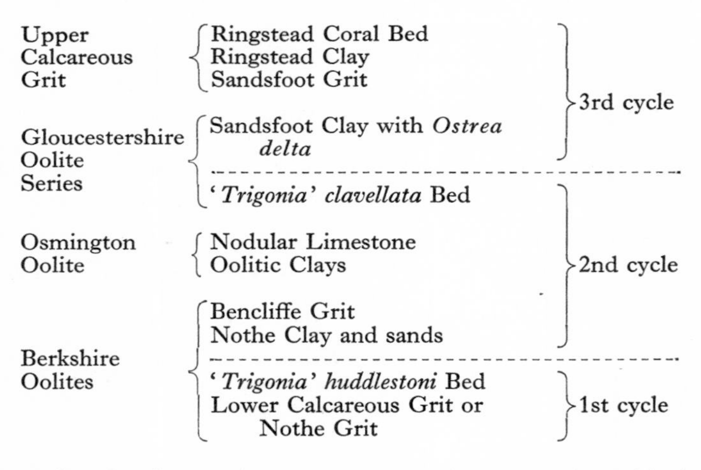

Upper Calcareous Grit	Ringstead Coral Bed Ringstead Clay Sandsfoot Grit	3rd cycle
Gloucestershire Oolite Series	Sandsfoot Clay with *Ostrea delta*	3rd cycle
	'*Trigonia*' *clavellata* Bed	2nd cycle
Osmington Oolite	Nodular Limestone Oolitic Clays	2nd cycle
Berkshire Oolites	Bencliffe Grit Nothe Clay and sands	2nd cycle
	'*Trigonia*' *huddlestoni* Bed Lower Calcareous Grit or Nothe Grit	1st cycle

Three cycles of sedimentation are present, each commencing with the deposition of clays (the Upper Oxford Clay in the first cycle), but the bulk of the succession is calcareous although no reefs occur in this southernmost area. Where beds of Corallian facies reappear in Wiltshire, for they are overstepped by the Cretaceous northwards away from the Dorset coast, certain differences are apparent. Reefs are developed in the Berkshire Oolite Series in north Wiltshire, in a belt extending for 50 miles as far as Abingdon in the

Osmington Oolite Series, as well as being locally developed in the Gloucestershire Oolite Series. Seventeen species of coral are recorded. By this time the Mendips and Moreton Axes seem to have been ineffective, though as they lie to the west of present outcrops evidence is slight. New axes had developed separating basins of subsidence. Over the Purton Axis beds as high as the Nothe Clay are absent while between this axis and the Oxford Axis lay a basin of subsidence. In this reefs developed at the '*Trigonia*' *huddlestoni* horizon. In the region of Oxford three facies have been described in the Osmington Oolite by Arkell. Between the reefs with a fauna of corals in the position of growth, were channels in which rubbly limestones were deposited, while in a deep channel south of Oxford the Littlemore Clays were laid down. The higher Corallian beds are absent in this area.

It is of interest that coral reefs occurred as far north as latitude 54°N. in Oxfordian times, and as far as 58°N. in the succeeding Kimmeridgian times, while at the present day reefs are restricted to a belt within 30° of the equator. A much warmer climate for Britain in Upper Jurassic times is inferred if corals required a mean sea temperature of about 70°F. as at present. Conditions could be met if the equator was situated at about latitude 46°N. in Jurassic times or alternatively if the world climate was warmer. The latter seems unlikely from the world distribution of Jurassic reefs.

Eastwards from Dorset the Corallian has been located in deep bores in Kent and Sussex where it attains its greatest thickness. Here the thick coralline limestones may have originated as fringing reefs near the southern flank of the London-Belgian landmass. Coralline development can generally be related to the margins of the sea, occurring also on the west flank of the East Anglian Massif and on the north flank of the Market Weighton Axis while Dorset, further from the Upper Jurassic shores, is devoid of true reef limestones.

North of the Oxford Axis the corallian facies passes into a deep-water clay facies, the Ampthill Clay, which extends northwards flooring part of the Fenland country where it has been proved in boreholes, with the exception of one boring at Upware which located an isolated reef. North of Market Weighton strata of this age are again of calcareous Corallian facies surprisingly closely resembling the beds of southern England in both succession and fauna. They are exposed in the Tabular, Hambleton and Howardian Hills as well as in coastal sections north of Filey. The succession in Yorkshire is:

Upper Calcareous Grit (45–80 ft) = Upper Calcareous Grit and Upper part of Gloucestershire Oolite

Osmington Oolite Series, with reefs at several horizons (50–90 ft)

Middle Calcareous Grit (0–80 ft) = Berkshire Oolite

Hambleton Oolite, with channelled upper surface (0–120 ft)

Lower Calcareous Grit (including beds transitional from Oxford Clay) (70–200 ft)

The typical Kimmeridgian succession is exposed in a long coastal section in Kimmeridge Bay, Dorset, where dips are low and only minor faulting has occurred.

The Kimmeridgian marked a return to a clay facies with conditions of deposition which closely resembled those prevailing during the deposition of the Oxford Clay. The Kimmeridge Clay comprises clays, shales with some cementstones and siltstones but is remarkably uniform in lithology despite great variations in thickness. Variations in lithology do occur. Locally the Abbotsbury Iron Ore is developed at the base north-west of Weymouth, while further north a sandy facies developed towards the top of the succession in Wiltshire, Oxfordshire and Buckinghamshire. Oil shales are found in Lincolnshire and in Dorset while oil is still obtained by pumping near the village of Kimmeridge in Dorset. The thickness of the Kimmeridgian ranges from over 1,600 feet in Dorset thinning steadily northwards so that in Oxfordshire it is only 300 feet, in the Fens no more than 120 feet, and it is absent over the Market Weighton Axis. The latter was the only stable axis operative by this time. North of this the Kimmeridgian thickens up to 500 feet and it floors the Vale of Pickering where it is obscured by thick drift. The fauna is chiefly molluscan, ammonites and lamellibranchs are abundant, but large saurians have been found.

In Scotland Upper Jurassic rocks are present, chiefly in Sutherland, but small exposures of strata from Callovian to Kimmeridgian age in Skye and Kimmeridgian in Mull prove the extension of Upper Jurassic seas on the west side of the Scottish Highlands. The Sutherland succession which follows the Brora Coal is marine. Again it differs considerably from succession in England and lithological divisions do not coincide with Stages. The succession is-

Kimmeridgian	Kimmeridge Beds
Lower Oxfordian and Upper Callovian	Brora Arenaceous Series
Middle Callovian	Brora Argillaceous Beds Brora Roof Bed (shelly sandstone)

The Jurassic rocks are faulted against Precambrian, Old Red Sandstone and granite. They may have been deposited off a faulted shoreline and certainly a submarine fault scarp must be postulated for there is abundant evidence of submarine slumping of shallow-water shelly deposits to form boulder beds in the contemporaneously deposited deeper water argillaceous beds. The structures of these slumped beds have recently been described in great detail and the suggestion that it was earth tremors which initiated the sliding of the shallow-water deposits seems probable. Sedimentary structures establish that material was transported from the north-west. Younger Mesozoic rocks than the Kimmeridgian are not found on this eastern side of Scotland although Cretaceous rocks are present in the glacial drift and beneath the North Sea.

Late Jurassic rocks of southern England

Major changes in geography took place at the end of the Kimmeridgian resulting in the recession of the sea from all except the south of England. There sedimentation continued but the clays and shales with occasional bituminous beds of the Kimmeridgian grade upwards into the sandier clays, marls and sandstones of the lower division of the Portland. Elsewhere the Kimmeridgian is overlain unconformably by considerably younger beds, chiefly of Cretaceous age. Evidence of yet wider palaeogeographical changes is revealed by faunal changes, for the Kimmeridgian ammonites of Russian affinity died out and in Portlandian times were replaced by large forms such as *Glaucolithes* in the Portland Sand and *Titanites* in the Portland Stone. Thus the elevation of much of northern Britain coincided with the separation of the Franco-British region of deposition from the Russian seas.

The Portland Beds take their name from the Isle of Portland where they outcrop extensively, although they attain an even greater thickness in the Isle of Purbeck. Northwards they are obscured by the Cretaceous overstep but reappear in the Vales of Pewsey and Wardour where they are greatly attenuated. Similarly, the Portland Beds in south-east England are known, from borehole information, to thin northwards towards the London-Belgian Ridge. The lower division, the Portland Sand, is largely argillaceous or marly but with some harder bands of sandstone and cementstone and is only well seen in certain outcrops below the Portland Stone, for example in Hounstout Cliff, Dorset. The higher division, the Portland Stone, often a sandy limestone and containing much chert in the lower beds, has been world famous as a building stone since the time of Wren. It has been obtained by mining as at the Tilly Whim 'Caves' just south of Swanage and it is still extensively quarried in both Portland and Purbeck. As well as the variations in thickness mentioned there are lateral variations in lithology. The cherty series is thicker in the Isle of Portland than further to the east, while the porcellanous limestone at the top of the sequence in Purbeck, known as the Shrimp Bed is not found in Portland. There the uppermost bed is the rubbly Roach. However, a succession, with typical lithologies, is:

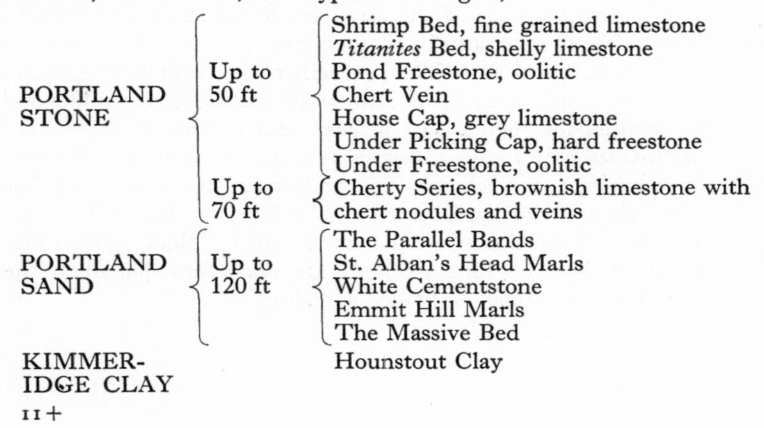

Division	Thickness	Beds
PORTLAND STONE	Up to 50 ft	Shrimp Bed, fine grained limestone
		Titanites Bed, shelly limestone
		Pond Freestone, oolitic
		Chert Vein
		House Cap, grey limestone
		Under Picking Cap, hard freestone
		Under Freestone, oolitic
	Up to 70 ft	Cherty Series, brownish limestone with chert nodules and veins
PORTLAND SAND	Up to 120 ft	The Parallel Bands
		St. Alban's Head Marls
		White Cementstone
		Emmit Hill Marls
		The Massive Bed
KIMMERIDGE CLAY		Hounstout Clay

The Portland Beds have also been designated by letters ranging from the highest Portland Sand, Bed A, to the Shrimp Bed, Bed V. Outcropping at numerous places along the east Dorset Coast, the Portland Stone is responsible for spectacular coastal scenery. The low-dipping Portlandian makes the steep cliffs of Dungy Head and St. Alban's Head, while the near-vertical Portland Stone is breached at Lulworth Cove and Durdle Door.

In the Vale of Wardour the Portland Sand is about 60 feet and the Portland Stone 50 feet. While the Cherty Series is well developed, the higher beds have been worked for building stone. Further north in the Vale of Pewsey the Portlandian is no more than 40 feet thick and outcrops are small. East of Oxford the Portlandian diminishes to 24 feet and the most northerly outcrop known is near Leighton Buzzard. Although in this area the Kimmeridgian is locally sandy near the top, reflecting the proximity of the East Anglian Massif, the Portland Sand is argillaceous with pebble beds (Upper Lydite Bed) and derived Kimmeridgian ammonites proving a non-sequence. Although of littoral facies, the Portlandian is transgressive. The Portland Stone of this area is partly sandy.

The Purbeckian strata, splendidly exposed from Swanage southwards in the coastal section of the Isle of Purbeck, were deposited over an area similar to that of the Portlandian (Fig. 14.1), although conditions of deposition were markedly different since ammonites are absent and marine macrofossils are found only at a few horizons. Clearly therefore geographical changes took place at the end of the Portlandian, reducing the Portlandian Gulf to an area of generally non-marine deposition. Fossil soils testify to its emergence at times although the occasional marine beds indicate that the sea invaded this area of lakes and mudflats. Freshwater limestones, marls and evaporite beds are present. The main marine horizon which occurred, in Middle Purbeck times, known as the Cinder Bed, is now accepted as the Jurassic-Cretaceous boundary for it may coincide with a widespread marine invasion which elsewhere initiated the Cretaceous. The Purbeckian cannot readily be related to the ammonite zones of the Upper Jurassic (Tithonian) of southern Europe or of Russia (Volgian). Some European geologists do not recognize the Purbeckian Stage although the rocks of this facies are found in western Europe. In Britain the 400 feet of strata can be divided into three faunal divisions or zones based on the ostracod faunas. These, together with freshwater molluscs such as *Viviparus* and *Unio*, are the most important fossils but crocodiles and dinosaurs are occasionally found. The marine beds include molluscs—the Cinder Bed is crowded with *Liostrea*, *Isognomon* and *Protocardia* as well as the rare echinoid *Hemicidaris purbeckensis*. This widespread horizon is found as far north as the Vale of Wardour. Like the Portlandian, the Purbeckian thins rapidly northwards and small outliers occur as far north as Buckinghamshire. In the Swindon Outlier the Purbeck Beds are unconformable upon the Portland Stone. The Purbeckian is divided as follows:

	Zone
Upper Purbeck	*Cypridea setina* (*C. punctata* and *C. ventrosa* also present)
Middle Purbeck	*Cypridea granulosa* (*Metacypris forbesi* also present)
Lower Purbeck	'*Cypris*' *purbeckensis* (*Candona ansata* also present)

The Lower Purbeckian includes tuffaceous limestones (the Caps) at the base overlain by the Broken Beds which are brecciated limestones. These are well seen at Durlston Head but better exposed near the entrance to Lulworth Cove where the structures have been described in detail. Brecciation seems probably largely due to tectonic deformation and not to volumetric changes on diagenesis of the evaporites, although diagenetic changes certainly occurred. Gypsum is present in the thick shales, with argillaceous limestones, which occur above the Broken Beds.

The basal bed of the Middle Purbeckian is called the Mammal Bed and has yielded pre-marsupial bones. It is followed by the freshwater limestones known as the Lower and Upper Building Stones with a non-marine fauna and separated by the Cinder Bed, so called from its appearance due to the crowded mollusc shells. It is the most conspicuous and widespread of the marine horizons in the Purbeckian. However, study of microfauna reveals that in Purbeck times marine incursions occurred more frequently than had formerly been supposed. The remainder of the Purbeckian succession is described in the next chapter.

Although the Purbeckian thins northwards like the Portlandian, it thickens eastwards attaining a maximum in Kent and is the only Jurassic stage to outcrop. Occurring in three faulted inliers along the axial region of the Wealden Anticline, about three-fifths of the succession is exposed and it is known from many boreholes. The Lower Purbeck includes worked beds of anhydrite originally deposited in hypersaline lagoons and in the Middle Purbeck the Cinder Bed is again found.

Jurassic fossils

In the generally shallow epicontinental seas of the Jurassic which spread northwards from the Tethyan geosyncline neritic forms are abundant. Molluscs are especially important and the zonal system based on ammonite faunas has been given (p. 287–8). The history of the ammonites is complex for, in addition to forms such as the Phylloceratids and Lytoceratidae which are present from Liassic times onwards (and these families persist into the Cretaceous), there was repeatedly the entrance of new forms replacing others which died out. Some are of very limited stratigraphical range, for example the Psiloceratids which appeared, rapidly diversified and died out within the Hettangian. They were followed in the Lower Sinemurian by the Arietids (*Arietes*, *Asteroceras*, etc.) and in turn the Echioceratids and Deroceratids in the Upper Sinemurian. Polymorphoceratids and Liparoceratids dominate in the Lower Pliensbachian and Amaltheids in the Upper (= Middle Lias). Hildoceratids are common in the Whitbian. A second major influx of new families took place in Inferior Oolite times, though Sonninians such as *Dactylioceras* occur first in the Upper Lias. The next major influx was within the Cornbrash at the beginning of Callovian times and forms such as *Quenstedtoceras* and *Kosmoceras* appeared and continued into Oxfordian times. The exceptionally large ammonites of late Jurassic times have been mentioned.

Of the other mollusca belemnites were common, as were gastropods. The earliest freshwater gastropods are known from the Jurassic. Many new genera of lamellibranchs made their appearance such as *Trigonia*, of world-wide distribution, while oysters (especially *Gryphaea*) were abundant. Elongate forms such as *Modiolus* were common in limestone environments while *Avicula*, *Steinmannia*, etc., occurred in argillites.

Coelenterates, chiefly scleractinian corals, which became common in Triassic times became abundant especially in the Corallian facies of the Upper Jurassic. Hydrozoans and sponges also are found in the reefs. Brachiopods were abundant. They were often restricted geographically and their migrations have been worked out in considerable detail. Echinoids of both the regular and irregular kinds were abundant, *Cidaris* and *Acrosalenia* of the former, and flat forms of the latter such as *Clypeaster*. Both stalked and free-swimming crinoids were common, including the distinctive *Pentacrinus*, while at some horizons Ophiuroids are present.

Among vertebrates primitive bony fish as well as sharks occurred. Marine saurians have been referred to. The commonest land-living vertebrates were, of course, reptiles, and the dinosaurs of both carnivorous and herbivorous types abounded, some growing to great size. Flying reptiles, and in south-west Germany the earliest known birds, occurred. Although four orders of mammals were present by late Jurassic times they were still relatively insignificant numerically.

Plant fossils have been described from the deltaic facies and comprised conifers and cycads. The flora appears to have been generally similar over the whole world.

14 The Cretaceous System

Classification

The Cretaceous System takes its name from the thick development of chalk of Upper Cretaceous age. However, the earliest beds of the Cretaceous were laid down in the same general area as were the higher Jurassic beds and were initially of similar facies. During Cretaceous times spectacular palaeogeographical changes took place.

The base of the Cretaceous, taken for many years as the junction between the Purbeckian and the Wealden Beds (a junction seldom well exposed), is now taken as a rather earlier major marine incursion which found its expression in southern Britain as the readily recognized Cinder Bed of the Middle Purbeckian. Further north it is represented by a nodular bed between the Lower and Upper Spilsby Sandstone. The problems of defining the top of the Cretaceous by fixing a Cretaceous/Tertiary boundary are frustrated in Britain by the absence of both the higher beds of the Chalk and the earliest Tertiary. Here, a major hiatus occurred so that the Tertiary beds of Eocene age rest unconformably on eroded Chalk. The Cretaceous is divided into the Lower and Upper and these are each divided into six stages although it is common practice to combine some of these as in Table 14.1.

Early Cretaceous geography and conditions of deposition

The Upper Purbeck Beds, shales with fibrous calcite ('beef') and thin fossiliferous freshwater limestones known as Purbeck Marble—still quarried as ornamental stone in the Swanage district—were deposited in the same general environment as the Lower and Middle Purbeckian strata. The Upper Purbeckian, both in Dorset and in the Wealden area, grades upwards into the Wealden Beds which were also laid down in a predominantly non-marine environment. However, the Wealden Beds are of a more deltaic character

TABLE 14.1

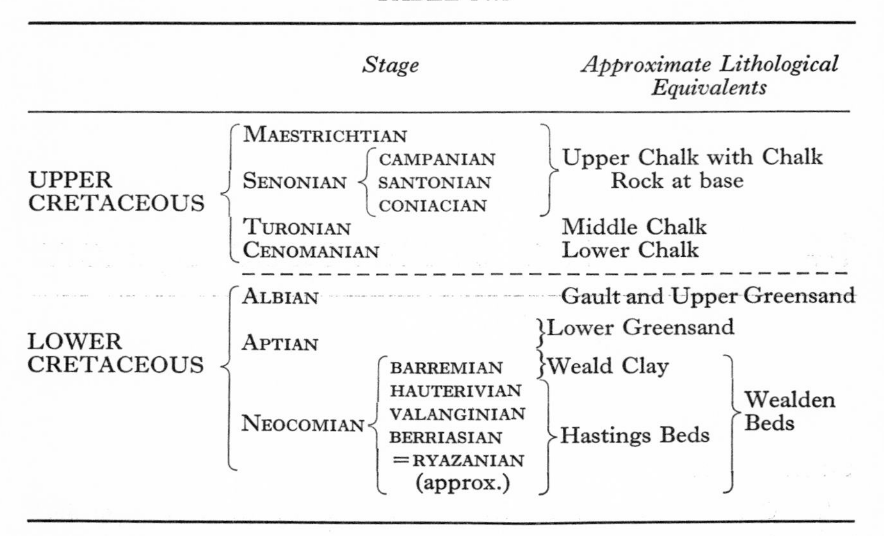

	Stage		*Approximate Lithological Equivalents*	
UPPER CRETACEOUS	MAESTRICHTIAN		Upper Chalk with Chalk Rock at base	
	SENONIAN	CAMPANIAN		
		SANTONIAN		
		CONIACIAN		
	TURONIAN		Middle Chalk	
	CENOMANIAN		Lower Chalk	
LOWER CRETACEOUS	ALBIAN		Gault and Upper Greensand	
	APTIAN		Lower Greensand	
	NEOCOMIAN	BARREMIAN	Weald Clay	Wealden Beds
		HAUTERIVIAN	Hastings Beds	
		VALANGINIAN		
		BERRIASIAN =RYAZANIAN (approx.)		

indicating, probably, rejuvenation of the London-Belgian Ridge, which comprised uplands over 1,000 feet high forming a source of supply of detritus. This ridge separated the southern province of dominantly non-marine deposition from the subsiding basin of marine sedimentation which included part of Norfolk and Lincolnshire, although these had been temporarily united in Cinder Bed times (Fig. 14.1). The Wealden Beds overlap the Purbeckian on to older strata in northern France while in Germany they pass laterally into contemporaneously deposited marine beds. The most typical development of the non-marine facies occurs, of course, in the Weald. The deposits there were laid down in a subsiding area to the south of the London-Belgian Ridge, the geography of which has been worked out in detail from the provenance of pebbles. These indicate an upland composed of Upper Palaeozoic rocks with, undoubtedly, outcrops of Upper Jurassic (Fig. 14.2). The Wealden Beds, formed as silts, gravels and clays, were deposited partly as alluvial flats and as deltas built out into a lake or lagoon (although the concept of 'The Wealden Lake' was long ago abandoned in the belief that the Wealden area was not completely isolated from the Tethyan sea across southern Europe). They are cyclic sediments; the coarsely cross-bedded sands show evidence of transgression by lacustrine sediments which commenced with a pebble bed and graded into finer silts and sandstones of the delta complex. This cyclic character is conspicuous in parts of the succession and appears to reflect eustatic changes. Megacyclothems have also been defined. Four in number, they are essentially alternations of arenaceous and more argillaceous strata. The Wealden Beds accumulated to a great thickness in this slowly downwarping basin, perhaps over 2,500 feet (Table 14.2).

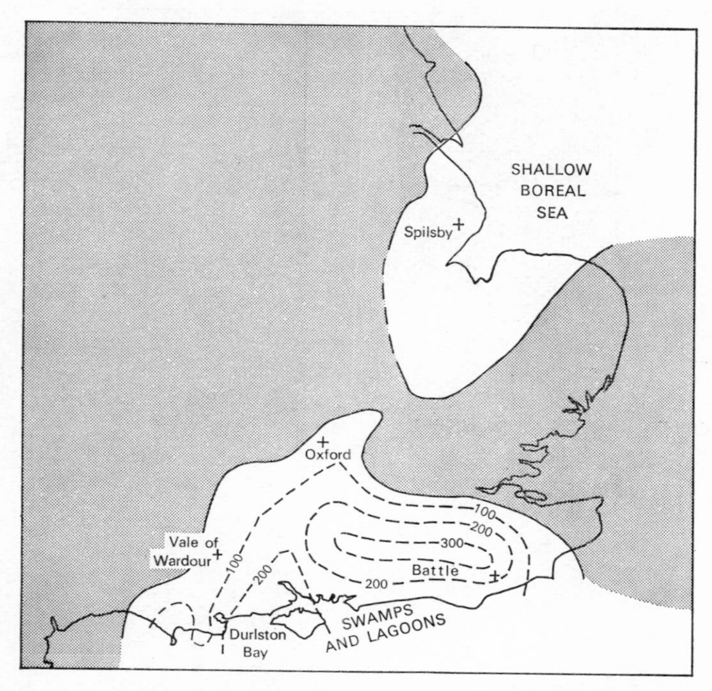

Fig. 14.1 Palaeogeography immediately preceding Cinder Bed times, the beginning of the Cretaceous Period, with isopachytes for the youngest Jurassic rocks (Basal Purbeckian to Cinder Bed). (Based on R. CASEY, 1963, *Bull. S.-East Un. scient. Socs.*, **117**, 1; isopachytes from F. HOWITT, 1964, *Q. Jl geol. Soc. Lond.*, **120**, 77.)

The Wealden of south-east England

TABLE 14.2

	Weald Clay including Horsham Stone, up to 1,500 ft	
Hastings Beds	Upper Tunbridge Wells Sand	400 ft
	Grinstead Clay	
	Lower Tunbridge Wells Sand	
	Wadhurst Clay	100–230 ft
	Ashdown Sands with Fairlight Clay	500–700 ft

The Ashdown Sands, named from the Ashdown Forest area, are fine grained sandstones and siltstones of deltaic origin. Near Hastings they pass into the Fairlight Clays which were laid down in front of the delta. Both the Ashdown Sands and Fairlight Clays are well seen in the coastal section though the Ashdown Sands are more argillaceous than in inland exposures.

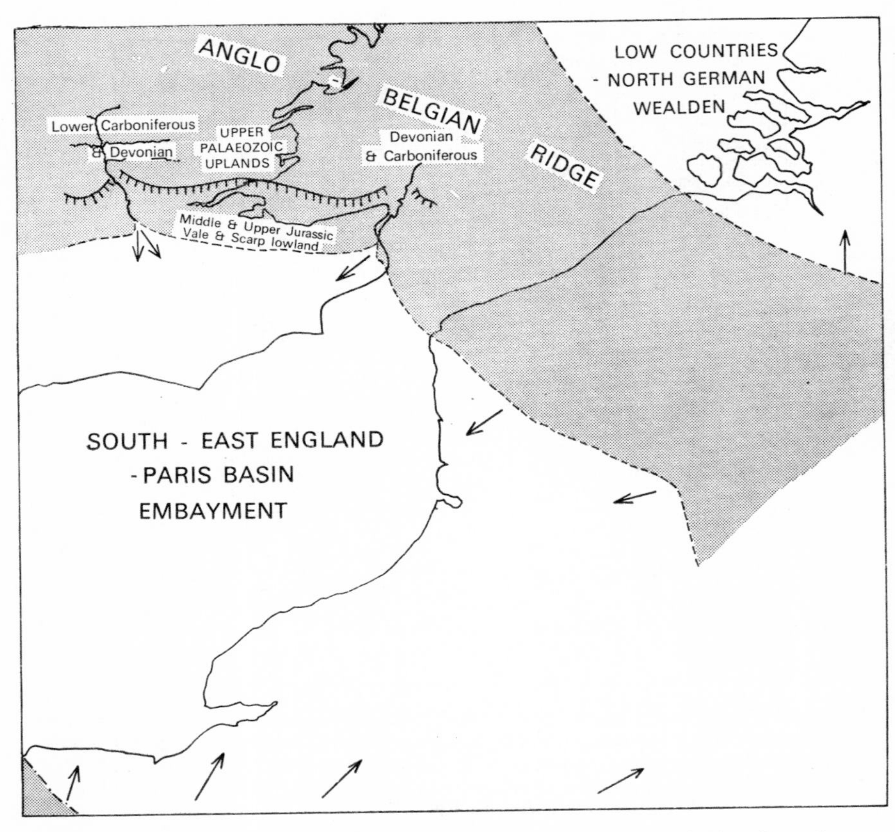

Fig. 14.2 Wealden palaeogeography (Berriasian-Valanginian). (Based on several sources including P. ALLEN, 1954, P. KAYE, 1966 and J. F. KIRKALDY, 1963.)

However, they exhibit the typical cross-bedding and festoon bedding, ripple markings and rain pitting, all of which confirm their deposition in shallow water. Lignite formed from driftwood and rather rare and poorly preserved fossils (except for good plant fossils in the Fairlight Clays) also testify to the deltaic environment. The Wadhurst Clay comprises chiefly shales and mudstones of lacustrine origin deposited on top of the deltaic earlier Wealden Beds. Calcareous siltstones also occur, reflecting periods of transportation of coarser material—and these currents gave rise to flute and groove casts in the deposited sediments. The calcareous siltstones, known as Tilgate Stone, show ripple marks and rain prints and the casts of dinosaurs' footprints including those of *Iguanodon*. Clay ironstones occur in the Wadhurst Clay and were, from the Iron Age until the late Middle Ages, the main source of iron ore in Britain which was smelted using charcoal from the then heavily forested

Wealden area. Freshwater lamellibranchs and ostracods indicate a generally brackish environment and plant fossils are common at some horizons, although marine microplankton has also been found. The Tunbridge Wells Sand represents a return to conditions of deposition closely resembling those of the Ashdown Sand, fine silts and sands being laid down as deltas advancing into the Wealden lakes. Except in the eastern part of the Weald, the Tunbridge Wells Sand includes the Grinstead Clay. This is largely argillaceous but with a thick calcareous Tilgate Stone (like those of the Wadhurst Clay) from which the first *Iguanodon* was discovered. The Grinstead Clay represents only a temporary break in deltaic deposition and the Upper Tunbridge Wells Sand, though poorly exposed, appears to closely resemble the Lower.

Towards the end of the deposition of the Hastings Beds the uplands to the north had been greatly reduced by erosion and the highest member, the Upper Tunbridge Wells Sand, grades upwards into the Weald Clay. This formation is much more argillaceous and consists predominantly of silty greenish-grey clays, although brown and reddish clays occur, as well as some beds of siltstone and occasional shelly limestone—the '*Paludina* Marble'. The resemblance to the clays of the Hastings Beds is close but, in late Weald Clay times, a transition to marine conditions is indicated by the replacement of the typical non-marine fauna by marine lamellibranchs in the highest Wealden Beds. Cyclothems have been described in the Weald Clay of the Maidstone area but in general the evidence is of a singularly static environment of deposition for, despite some variations in lithology, there is little or no evidence of emergence. The high percentage of silt in the clays has led to a comparison with some modern pro-delta sediments. The sandstones within the Weald Clay are generally thin and have sharply defined bases but grade upwards into siltstone and clay. Only the Horsham Stone attains any considerable thickness.

The Weald Clay is poorly exposed in south-east England giving rise to a low-lying belt surrounding the High Weald of the Hastings Beds. The best exposures are seen in brickwork claypits. Westwards the Weald Clay thickens from only 400 feet at Hythe to 1,500 feet near Guildford.

West of the Wealden area

In Dorset, in Swanage Bay, the Wealden is only slightly less well developed than in the Weald, but it decreases rapidly further westwards, becoming more arenaceous, to half the thickness at Worbarrow Bay and to a mere 350 feet at Upwey near Weymouth. The Wealden Beds are also well seen in the Isle of Wight in the anticlinal cores. In these westerly outcrops the lithological divisions of the Wealden area cannot be recognized nor is there evidence of the megacyclothems. The Wealden is divisible only into a lower dominantly arenaceous division and an upper argillaceous division, the Wealden Shales.

Although the sea had retreated from much of Britain at the end of Kimmeridgian times, sedimentation appears to have been essentially continuous in south-east England and Wessex during late Jurassic and Lower Cretaceous times. However, further north occur deposits which rest unconformably on the Kimmeridge Beds but which were clearly deposited prior to the widespread Upper Cretaceous marine transgression. These beds have been the subject of great controversy. There are, in Buckinghamshire and Wiltshire lying to the west of the Lower Greensand, ferruginous beds called the Whitchurch Sands. Though resembling the Wealden Beds of Dorset, and they have been referred variously to the Lower Greensand, the Wealden, and the Portland Sand in the past, faunal and lithological evidence infer their equivalence to the Cinder Bed of the Middle Purbeckian.

The Northern Basin

In Lincolnshire and Norfolk an embayment of the sea was separated from the southern area of non-marine deposition by the London-Belgian Ridge. In this subsiding basin a thick succession of marine Lower Cretaceous beds accumulated (Table 14.4). The lowest formation, the Spilsby Sandstone with a basal nodule bed, rests unconformably on the Kimmeridge Clay, but the hiatus must have been of short duration for sedimentation recommenced before the close of the Jurassic. The Lower Spilsby Sandstone is equivalent to the lower part of the Purbeck since a nodule bed which occurs between the Lower and Upper Spilsby Sandstone is accepted as the equivalent of the Cinder Bed and Whitchurch Sands. All are products of the same marine incursion. It was brief, for throughout Wealden times (Ryazanian to Barremian and perhaps Lower Aptian) the sedimentary basins remained separate and the Wealden area remained non-marine in character.

The Lower Greensand

Late in the deposition of the Weald Clay there is evidence of the transition through brackish conditions to marine as eustatic rise in sea level extended the Tethyan sea northwards into this area of lakes and deltas. The succeeding Lower Greensand marked the end of a long phase of non-marine sedimentation. Nevertheless the sea still lapped against the London-Belgian Ridge to the north and the Lower Greensand is known to be absent from north Kent and beneath the London Basin. There is in the Wealden area a slight disconformity at the base of the Lower Greensand, and eastern Kent was not initially submerged by the Lower Greensand sea.

The Lower Greensand is more arenaceous than the preceding sediments but it is seldom green although it is glauconitic. It comprises up to 800 feet of beds, attaining this maximum in the Isle of Wight, deposited in near-shore marine environments. It is divided on a lithological basis into four divisions which vary in thickness towards the London-Belgian Ridge as well as in an east—west direction. In addition, the lithological divisions have been shown to be diachronous.

The northern and southern provinces of deposition, separate in Lower Aptian times although marine sediments were deposited in both, became united in Upper Aptian times (Fig. 14.3) but essential differences remained between the deposits of the Northern Basin north of the Oxford Axis and those of the Southern Basin (Table 14.3).

TABLE 14.3

Stages	Zones	*Southern Basin* — *Isle of Wight (Vectian Prov.)*	*Southern Basin* — *Surrey/W. Kent (Wealden Prov.)*	*Northern Basin*
LOWER ALBIAN	*Mammilatum*	Carstone	Nodular Bed	Carstone
		------------ NS	------------ NS	Sutterby Marl with derived fossils of Upper and Lower Aptian Age
	Tardefurcata	Sandrock	Folkestone Beds	
UPPER APTIAN	*Jacobi*			
	Nutfieldensis	Ferruginous Sands	Sandgate Beds	
	Martinoides			
LOWER APTIAN	*Bowerbanki*		Hythe Beds	
	Deshayesi			
	Forbesi	Atherfield Clay	Atherfield Clay	
	Fissicostatus			
BARREMIAN		Wealden Shales	Weald Clay	Fulletby Beds

(NS = non-sequence)

The Southern Basin

In the Wealden Province shales and mudstones with iron concretions and occasional sandy beds form the lowest division, the Atherfield Clay. It is highly fossiliferous with many ammonites and lamellibranchs. Upwards the Atherfield Clay grades into the sandier Hythe Beds which vary very considerably both in thickness and lithology. Composed chiefly of glauconitic or ferruginous sands, rubbly sandy limestones known as Kentish Rag are locally developed. Chert is important in the western outcrops and so reinforces the Lower Greensand that it forms a higher scarp, at Leith Hill, than that of the Chalk Downs nearby. The fauna of the Hythe Beds includes echinoids which are common as well as ammonites, while occasional dinosaur remains have been found.

The Sandgate (and Bargate) Beds which follow are the most variable group of the Lower Greensand. They consist of glauconitic sandy and silty mudstones, with calcareous sandstone (the Bargate Beds) locally developed in the western Weald at their base. Derived Jurassic fossils have been found as well as the abundant indigenous fauna. The Folkestone Beds, quartzose sands,

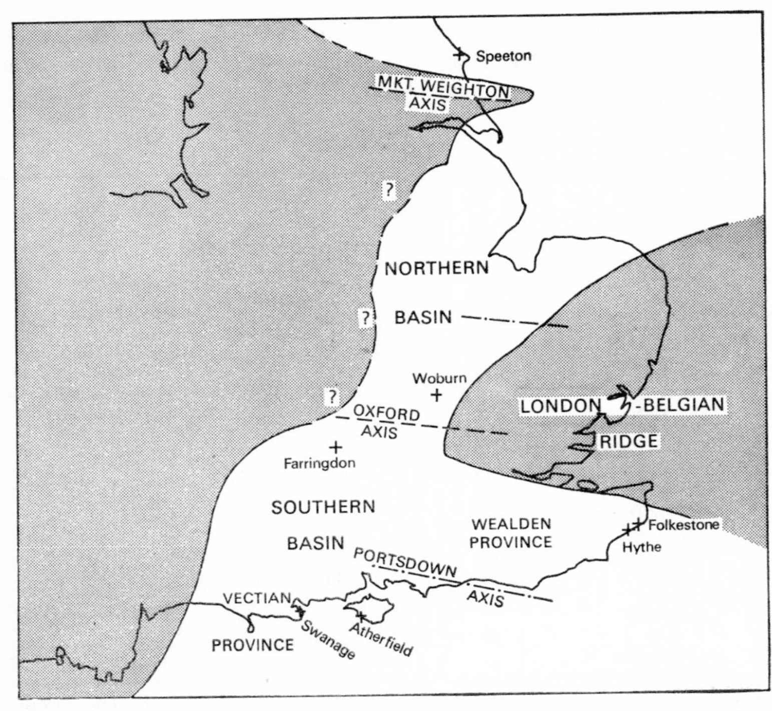

Fig. 14.3 Provinces of the Lower Greensand (Upper Aptian). More detailed palaeogeographical maps of Lower Greensand times are to be found in J. F. KIRKALDY, 1963, *Proc. geol. Ass.*, **74,** 127. (Simplified from R. CASEY, 1961, *Palaeontology*, **3,** 487.)

sometimes limonitic and often sparsely fossiliferous, complete the Lower Greensand succession. This last group is sometimes indistinguishable from the overlying Gault but faunal evidence places it in the Upper Aptian and Lower Albian. The cross-bedded sands include occasional dreikanter pebbles which have led to speculation on the origin of the Folkstone Beds, but there can be no doubt that with the rest of the Lower Greensand they are marine sediments.

In the Vectian Province, which includes outcrops in the Isle of Wight and Dorset, the Lower Greensand was divided over a century ago into 16 divisions. These conveniently group into 4, though they do not correspond to the four groups of the Wealden Province Lower Greensand. However, the Atherfield Clay (named from Atherfield in the Isle of Wight) is typically developed. The '*Perna*' Bed occurs at the base, as in the western Weald, while beds of concretions, the Crackers, mark the top. Corresponding to both the Hythe and Sandgate Beds there follows a thick series of Ferruginous Sands with some very distinctive fossiliferous beds. The Sandrock Beds, 100 feet or more of sands, are similar to the Folkestone Beds, their equivalent in the

Weald. The Carstones, brown pebbly grit, were laid down after a brief interval of non-deposition.

In this Vectian Province the Lower Greensand becomes attenuated westwards at a more rapid rate than the Wealden (Neocomian) Beds, although the diminution may be in part due to subsequent strike faulting. The sea spread rapidly throughout the Southern Basin initially, except over eastern Kent, but the Hythe Beds and the Ferruginous Sands represent a great influx of arenaceous material and a slight withdrawal of the sea, preceding the somewhat later major marine transgression of Middle Albian times. The attenuation of beds to the west in Wessex indicates greater stability of that area, and the so-called Punfield Marine Band of Swanage Bay has a distinctive fauna consequent upon its more brackish environment near to margins of the Lower Greensand transgression.

Western outcrops

Minor outcrops of Lower Greensand age are found in Berkshire. Among the most interesting is the Farringdon Sponge Gravel which has yielded an abundance of calcareous sponges. In Bedfordshire the Woburn Sands attain a thickness of 200 feet and overstep on to Ampthill and Oxford Clays. They include an indigenous fauna and many derived Jurassic fossils.

The Northern Basin

In north Norfolk and Lincolnshire a thick series of beds was laid down succeeding the Spilsby Sandstone in a subsiding basin bounded to the north by the still effective Market Weighton Axis. The succession beneath the Chalk (i.e. the complete Lower Cretaceous succession) is given in Table 14.4. Although this succession thickens southwards in Lincolnshire, the Lower Cretaceous succession in west Norfolk is thinner and these divisions cannot be recognized. There, beneath the Red Chalk, the succession is Carstone, Snettisham Clay and Sandringham Sands, the latter resting unconformably on the Kimmeridge Clay.

In Yorkshire, north of the Humber only thin pre-Chalk strata occur, the Red Chalk and Carstone—the latter resting unconformably on the Lower Lias. However, on the north side of the Yorkshire Wolds a narrow outcrop of clays is found and it extends to the coast near Speeton. These clays, some 300 feet thick and known as the Speeton Series, range in age from the Ryazanian to the Albian. Contemporaneous with the upper part of the Spilsby and the Tealby Series of Lincolnshire, the Speeton Series differs completely from those littoral arenaceous sediments and the Market Weighton Axis may have acted as a barrier preventing the spread northwards of terrigenous material. Certainly, the Speeton Series represents a very slow rate of accumulation since it spans almost the entire Lower Cretaceous. Despite this fact, sedimentation was occasionally interrupted by non-sequences and there is no sugges-

TABLE 14.4

		Approximate equivalents in Southern England
UPPER ALBIAN	Red Chalk	Gault and Upper Greensand
MIDDLE ALBIAN	Carstone, sandstone and sandy marl	
LOWER ALBIAN	Sutterby Marl, pyritous marl—sandy at top	Lower Greensand
APTIAN	Fulletby Beds, ferruginous sands and clay (Tealby Series)	
BARREMIAN	Upper Tealby Clay, Tealby Limestone and Lower Tealby Clay (Tealby Series)	Weald Clay (Wealden Beds)
HAUTERIVIAN	(Tealby Series)	Hastings Beds (Wealden Beds)
VALANGINIAN	Claxby Ironstone and Hudleby Clay (Tealby Series)	Hastings Beds (Wealden Beds)
RYAZANIAN-PURBECKIAN	Upper Spilsby Sandstone	Upper Purbeck Beds

tion that it took place in deep water. On the other hand, littoral deposits are unknown.

A basal coprolite bed containing derived Jurassic ammonites rests unconformably on the Kimmeridgian, and the succeeding beds were laid down in shallow water for current action affected them. Gentle downwarping of the

TABLE 14.5

		Yorkshire		*Lincolnshire*
LOWER CENOMANIAN / UPPER ALBIAN		Red Chalk		Red Chalk
MIDDLE ALBIAN			A.1	
LOWER ALBIAN	Speeton Series	Gault	A,2–4	Carstone
UPPER APTIAN	Speeton Series	*Ewaldi* Marl	A.5	Sutterby Marl
LOWER APTIAN	Speeton Series			
BARREMIAN	Speeton Series	*Brunsvicensis* Beds	B.1–5	Tealby Series
HAUTERIVIAN	Speeton Series	*Jaculoides* Beds	C.1–11	Tealby Series
VALANGINIAN	Speeton Series	*Subquadratus* Beds	D.1–8	Tealby Series
	Speeton Series	*Lateralis* Beds		Spilsby Series
RYAZANIAN	Speeton Series	Coprolite Bed	E	

The numbering of beds from the top of a succession downwards, as in this case, is quite unusual but it is a long established and highly satisfactory basis for the continued work on these beds.

basin led to the accumulation of clays in deeper water, though nodular beds infer periodic shallowing. The shales are often dark and pyritous but show a wide range of lithology and the higher part of the succession is lighter coloured and includes calcareous nodules. Ammonites are largely restricted to thin fossiliferous beds but belemnites are more uniformly distributed through the sequence. They form the basis of a fourfold division (Table 14.5). The bed overlying the Speeton Series, the *Ewaldi* Marl, which had generally been regarded as the approximate equivalent of the Sutterby Marl of Lincolnshire (indeed both are of Aptian age), possesses a quite different fauna and no species is common to both.

The Gault and Upper Greensand

The Weald

Over the whole of southern England a widespread marine incursion responsible for the deposition of the Gault Clay and the Upper Greensand occurred. It spread far beyond the confines of the Lower Greensand (Aptian-Lower Albian) basins, uniting them in the same facies, and spreading far to the west and north. For the first time during the Mesozoic era marine sedimentation spread across the London-Belgian Ridge. As a result, the Gault, which in the Wealden area rests on the Folkestone Beds of the Lower Greensand, northwards oversteps on to Jurassic rocks and beyond on to Palaeozoic rocks in north Kent and beneath London. The Gault in eastern Kent is entirely argillaceous and the classic exposure or type section at Folkestone is much confused by landslipping. Eastwards siltstones and sandstones become more important and in the western Wealden area they come to occupy the whole of the Upper Gault. Known as the Upper Greensand, they are a sandy facies of the Gault and the onset of arenaceous conditions was diachronous. The Gault and Upper Greensand together were deposited during the Middle and Upper Albian. This twofold division of the Gault-Upper Greensand is an oversimplification for, as well as sand and clay, a third lithology known as Malmstone occurs. This name is given to siliceous sandstone and calcareous sandstone which are well developed in the western Weald and in Wiltshire and Berkshire between the Gault Clay and the sands of the Upper Greensand. It dies out northwards. The Gault-Upper Greensand formations make generally low-lying ground between the escarpments of the Chalk and the Lower Greensand so that exposures are few. However, the Upper Greensand is seen in some of the sunken lanes.

The fauna of the Gault Clay is abundant, molluscs predominating. Ammonites, including the rapidly evolving Hoplitidae, are common and are the zone fossils but lamellibranchs and gastropods are also abundant, though they differ from the thick-shelled forms of the Greensand near-shore environment.

The Western outcrops

To the west of the Weald, in the Isle of Wight and Dorset, the Gault is sandier and it becomes increasingly difficult to establish a Gault-Upper

Greensand boundary. The transgressive nature of the Gault is clearly demonstrated by the westwards overstep on to successively older beds (Fig. 14.4). In the Isle of Wight and in Swanage Bay the Gault succeeds the Lower Greensand. In fact, the Carstone is a sandy Albian age deposit resting with non-sequence on the Sandrock Series of the Lower Greensand while grading upwards into the Gault. Traced westwards the Gault is found resting on the Forest Marble at Abbotsbury, on the Lias between Lyme Regis and Seatown and, further west, on the Trias. The Gault with the overlying well developed

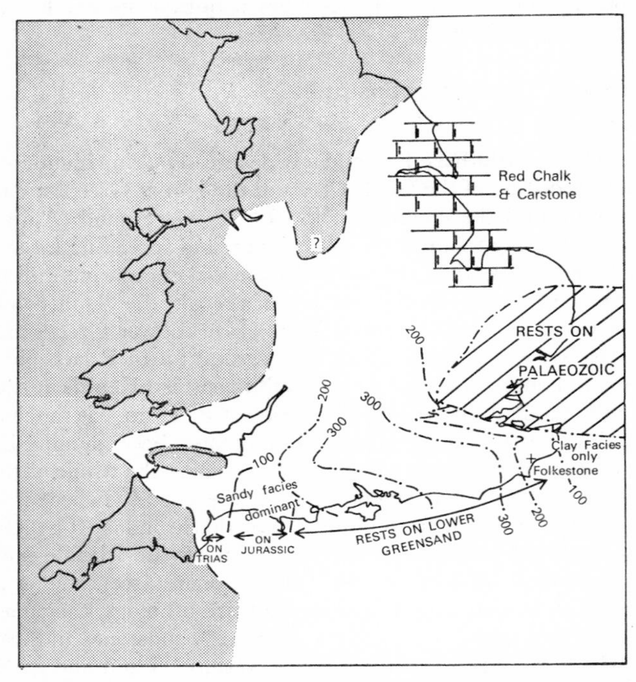

Fig. 14.4 The probable palaeogeography of Gault-Upper Greensand times (Middle-Upper Albian). (Isopachytes based on S. W. WOOLDRIDGE and D. L. LINTON, 1938, *Proc. geol. Ass.*, **49**, 264.)

Upper Greensand caps hills such as Black Ven, Stonebarrow and Golden Cap. The most westerly outcrops are the outliers of the Haldon Hills south-west of Exeter where the beds are entirely sandy. In Devon and Dorset the sandy facies is called the Blackdown Sands.

The Gault Clay west of the Chilterns is characteristically stiff dark clay varying in thickness but up to 230 feet. It rests on beds of Lower Greensand age where they are present, but in the Vales of Pewsey and Wardour oversteps on to Purbeck and Portland Beds. South-west of Cambridge pre-Albian uplift occurred so that the Gault, traced northwards, is seen to rest on Kimmeridgian, Ampthill Clay and Oxford Clay and then these beds in the reverse order. Late in Albian times this area was again elevated bringing the sea floor into a zone of tidal scour. This resulted in erosion of already deposited Gault-Upper Greensand Beds, and their contained fossils were rolled and concentrated in the slowly accumulating chalk marl. This unusual development, known as the Cambridge Greensand, although no more than a foot thick is of particular interest. Resting on eroded lower zones of the Gault, and filling hollows in the eroded surface, it includes pebbles and boulders of uncertain provenance, as well as derived fossils and phosphatic nodules (coprolites). The abundant fauna includes many species of ammonites and brachiopods as well as vertebrate remains.

Eastern England

The Gault, like earlier formations, is largely obscured by Fenland alluvium, but beds of this age appear in north Norfolk and are known from Lincolnshire and east Yorkshire. In Norfolk the sandy Carstone is followed by the Hunstanton Red Rock – Red Chalk. The latter is gritty at the base with similar heavy minerals to the underlying Carstone, although faunally it is quite different. It is essentially a compact red limestone. It is found northwards through Lincolnshire and Yorkshire—and its outcrop is continuous over the Market Weighton Axis. Overlying the Speeton Clay, it here extends into the Cenomanian, this rather exceptional environment persisting later than to the south. The red coloration is original but, since the abundant brachiopod-molluscan fauna proves a neritic marine sedimentary environment, it is assumed that the source of terrigenous material was lateritic. The position of the shoreline in Middle and Upper Albian times is uncertain except in southwest England, although it lay far to the west of present outcrops (Fig. 14.4). However, it was not until the Upper Cretaceous that sedimentation spread to Northern Ireland where sandy marginal facies, although glauconitic 'greensand', are of Cenomanian age.

The Upper Cretaceous

The Chalk is synonymous with the Upper Cretaceous beds in England. This unusually constant limestone sequence which outcrops extensively in the Downs and from the Dorset coast to the Yorkshire coast, forming the Chilterns and Yorkshire Wolds, generally gives rise to high ground (over 400 feet). It is relatively resistant to erosion compared with the majority of the Lower Cretaceous beds or the overlying Lower Tertiaries. As much as 1,500 feet thick, naturally lithological variations are present in the succession and formed the basis for the original division into Lower, Middle and Upper Chalk. These correspond closely to the stage names commonly used in Britain, Cenomanian, Turonian and Senonian, although the uppermost Chalk—found in Norfolk—is of Maestrichtian age (Table 14.6). Considerable variations in the thickness of Chalk now present partly reflect pre-Tertiary erosion (see p. 334) and recent erosion. There were, as well, original depositional variations in thickness. The Chalk overlapped beds of Albian age to the north and west but in the main outcrops of southern and eastern England it succeeds conformably the Upper Albian (Upper Greensand, Red Chalk, etc.) and there is a transition to the lowest argillaceous Chalk.

LOWER CHALK

The Lower Chalk is up to 200 feet thick in the Downs but rather less than this beneath the London Basin. Northwards it thins rapidly to no more than 60 feet in north Norfolk but in the Yorkshire Wolds reaches 150 feet. The lowest beds are locally cyclic, argillaceous bands alternating with purer bands. The insoluble material, as much as 50% in the lowest Chalk, decreases to no more than 10% towards the top of the Lower Chalk (compared with an average of about 5% insoluble material in the Upper Chalk). The lower zone, the *Varians* Zone, commences with the glauconitic 'Chloritic' Marl, sandy and nodular, which grades up into grey marly Chalk. The hard Totternhoe Stone marks the base of the upper zone, the *Subglobosus* Zone, and the Grey Chalk with many shell fragments follows. The upper part with common belemnites is called the *Actinocamax plenus* Subzone. Local variations in the Cenomanian beds of south Devon have been described in detail and related to contemporaneous uplift of minor periclinal axes.

MIDDLE CHALK

The lower zone, the *Cuvieri* or *Labiatus* Zone, of the Middle Chalk is marked throughout much of the outcrop by a harder basal bed, the Melbourne Rock. The remainder of the Middle Chalk is white, usually without flints except for the uppermost 30 feet (and in the Isle of Wight is completely without flints). The two zones, faunally distinctive, are lithologically similar. The fauna includes ammonites and echinoids, also common in the Lower Chalk, but brachiopods are predominant.

UPPER CHALK

The Upper Chalk attains a thickness of 1,300 feet in the Isle of Wight but is more complete, and perhaps even thicker, in north Norfolk. Its junction with the Middle Chalk is marked—in some areas—by another hard bed, the Chalk Rock, nodular in character and deposited in shallow water. The lower beds of the Upper Chalk contain the remarkable reussianum fauna, called after the ammonite *Hyphantoceras reussianum*, including a wide variety of molluscs. The fauna of the Upper Chalk as a whole is extensive and varied. Brachiopods are common, especially in the lower zones, and both regular and irregular echinoids are numerous. Of the latter, variations in *Micraster* and in *Echinocorys* have been shown to be of especial evolutionary interest. Ammonites are relatively few but lamellibranchs are common, although with the exception of *Inoceramus* none is of stratigraphical importance. The list of zones (Table 14.6) gives some guide to the variety of fossils present but corals, asterozoa, fish teeth and scales are not uncommon as well as many microfossils.

TABLE 14.6 Zones of the Chalk

Stages		*Zones*	
MAESTRICHTIAN		*Liostrea lunata*	
SENONIAN	CAMPANIAN	*Belemnitella mucronata*	
		Gonioteuthis quadrata	[*Inoceramus lingula*]
		Offaster pilula	
	SANTONIAN	*Marsupites testudinarius*	
		Uintacrinus socialis	
		Micraster coranguinum	[*Hagenowia rostrata*]
	CONIACIAN	*Micraster cortestudinarium*	
		Holaster planus	
TURONIAN		*Terebratulina lata*	[*Inoceramus lamarcki*]
		Orbirhynchia cuvieri	[*Inoceramus labiatus*]
CENOMANIAN		*Holaster subglobosus*	[*Holaster trecensis*]
		Schloenbachia varians	[*Holaster subglobosus*]

[ ] Alternative zone fossils used, especially in the Yorkshire Chalk.

Conditions of deposition

Undoubtedly the Cenomanian transgression, the spread of the Chalk Sea, was one of the most widespread incursions ever to take place in Europe. The greater part of Britain was inundated and evidence of shorelines is scanty even from the western outcrops but no doubt the Scottish Highlands were probably land as were the Ardennes and Brittany. The presence of littoral facies is rare. After the initial deposition of the lowest Chalk little terrigenous material was carried into the area of deposition, which led Bailey to suggest that waterless deserts bordered the Chalk Sea. The downwarping of the crust must have been widely uniform, though lateral variations do occur both in thickness of individual zones and in lithology. Shallower water conditions became established at times but there is no evidence that any area became emergent. The Chalk was laid down as a white calcareous mud and is very fine grained. Much attention has been given to the investigation of its mode of formation and possible origin but only recently has vital evidence, sedimentary and palaeontological, come to light. By a process of staining, biological structures (bioturbation) until now unsuspected have been revealed and infer deposition in shallow water. Once thought to be a foraminiferal limestone, perhaps analogous in origin to present-day oceanic foraminiferal oozes, it is known that foraminifera are not very common and nowhere exceed 15% of the rock. In many specimens they are absent. Believed to be largely an inorganic precipitate by some early workers, it has been found that precipitated aragonite, as found in present-day deposits, is rare. Chalk comprises many minute shell fragments (sometimes giving a 'gritty' texture) and microscopic and ultra-microscopic foraminiferal and coccolith fragments. The latter, revealed by electron microscope, make up the bulk of the fine matrix of the Chalk.

The evidence of the neritic benthonic fauna can be taken at its face value. The Chalk was a relatively shallow-water marine limestone deposited in water of, perhaps, a depth of 100 to 150 fathoms. Nodular beds represent shallower water deposition while marls indicate an increase in terrigenous material possibly consequent on the elevation of neighbouring land areas.

FLINTS

The origin of flints, such a characteristic and conspicuous feature of the Upper Chalk particularly, is still uncertain. Of silica, perhaps precipitated as gel, they may be an original depositional feature. However, they may have been formed subsequent to the deposition of the Chalk but before its consolidation, or they may be entirely secondary and have been formed much later by deposition from percolating ground-water, perhaps during the Tertiary Era. The absence of flints from the Lower Chalk where insoluble material is dispersed throughout the strata seems a highly significant fact, but is capable of more than one interpretation.

ERRATICS IN THE CHALK

The presence of boulders within the Chalk, especially in the absence of general sandy or detrital material, is puzzling. The majority of boulders come from the Cambridge Greensand at the base of the Chalk in Cambridge, but a considerable number have been found in Kent near Rochester and occasional erratic boulders are very widely dispersed through the Chalk. They represent an extremely wide range of petrological types, the commonest being vein quartz, chert and granite. Sedimentary rocks, including several different types of sandstone, volcanic rocks and rarely metamorphic rocks such as schist, have been described. Their provenance can seldom be given with certainty but many may have come from the west (from southwest England, north Wales and Shropshire). Some boulders are believed to have been carried from Sweden. The transport of boulders in the roots of floating trees has been suggested, and lignite has been found close to erratic boulders in Sussex, but many may have been carried as gastroliths (in the stomachs) of marine saurians and fish.

The Margins of the Upper Cretaceous Sea

The sea in Upper Cretaceous times spread far to the north and west, and in Northern Ireland glauconitic sands and chalk are preserved extensively beneath, and outcrop discontinuously around, the Antrim basalt plateau. The basal Cretaceous beds rest with great unconformity on strata of Dalradian, Triassic and Liassic age. The lower beds are glauconitic sands, sometimes known as the Hibernian Greensand, but the fauna includes numerous ammonites of the genus *Schloenbachia* suggesting an equivalence to the Chloritic Marl of southern England. These beds are undoubtedly younger than the Upper Greensand which lies beneath the Chalk in England and marine sedimentation did not spread into Ireland until Cenomanian times.

Initially sedimentation occurred in basins in south-east Antrim and in northern Derry. Deposition must have occurred in shallow water, and nodular beds occur and there is at least one break in sedimentation within the Cenomanian. The Turonian is scarcely represented by sediments but the sea spread back depositing glauconitic beds then chalk, the White Limestone of Northern Ireland. Zonally the most complete in the north at White Park Bay, higher zones overlap southwards and by *Mucronata* Zone times the Chalk sea covered all northern Ireland. The Chalk is known to be thickest west of Lough Neagh (over 400 feet) and the sea spread far to the south-west although the evidence is scanty. Chalk, forming the matrix of a breccia near Killarney in south-west Ireland, infers the presence of the Senonian sea and probably coincided with its maximum extent in Northern Ireland.

The evidence of the Cretaceous in western Scotland is derived from widespread scattered minor outcrops. Thickest in Morvern, where it is preserved beneath the Tertiary lavas—as in Antrim—a similar history can be deduced. Cenomanian age greensands lie unconformably on Lias or older rocks and are followed, after the Turonian period of non-deposition, by thin Senonian age Chalk. The pre-Chalk beds are partly glauconitic sands with calcareous beds, but pure white sandstone of Loch Aline (Morvern) is well known as a glass sand. The succession is generally similar in Mull, while less complete successions of Cretaceous are known from small outcrops in Skye, Raasay and Eigg. The minor outcrops of hardened chalk preserved by down faulting in a volcanic vent in Arran are well known. In eastern Scotland no Cretaceous rocks occur *in situ* but flints, sometimes with fossils, confirm the derivation of gravels from adjacent Cretaceous beds which floor the North Sea.

Cretaceous fossils

In the Lower Cretaceous and the Lower Chalk cephalopods are highly important in beds of marine facies and are used to zone the Speeton Series, Gault and lowest Chalk. In higher strata of the Chalk they are less common. In addition to representatives of long-ranged families such as the Phylloceratids which survived from the Trias, there are important groups such as the Hoplitidae which evolved rapidly and diversified during the Cretaceous Period. Unusually coiled ammonites are known such as *Turrilites* and those showing tendency to uncoiling, *Crioceras*, *Hamites* and the almost straight *Baculites*. As well as ammonites the dibranchiate cephalopods are stratigraphically useful.

Of the other molluscs gastropods are numerous in the Cretaceous but are stratigraphically unimportant. Lamellibranchs are abundant, especially oysters, and the coral-like Rudists are important, forming reef environments in Tethys.

Alcyonarian and scleractinian reef-building corals are not abundant in the British Cretaceous, in fact reefs are rare. On the other hand, solitary scleractinians are common. The abundant benthos also includes many calcareous sponges, and siliceous sponge remains and (especially in the Middle Chalk) brachiopods, chiefly terebratulids and rhynchonellids. The best known echinoderms are the irregular echinoids which are used as zone fossils of the Chalk, but regular echinoids, asterozoans and stalked crinoids are all common.

Much of the Lower Cretaceous is non-marine and the invertebrate fauna, seldom abundant, consists of freshwater gastropods and fresh- to brackish-water lamellibranchs. Land vertebrate fossils are chiefly dinosaurs, both herbivores and carnivores, and the large flying types (Pterosaurs) are occasionally found. The relatively uncommon mammals had reached the placental stage of development.

While the Cretaceous fauna, both vertebrate and invertebrate, is characteristically Mesozoic, great changes took place at the end of the Period and many forms do not survive beyond the Cretaceous. However, changes in the flora occurred much earlier. The plant fossils of the Early Cretaceous (the Wealden Beds) are chiefly conifers, cycads and ferns, and in general aspect the flora resembles that of the Jurassic. From mid-Cretaceous times onwards major changes in the flora became apparent and angiosperms oust the characteristically Mesozoic flora. Of course, non-marine strata of Middle Aptian age and later are not found in Britain and knowledge of the Upper Cretaceous flora is based on finds from far to the north in east Greenland and Spitzbergen.

VI The Cainozoic Era

15 The Tertiary System

Post-Cretaceous time has been divided into the Tertiary and Quaternary Eras, although the latter represents a mere 2 million years and the Tertiary is of very much shorter duration than the Mesozoic or Palaeozoic Eras. The Tertiary is best considered as a Period, following the lead of the American Commission on Stratigraphical Nomenclature of 1961.

A fivefold division of the Cainozoic (= Tertiary + Quaternary), proposed over a century ago by Lyell and based on the percentage of still living invertebrate forms, has been widely used, but with amendments such as the recognition of the Palaeocene. There is some merit and convenience, when dealing with strata of this age in western Europe, in grouping these divisions into the Palaeogene and Neogene as in Table 15.1.

TABLE 15.1

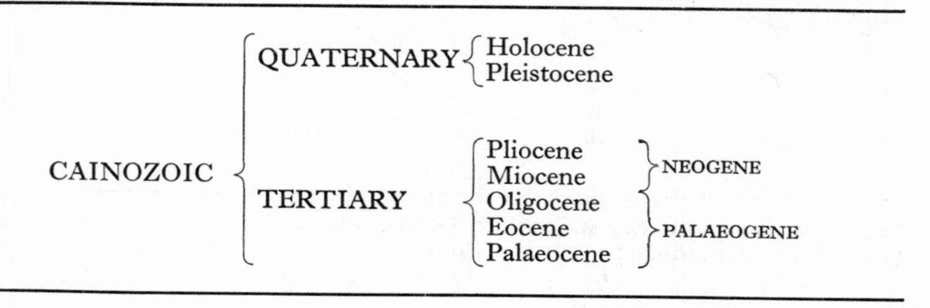

CAINOZOIC	QUATERNARY	Holocene	
		Pleistocene	
	TERTIARY	Pliocene	NEOGENE
		Miocene	
		Oligocene	PALAEOGENE
		Eocene	
		Palaeocene	

There is some justification for the elevation of the brief Quaternary to the rank of Period since there exists a vast amount of data, including that of the evolution of Man. It seems preferable to its inclusion in the Tertiary extending the latter to be synonymous with the Cainozoic (as done by some authors).

The Palaeogene

The Lower Tertiary unconformity

Great changes in geography marked the close of the Cretaceous Period. The Chalk sea withdrew from northern Europe and, in Britain, a long period of erosion occurred before the deposition of the earliest Tertiary beds. Despite the removal of great thicknesses of Chalk (over 500 feet in the region of

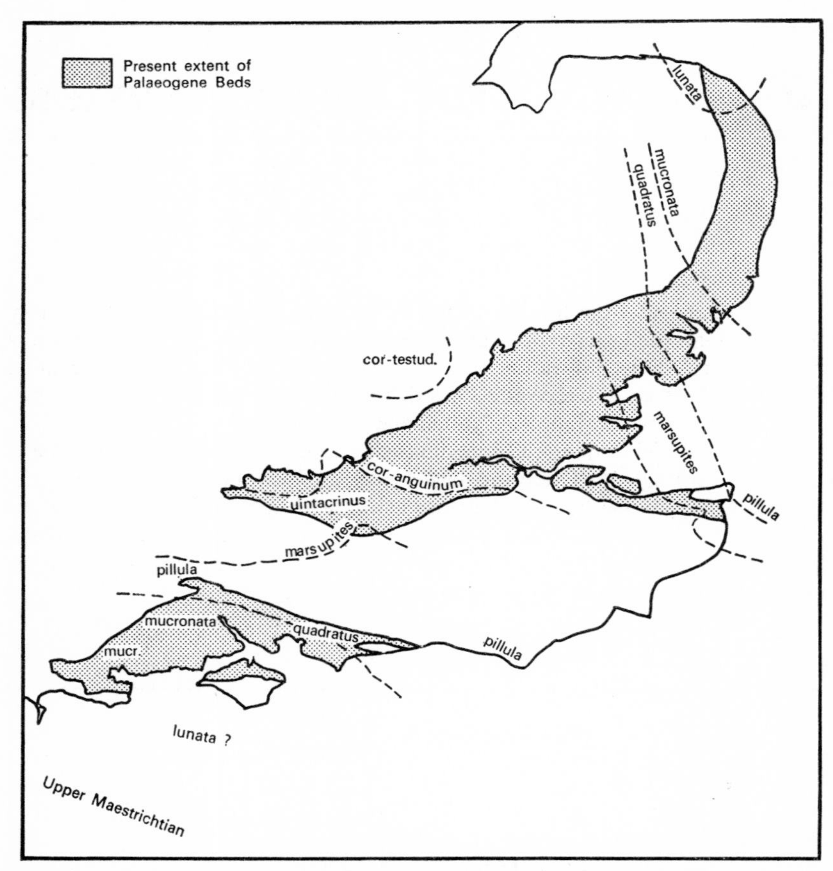

Fig. 15.1 Zones of the Chalk which lie beneath the Lower Tertiary (Palaeogene) Beds of south-east England. (From D. CURRY, 1965, *Proc. geol. Ass.*, **76,** 151.)

the London Basin) representing several zones (Fig. 15.1), the earth movements which caused the withdrawal of the Cretaceous seas and uplift of Britain and north-west Europe were slight compared with the Alpine Orogeny of Miocene times. As a result, although the lowest Tertiary beds rest unconformably on different zones of the Chalk, the discordance of dip is frequently slight. Sedimentation returned to south-east England with the formation of the Anglo-Parisian Cuvette in which were laid down beds of marine facies marginally interdigitating with deposits of continental facies. These strata are now preserved in four synclinal structures: the London Syncline (or 'Basin'), the Hampshire Basin, the Paris Basin and the Belgian Basin.

The crustal instability preceding the Miocene Orogeny gave rise to cyclic sedimentation. A concept of megacycles, first described by Stamp, neatly summarizes and explains in simplified form the interdigitation of beds of different facies (Fig. 15.2). It follows that it is impossible to give a completely typical succession for the Palaeogene of southern England because of the lateral variations in thickness and facies of beds laid down contemporaneously. A wide variety of lithologies is represented, although limestones are rarely present. The beds are usually soft and poorly cemented, bearing a general resemblance to the Lower Cretaceous beds of south-east England.

Despite differences in the successions found at different localities, the same general sequence of events, giving rise to the same major cycles of sedimentation, can be demonstrated from each of the four structural basins where Palaeogene strata are now preserved. This, together with similar faunal sequences, illustrates that the beds of the four basins were originally laid down in the one cuvette of deposition. Due to their positions relative to the margins of the cuvette, the London and Hampshire Basins differ and, in fact, closer comparison can be made between the London and Belgian Basins. The earliest transgression which reached the eastern part of the London Basin did not reach the Hampshire Basin. On the other hand, the succession is much more complete in Hampshire and the Isle of Wight since in the London Basin much of the sequence has been removed by post-Palaeogene erosion (Table 15.2).

TABLE 15.2

	'Stages'		*Formations*	*Megacycles*	
OLIGOCENE	RUPELIAN [STAMPIAN]		Upper Hamstead Beds	7	Absent from the London Basin (7–5)
	LATTORFIAN [SANNOISIAN]		Lower Hamstead Beds Bembridge Marls Bembridge Limestone Osborne Beds Upper and Middle Headon Beds	6	
EOCENE	BARTONIAN		Lower Headon Beds Barton Beds	5	
	AUVERSIAN LUTETIAN		Upper Bracklesham Beds Lower Bracklesham Beds	4	
	CUISIAN YPRESIAN		Lower Bagshot Beds London Clay	3	
	SPARNACIAN	LAN-DENIAN	Woolwich and Reading Beds	2	
	THANETIAN		Thanet Beds	1	

[1–7 = cycles of sedimentation]

Lower Tertiary of the London Basin

The earliest Tertiary beds found on the continent of Europe, the Danian and Montian which together form the Palaeocene (though some authors consider the Danian late Cretaceous), were never deposited in the British area. Sedimentation commenced with the inundation by the sea of the eastern part of the London Basin in Landenian times, depositing the Thanet Beds. Over the eroded Chalk surface spread a basal bed of rolled flint pebbles which was followed by beds chiefly of glauconitic sands and silts. The Thanet Beds are well exposed in the north Kent coast, including the Isle of Thanet where they attain a thickness of approximately 100 feet. They thin steadily westwards forming a narrowing outcrop along the north side of the North Downs as far

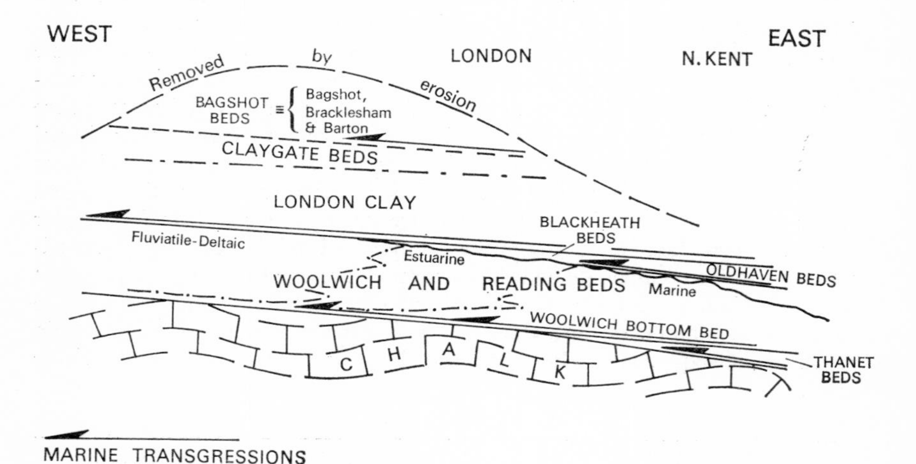

Fig. 15.2 Cycles of sedimentation in the Lower Tertiary beds of the London Basin. (Based partly on S. W. HESTER, 1965, *Bull. geol. Surv. Gt. Brit.*, **23,** 117.)

as East Clandon in Surrey. Their western extension as far as Weybridge and Sunbury has been proved in borings, but they are overstepped by younger beds. Contained fossils are all marine, chiefly poorly preserved molluscs and foraminifera. The latter include forms indicating deposition in water probably less than 150 feet in depth. Although not extensive (Fig. 15.3) the Thanet Beds have a number of characters typical of most of the marine transgressions of the Palaeogene, namely the basal pebble bed resting on a surface of marine erosion succeeded by beds which become thinner and sandier westwards.

The Woolwich and Reading Beds succeed the Thanet Beds in the eastern part of the London Basin but overstep westwards to rest on eroded Chalk in the west of the London Basin and in the Hampshire Basin. The Chalk surface bored by a marine worm testifies to the western extent of the marine transgression of the Woolwich Bottom Bed. The Woolwich and Reading Beds are a variable group of sands and clays, the Woolwich Beds facies is marine in the east of the London Basin grading westwards into estuarine or brackish beds. Yet further west gradation takes place into fluviatile and deltaic deposits of the Reading Beds facies. The nomenclature is thus somewhat confusing since three facies are present but two names, Woolwich and Reading, are employed. The Woolwich facies includes a freshwater shell bed. The

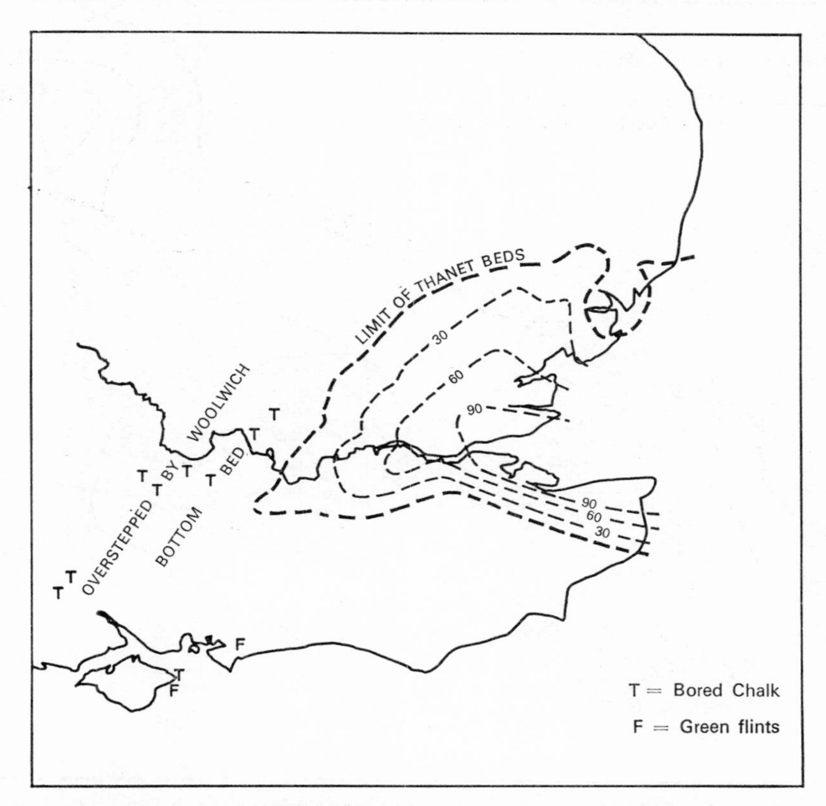

Fig. 15.3 The extent of the Thanet Beds marine incursion in south-east England and its overstep by the Woolwich Marine Bottom Bed. (After S. W. HESTER, 1965, *Bull. geol. Surv. Gt. Br.*, **23,** 117.)

Reading Beds facies is the most extensive of the three in the London Basin (Fig. 15.4). In the Hampshire Basin the westward gradation is seen, although much of the area is of Reading Beds facies, for in Sussex beds of freshwater or brackish facies are found. However, as beds of fluviatile origin are so widespread here they are commonly referred to simply as the Reading Beds. Variations in thickness of the Woolwich and Reading Beds are somewhat complex and they attain a maximum in the Southall-Brookwood area west of London and near Chichester.

After the initial Landenian marine transgression which deposited the widespread Woolwich Marine Bottom Bed the three facies discussed above existed simultaneously. However, it is probable that marine offlap continued, resulting in the spread of the fluviatile and deltaic facies. The marine facies may not have persisted throughout the Landenian even in the east of the London

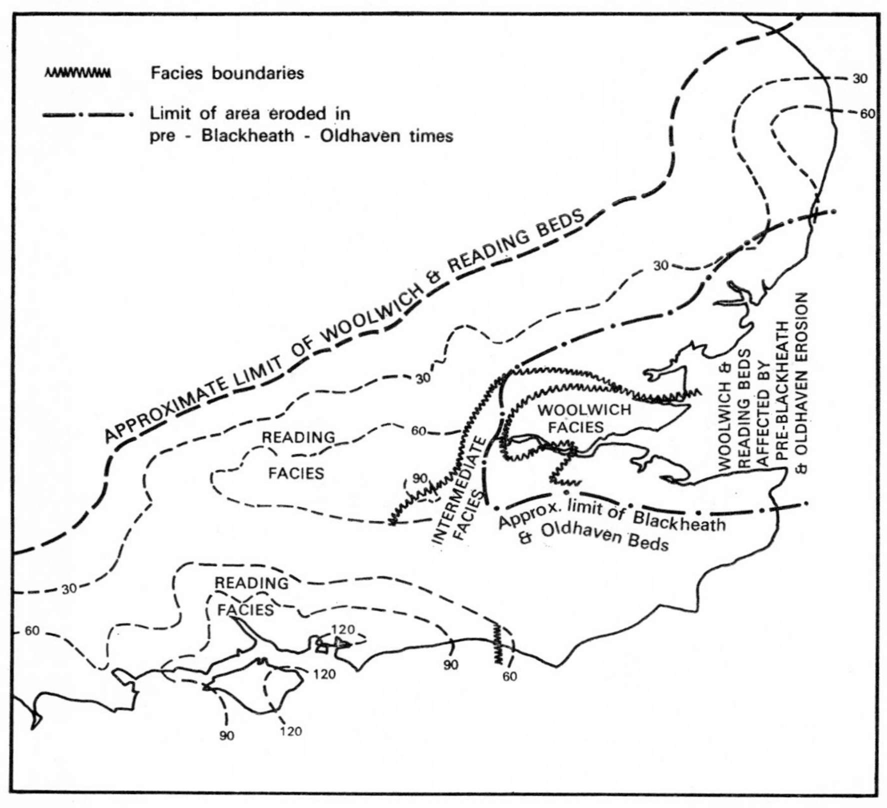

Fig. 15.4 The distribution of facies of the Woolwich and Reading Beds in south-east England. (Redrawn from S. W. HESTER, 1965, *Bull. geol. Surv. Gt. Br.*, **23,** 117, by permission of the Director, Institute of Geological Sciences.)

Basin, but the evidence is obscured as the higher beds were removed by erosion preceding the deposition of the Blackheath and Oldhaven Beds, discussed below.

Boulders, some measuring many feet, of cemented sandstone known as 'sarsens' are found scattered both in the present area of Tertiary outcrops and beyond. Some have been derived from the Reading Beds, others from higher horizons including the Bagshot Sands, while similarly cemented pebble beds are known as Hertfordshire Puddingstone. The Sarsen Stones of Stonehenge are believed to have been taken there from the Marlborough-Newbury area.

In the London Basin a further marine transgression laid down the Oldhaven Beds of north Kent and their equivalent further west, the Blackheath Beds. They rested on eroded Woolwich beds generally but southwards are found resting on Chalk, either as a result of southwards overstep (formerly cited as evidence of an already elevated Wealden Island) or more probably due to deposition in deeply cut estuarine channels. The fauna of these beds is marine in the eastern outcrops but west of Chatham brackish fossils appear and the proportion of brackish forms increases westwards. The pebbles may therefore also have been derived from the west and not from the Wealden Island postulated by Prestwich over a century ago. The concept of this island now finds little favour and there is no direct evidence of the elevation of the Wealden Axis in pre-Miocene times.

The Ypresian cycle commenced with the basement bed of the London Clay. The base of this cycle, like almost all others, is transgressive and cannot be taken as a true time-plane. Undoubtedly all lithological changes and facies boundaries in the British Palaeogene are diachronous. Fossiliferous horizons are probably of less than usual value in defining time-planes because of the relationship of faunas to facies and the interdigitation of marine and non-marine facies. The London Clay spread far to the west from the North Sea Basin and faunas infer a south-westerly connection with Tethys as well (Fig. 15.5). Certainly, it represents the greatest marine extension in southern England during the Tertiary Period. Typically a stiff blue clay, weathering brown, it is remarkably uniform in lithology, though it becomes sandier towards the west. Even so the middle beds of the London Clay remain more argillaceous and foraminifera confirm that they were laid down in deeper water than the lower or higher beds. The sandy upper beds which occur both in Essex and west of London are called the Claygate Beds which in Surrey consist of rapid alternations of sand and clay. In the Hampshire Basin the uppermost sandy beds of the London Clay are scarcely distinguishable from the overlying Bagshot Sands.

Molluscs are numerous in the London Clay but only at some localities and it often lacks fossils, suggesting that many have been destroyed resulting in the production of selenite crystals which are common. However, the lowest 50 feet of the London Clay may have been deposited in conditions too unaerated to permit molluscan life. Attempts have been made to zone the London Clay by means of both foraminifera and ostracoda and it has recently been pointed out that these fossils fail to substantiate the view that the London Clay transgression was diachronous. Up to 600 feet thick in the eastern part of the London Basin, the London Clay thins to 300 feet at Windsor and even less further to the west, a trend also seen in the Hampshire Basin where it thins from 300 feet in the eastern Isle of Wight to 80 feet in Purbeck. However, the position of the shoreline is not known with certainty. The remarkable flora of the London Clay, over 500 species of plants known chiefly from fruits and seeds, has been the subject of much published work.

It infers a land with tropical rain forest adjacent the London Clay sea, although the marine molluscs of the latter include forms which have a boreal affinity.

In the London Basin extensive outcrops of Bagshot Sands overlie the London Clay, giving rise generally to heath and woodland including Bagshot

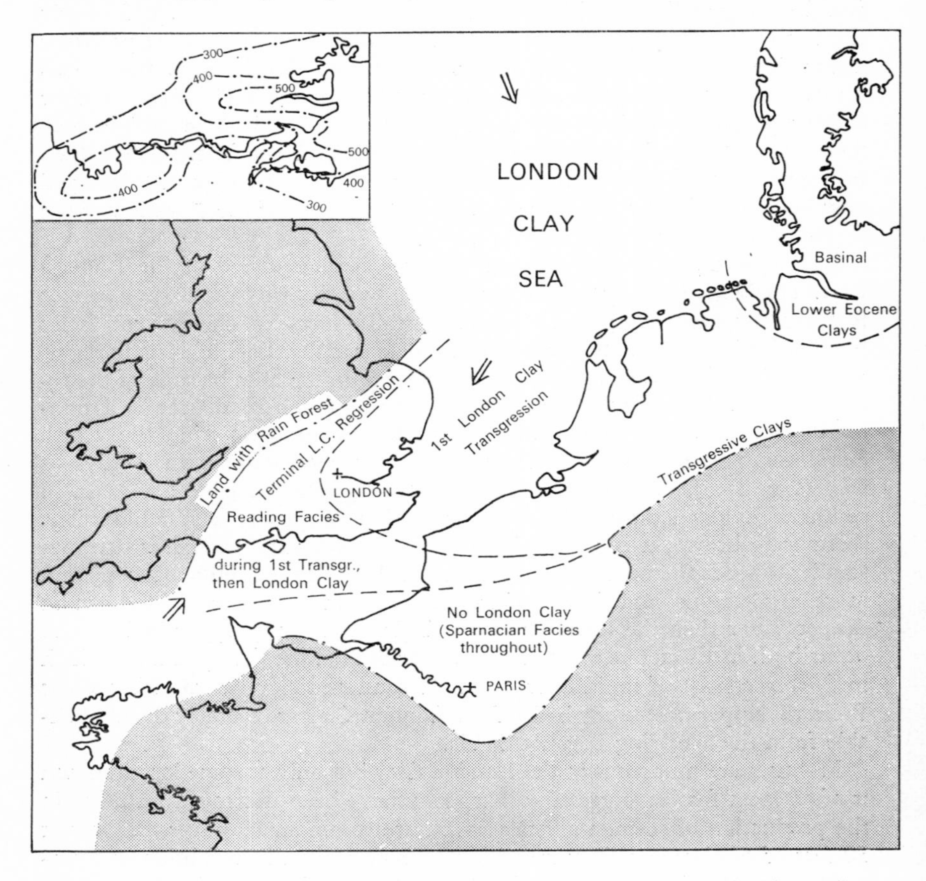

Fig. 15.5 The palaeogeography of the London Clay Sea. *Inset:* Isopachytes for the London Clay of the London Basin. (From A. G. DAVIES and G. F. ELLIOTT, 1958, *Proc. geol. Ass.*, **68,** 255. Isopachytes after S. W. WOOLDRIDGE and D. L. LINTON, 1938, *Proc. geol. Ass.*, **49,** 264.)

Heath and Hampstead Heath. The outcrops run east-north-eastwards, preserved in the axial region of the syncline. These sands, like the sandy Claygate Beds which they succeed, were laid down as a consequence of the shallowing of the London Clay Sea. They are poorly fossiliferous but sufficient evidence is available to demonstrate that, in the Windsor to Newbury area, beds as late as Bartonian age are present. However, the usage of either the name Bracklesham or Barton Beds for parts of this sequence of sands implies a certainty of correlation with those beds of the Hampshire Basin which is not warranted.

The Hampshire Basin—the higher Palaeogene Beds

In the Hampshire Basin sandy beds overlying the London Clay are well developed, for example in the Isle of Wight where they are called the Bagshot Sands and in the Isle of Purbeck where they are known as the Redend Sandstone. In the Bournemouth area the Bagshot Sands are deltaic and the higher part is called the Bournemouth Freshwater Beds. Interbedded clays yield plant fossils. In Dorset the Bagshot Beds overlap on to the Chalk and include pebbles of Purbeckian chert, quartz and Cornish rocks.

Succeeding the Bagshot Beds are some 600 feet of clays and glauconitic sands at Bracklesham Bay, Sussex, and in the eastern Isle of Wight, known as the Bracklesham Beds. They include the earliest appearance of the foraminifer *Nummulites* in Britain. *N. laevigatus* occurs in the Lower Bracklesham Beds (and has been found in the London Basin) and *N. variolarius* in the Upper Bracklesham Beds (though their ranges overlap). In fact, the Bracklesham Beds have an extensive marine fauna, chiefly of molluscs, over 500 species being present. Thirty of the genera are not found in the London Clay. Westwards the facies changes rapidly and in the western Isle of Wight the Bracklesham Beds are estuarine and sandy with thin beds of lignite. They are brightly coloured and, with the Bagshot Sands, outcrop at Alum Bay. In the Bournemouth area the Bournemouth 'Marine Beds' of this age are transitional between the Bournemouth Freshwater Beds below and the Hengistbury Beds, with a marine fauna, above.

An expansion of the sea followed the deposition of the Bracklesham Beds and the Barton Beds succeed without interruption in the eastern Isle of Wight but in the west have a basal pebble bed. The type section is at Barton on the Hampshire coast. The Barton Beds are clays and sands with an abundant marine—chiefly molluscan—fauna. The middle beds are especially fossiliferous. However, in late Barton times a transition to a freshwater environment occurred and the uppermost beds include fossils indicating brackish conditions. The overlying Lower Headon Beds were laid down in freshwater lakes.

The Lower Headon Beds are regarded by some authors as of Oligocene age but it seems logical to regard the following marine transgression as marking the base of the Oligocene. This transgression deposited the Middle Headon Beds—commencing with the Brockenhurst Beds, which were deposited east of a line from Fordingbridge to Ventnor and have a marine fauna including corals. The estuarine '*Venus*' Beds follow and then some 500 feet of essentially continental facies beds, which have been divided into the Middle and Upper Headon Beds, the Osborne Beds, the Bembridge Beds and the Lower Hamstead Beds. The persistent Bembridge Limestone is freshwater.

An outlier of this bed caps the small Creechbarrow Hill west of Corfe, Dorset, and apart from this beds younger than the Barton Beds are restricted to the Isle of Wight.

The Bembridge Limestone is followed by an oyster bed laid down by a marine transgression of limited westerly extent and short duration (like that of the Brockenhurst Beds), and continental conditions of deposition quickly returned. However, the Palaeogene succession is completed by the representative of yet one more marine incursion, the Upper Hamstead Beds—clays of Rupelian age.

BOVEY TRACEY

A unique outlier of Tertiary strata is found occupying an elongate basin to the east of Dartmoor lying between Bovey Tracey and Newton Abbot in south Devonshire. At least 650 feet of beds have been proved by borings near the centre. Although gravelly and even coarsely pebbly near the margins, the beds are chiefly clays with interbedded lignites, some several feet thick. The clays are kaolinitic ball-clays and pipe-clays, the material derived from the adjacent granite. It is clear that the sediments were laid down in a lake and the lignites formed from rafted logs. The trees were chiefly the conifer *Sequoia*, no longer found in Europe, *Cinnamon* and *Nyssa*. No fauna has been found and evidence of age is not entirely conclusive but the beds are now generally referred to the Oligocene and may be contemporaneous with the brown coals of late Oligocene age in Germany. Similar minor outcrops of clays with thin lignitic bands in north Devon and Pembrokeshire have not been dated but may be of Palaeogene age as may a similar basin proved by geophysical and borehole evidence in Harlech Bay (beds overlying Upper Lias in the Mochras Bore).

The Tertiary Igneous Province

While southern Britain was reinvaded by successive marine transgressions in Lower Tertiary times, northern Britain had become part of a large province of igneous activity, the Thulean or North Atlantic Igneous Province. This extended far beyond the British Isles to include the Faroes, Iceland and Greenland. In Britain the igneous rocks, extensive lava flows and intrusions, are found chiefly in the Inner Hebrides, the west of Scotland and in Northern Ireland (Fig. 15.6). However, dykes extend far from the igneous centres, some crossing northern England, while the Lundy Granite forming an island in the Bristol Channel, and isotopically dated as 50 m.y. old, is the most southerly expression of Tertiary igneous activity. Recent work on the palynology of sediments which immediately underlie the earliest volcanic rocks suggests that vulcanicity began in late Oligocene times in Skye, but perhaps as early as the Eocene in some parts of the province. Isotopic dates of lavas in Mull and Antrim of 74 m.y. infer an Eocene age or even earlier. In Iceland vulcanicity persists to the present day; however, the last vestiges of activity ceased long ago in the British area.

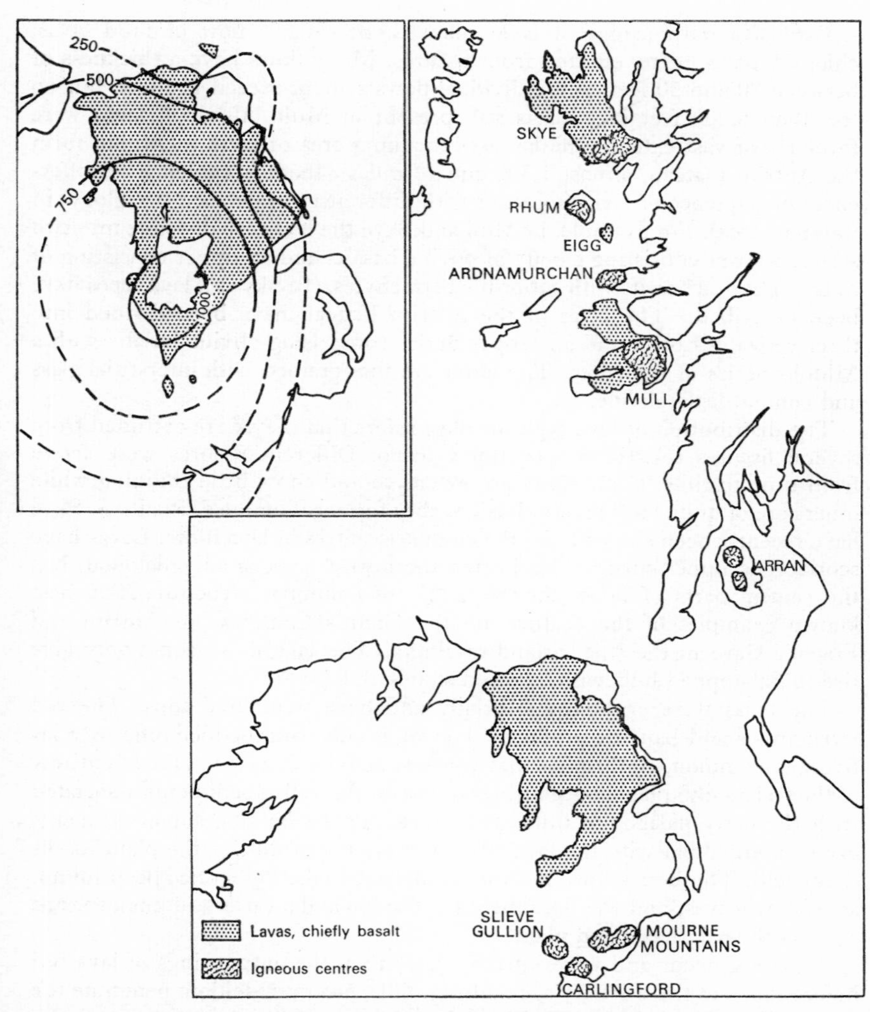

Fig. 15.6 Tertiary igneous centres and lavas of the Thulean Province. *Inset:* Isopachytes for the Lower Lavas of Antrim. (Outcrops based on Crown Copyright Geological Survey Maps, by permission of the Director of the Institute of Geological Sciences. Inset from S. I. TOMKEIEFF, 1964 Petrochemistry and Petrogenesis of the British Tertiary Igneous Province, p. 327. In *Advancing Frontiers of Geology and Geophysics*).

The lava flows

The earliest volcanic activity was, in northern Skye, the explosive eruption of basic tuffs. These palagonite tuffs accumulated in shallow water for they show evidence of rapid cooling and are associated with plant-bearing shales and flags.

Extensive outpourings of lava followed, flow upon flow of fluid lavas, chiefly basalts, being emitted from fissures. Many flows have a thickness of between 20 and 50 feet, and individual flows seldom exceed 100 feet, but no less than 6,000 feet of lavas is still present in Mull. Although lavas were formerly of vast extent—the largest remaining area of lavas in Britain form the Antrim Plateau, almost 1,500 square miles—there is evidence of differences of sequence of lavas (together with differences in detailed petrology) in different areas. For example, in Mull and Skye the lavas are divisible into two suites, a lower consisting chiefly of olivine basalts and an upper consisting of mugearites and basalts with subordinate trachytes. (In Skye six lava types have been described.) The lavas of the Antrim Plateau have been divided into three series, a Lower and an Upper Series comprising olivine basalts with a Middle Series of tholeiites. The latter are fine grained with interstitial glass and contain little olivine.

The distribution of lava types in Skye infers that they were extruded from several fissures related to a central volcano. Different fissures were active from time to time. Some flows are extensive and show little thinning while others are of quite local extent. Basic bodies formerly mapped as sills in Skye have recently been shown to be the compact centres of lava flows. Lavas have scoriaceous upper surfaces (and often the lowest layer is amygdaloidal) but the central parts of flows are frequently of columnar structure. The best known examples of this feature are the Giant's Causeway of Antrim and Fingal's Cave in the small island of Staffa. The lava flows commonly give rise to flat-topped hills with terraced slopes.

The lavas were erupted subaerially and have weathered tops. The red ferruginous and bauxitic product is known as bole. Interbedded tuffs are relatively uncommon, confirming that volcanic activity consisted almost entirely of the (relatively) placid outpourings of basalt. As well as sediments associated with the early palagonite tuffs, and preceding the basalts, some sediments occur interbedded with the lava flows and are important for the plant fossils they yield. The best known are the Mull Leaf Beds. Pollen has been found, as well as leaves, and the flora included *Ginkgo* and a variety of angiosperms such as the oak, hazel and plane.

Basic sills occur and were intruded later than the outpourings of lava but before much of the Tertiary age faulting. Sills, however, seldom penetrate the lavas and were chiefly intruded into the mesozoic sediments.

The Central Igneous Complexes

In Scotland central igneous complexes are found in Skye, Rhum, Ardnamurchan, Mull and Arran. Three are present in Ireland at Carlingford, Slieve Gullion and the Mourne Mountains.

A central igneous complex comprises a large number of igneous bodies of widely differing structures and petrology representing a complex igneous history. In most cases the 'focus' of activity migrated, so that two or three distinct centres are recognized, and the concentric intrusions of the later centre (ring dykes, cone sheets, etc.) cut those of the earlier. The intrusions can be classified according to their form. There are arcuate or circular steeply dipping bodies known as ring dykes. Ring dykes may be separated by narrow screens of the country rock into which they have been intruded. Some granite boss-like plutons may have also been emplaced as a result of cauldron subsidence. Other important plutonic bodies are formed of layered basic rocks, for example the Cuillins gabbros of Skye. Associated with many of the central igneous complexes are vast numbers of dolerite cone sheets. The plutonic history is now believed to have been prolonged and not entirely later than the basaltic lava flows. Even so, the isotopic age of 70 m.y. for the Mourne Mountains granite (which places it in the late Cretaceous) presents a problem. Dykes are incredibly numerous and dyke intrusion extended over a lengthy period. Tending to radiate from the igneous centres, the north-west—south-east trend predominates. The emplacement of dykes poses a problem of displacement for in some areas the dykes locally make up as much as a fifth of the width of exposed outcrops. They are mostly of olivine dolerite or teschenite but there are, as well, smaller numbers of felsite dykes and some of pitchstone. The latest dykes cut all other intrusions and were the last phase of activity. Frequently, dykes can be seen to cut the basaltic lava flows so they clearly were not feeders for the lavas now seen, although they may have fed higher flows since removed by erosion.

The igneous history, or chronology, of most of the centres of activity has been worked out in very great detail. The literature extends back over half a century but important work is still being done. Naturally, it is impossible to review this extensive data in the space available here but the well known chronological series of events in Mull is summarized. Into a terrain of Moinian rocks overlain by Mesozoic strata, granophyres were intruded along the margin of a caldera. Explosion vents were produced and then a vast thickness of cone sheets was intruded. Next, gabbro-eucrite masses were intruded at the margins of the caldera, followed by another phase of explosive activity which produced vents. A further set of cone sheets was intruded and then, locally, intrusions of olivine-gabbro. Now the centre of activity shifted north-westwards with the production of a later caldera. The history of this centre is not quite so complex but commenced with the intrusion of thick ring dykes, followed by the intrusion of cone sheets and then large granophyre bodies. Dykes are numerous and there are other intrusions which cannot be fitted into this sequence through lack of evidence.

The other central igneous complexes each have particular characters of interest. Ardnamurchan has three centres of activity of which the most easterly was the earliest, the most westerly followed and activity finally centred between these two. The igneous intrusions of this last centre have obliterated considerable parts of the intrusions of the earlier centres. Arran has a northerly granite mass and to the south a later central ring complex. In Skye the plateau basalts are extensive in the north but to the south of them the plutonic mass of layered gabbros form the Cuillins. Their intrusion was followed by ring faulting and further intrusions, the arcuate acidic rocks of the

Western Red Hills Centre and the complex intrusions of the Eastern Red Hills, the chief of which is a granophyre boss.

The Miocene Orogeny

No sediments of Oligocene age younger than the earliest Rupelian (Stampian) age are known from Britain, although sedimentation continued for some considerable time in the Anglo-Parisian Cuvette and persisted into the Miocene in the Aquitaine Basin. Sediments of Miocene age are absent from Britain and during this period major palaeogeographical changes took place in Europe. The geosynclinal deposits of Tethys were folded, thrust and piled into nappes to a height of 10 miles to form the Alps and contemporary mountains. This major, and most recent, orogeny to affect Europe was prolonged, and the instability of the crust through Palaeogene times might be considered to be due to movements which were precursors of the Alpine Orogeny. Its effects, in Britain, were to institute an interval of non-deposition which lasted perhaps for 20 m.y. Despite the geographical changes which this infers, the structures produced in Britain were not complex and regional metamorphic effects are unknown.

Folding in southern England, with an approximately east—west trend, is not intense although the principal structures are large. The monoclinal fold of the Isle of Wight has an amplitude of the order of 20,000 feet. Some overturning of beds occurred locally and minor thrusts are known, but the effects of the orogeny in Britain chiefly were the production of gentle folding in southern England and, further north, tilting of strata and block faulting. The dominant strike of the Mesozoic rocks of the main outcrop from Dorset to Yorkshire is a feature of Miocene age.

Britain, as a result of Miocene uplift and folding, began to take on its present physiographic form, although there have been innumerable modifications resulting from changes in sea level, widespread Pleistocene deposition and prolonged erosion. Nevertheless the drainage patterns of the Wealden area, southern England, the north Pennines and the Lake District were all initiated by Miocene uplift. The consequent drainage pattern developed on the eastwards-tilted Mesozoic rocks included the proto-Thames and proto-Trent.

East—West fold structures

In southern England the main anticlinal axes, the Wealden and the Weymouth-Purbeck-Isle of Wight (with their respective counterparts across the Channel as the Pays de Bray and Pas de Calais Anticlines), are the most conspicuous major structures. In the Isle of Wight the fold is steeply monoclinal, the near vertical Chalk of the 'middle limb' making a central topographic ridge across the island, while in the south of the island the beds dip southwards at no more than 5°. The fold really comprises two anticlinal folds en echelon. In Purbeck the monoclinal fold is evident, although strike faulting within the

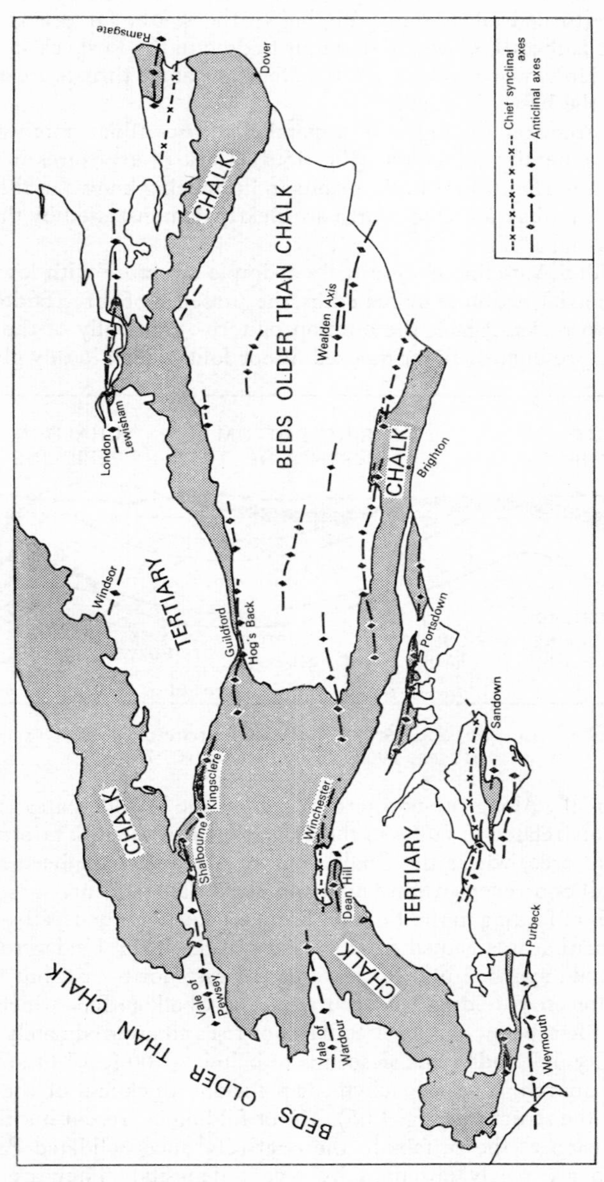

Fig. 15.7 Miocene fold structures of south-eastern England.

Chalk of Ballard Down complicates the structure. Westwards the Cretaceous beds are overturned and steeply dipping to the south, for example on the north side of Lulworth Cove. Here minor folding in the Purbeck Beds is well seen in Stair Hole, while west of Durdle Door low angle thrusts are eroded by the sea near the base of the cliff.

The axial trend of folding here is parallel to pre-Albian folding, though axes are not generally coincident (Fig. 15.8). Lack of exposures in the Tertiary beds of the Hampshire Basin results in little being known of the detailed structure, but fold axes further north are clearly demonstrated by the pattern of outcrops (Fig. 15.7).

The Wealden Anticline is essentially a simple structure with low regional dips and the axial region is indicated by the presence of three faulted inliers of Purbeckian rocks. Again, the outcrop pattern—especially of the Chalk—indicates the presence of superimposed minor folds which locally give rise to

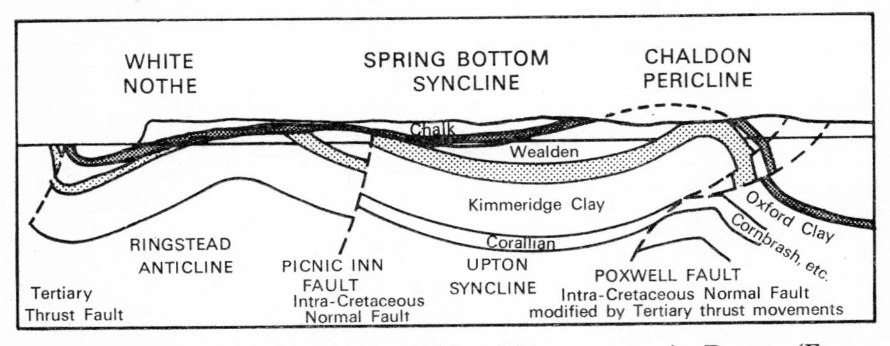

Fig. 15.8 The Miocene and pre-Albian fold structures in Dorset. (From W. J. PHILLIPS, 1964, *Proc. Geol. Ass.*, **75**, 373.)

dips of 20 to 30°. Along the northern margin of the Wealden area are monoclines probably related to faults in the Palaeozoic basement. The structure in the Guildford area, where the Chalk outcrop narrows, have been described in great detail and reversed dip faults and strike faults occur.

The details of folding in the London Basin are known imperfectly, although useful information was gained from the excavation of the Underground railways. The main syncline plunges towards the east-north-east, but not uninterruptedly for cross-folding has occurred. The Chalk surface which lies at a depth of 400 feet below sea level at Bagshot rises to approximately sea level under the City of London and descends to below −400 feet in the region of the Thames mouth. The isopachyte map for the thickness of the London Clay reflects the structure (Fig. 15.5). Minor folding is present but finds little or no expression at the surface in the relatively unconsolidated Palaeogene rocks (which are partly obscured by recent deposits). They are of small amplitude, no more than 200 feet, compared with the minor folds of the Wealden area which have an amplitude exceeding 600 feet. Undoubtedly this difference reflects the greater depth of the basement in the Wealden area.

Faulting

Many large faults in Britain are known to have had a long and complex history but their most recent movement can be dated only as post-Triassic (since younger strata are not present). This post-Triassic throw, often a very large vertical displacement, is probably of Miocene age since this was the only epoch of widespread orogeny and uplift to occur in post-Triassic times. Important faults with a large displacement are the Church Stretton Fault, the boundary faults of some Midlands Coalfields and the faults bounding the North Pennine Fault Block. The faulting in coalfields is particularly well documented from several sources, surface mapping, borehole information and underground workings. While much of this faulting occurred as a result of the Variscan Orogeny, many faults can be shown to displace the cover rocks in concealed coalfields and partly concealed coalfields. These faults are clearly post-Permian or post-Trias and by inference may be of Miocene age.

Undoubtedly the South-West Peninsula of England lay in line with the Miocene folding which affected southern and south-eastern England. It is impossible to completely disentangle the effects of the Variscan and Miocene Orogenies here and there may have been post-Jurassic-pre-Cretaceous folding too. It seems likely that the relatively rigid Palaeozoic rocks were little affected by folding of Miocene age, but a system of wrench faults has been described and these are believed to be attributable to Miocene age stress (Fig. 10.1).

Tertiary fossils

Marine faunas of Tertiary age are prolific and cannot be adequately summarized in a few pages. However, two important characters are outstanding, the dominance of molluscs and the profusion of large foraminifera.

As a result of major changes in late Cretaceous times and at the end of the Cretaceous, which included the extinction of the ammonites and the great reduction in numbers and variety of brachiopods, the commonest marine invertebrates of the Tertiary seas were lamellibranchs and gastropods. The latter have proceeded from strength to strengh, with new families and genera constantly appearing and few dying out so that they are very numerous in the Tertiary—as they are in modern seas. The presence of ornate genera like *Murex* and *Voluta* probably indicates that Tertiary seas were warm, and *Cerithium* and *Cypraea* lived in relatively shallow water. Lamellibranchs appear to have taken little heed of the Mesozoic/Tertiary junction and many genera of Mesozoic lamellibranchs continue into the Tertiary. Oysters, mussels (*Mytilus*) and cockles (*Protocardia*) abound and, in the Palaeogene, fossil wood is frequently bored by the lammelibranch *Teredo* (ship-worm). Echinoderms of many kinds are common in the Tertiary, regular and irregular echinoids occupying their different ecological niches. Corals too are common, both Scleractinians and Alcyonarians, although reefs are absent from the British Tertiary. Cephalopods are unimportant as Tertiary fossils as a result of the reduction in the shell or its absence in many forms. Arthropods are not common from Tertiary marine beds except for crabs and barnacles but they form an important part of terrestrial faunas. Freshwater faunas were extensive but chiefly comprise lamellibranchs and gastropods. While land vertebrates have been used to zone the Tertiary in other parts of the world, in the marine strata the large foraminifera such as *Nummulites* provide the best means of zoning.

16 The Pleistocene

Pliocene-Pleistocene sediments

Marine deposits of Pliocene and Lower Pleistocene age are of restricted area and are found chiefly in East Anglia where they clearly pre-date the glacial and interglacial deposits described later in this chapter. The beds are listed in chronological order in Table 16.1 which includes minor outcrops from Cornwall and Kent.

TABLE 16.1

LOWER PLEISTOCENE	CROMERIAN	Cromer Forest Bed Series
	ICENIAN	Weybourne Crag Chillesford Beds Norwich Crag
	LUDHAMIAN	Red Crag
PLIOCENE	GEDRAVIAN	Coralline Crag and Lenham Beds ? St. Erth Beds

The St. Erth Beds of Cornwall are marine clays with a considerable fauna of molluscs and foraminifera and were laid down in shallow, but not necessarily littoral, conditions. They now occur some 100 feet above sea level. Some doubt remains about their age which has been variously placed between the Miocene and the Pleistocene, and a recent suggestion (1965) is that they are referable to the Cromerian of Lower Pleistocene times.

The Lenham Beds are ferruginous sands, preserved in solution pipes near the summit of the North Downs in Kent at a height of about 600 feet above sea level, and they were deposited near the margin of the North Sea Basin. Similar beds are known from the South Downs and from Holland. Perhaps correlatable with the Coralline Crag, they are of late Pliocene age and the earliest marine strata laid down in eastern England after Miocene uplifting and tilting. In the North Downs they are associated with a plane of marine erosion.

In East Anglia the Pliocene-Lower Pleistocene deposits are of considerable area, extending from near Harwich to the north Norfolk coast. They rest on London Clay in the east but overstep on to Woolwich and Reading Beds and Upper Chalk, for their deposition was preceded by a lengthy interval of erosion. Conventional stratigraphical principles cannot be widely applied to these beds since successively younger beds do not normally overlie older beds except locally. The earliest beds, the Coralline Crag, are found between Orford and Aldeburgh while the Red Crag spread beyond and occurs to the west. The still later Norwich Crag, the most extensive of these deposits, occurs to the north of the earlier deposits. The Crags are shelly sands laid down in sublittoral conditions, at a depth of 50 fathoms or less in the case of the Coralline Crag. However, there is a scanty evidence of the actual positions of shorelines. The Coralline Crag is about 100 feet of yellow sands with abundant shell fragments. The fauna is varied including lamellibranchs, gastropods, brachiopods and many species of Bryozoa from which the crag was erroneously named. Corals do occur but are not numerous.

If the Lenham Beds are equivalent in age to the Coralline Crag there has been considerable tilting of eastern England (deduced from their present heights above sea level), reviving a theory put forward over half a century ago. Certainly, the Red Crag, ferruginous cross-bedded shelly sands, is banked against the Coralline Crag. However, the Red Crag does, more generally, rest on eroded London Clay. The red coloration is apparently of secondary origin. A threefold division of the Red Crag has been recognized, the divisions succeeding each other northwards, confirming a shifting shoreline. Eoliths, once thought to be among the earliest artifacts, are found in the Red Crag (in the Stone Bed at its base) together with pre-Chellean implements, chiefly rostrocarinate flints. However, some authorities regard all pre-Cromer Forest Bed flints as merely simulating artifacts. The Red Crag includes an abundant, diverse, shallow-water marine fauna. The Norwich Crag thickens northwards to as much as 150 feet. Varying considerably in lithology it includes gravels, sands and laminated clays and is locally highly fossiliferous.

The fossils of the Crag deposits have a twofold importance, as indicators of climate and of relative age of the beds. The younger beds have a lower percentage of extinct forms, the figure falling from almost 40% in the Coralline Crag to as little as 10% in the Norwich Crag, confirming the field evidence of relationships. Accompanying the reduction in the percentage of extinct forms there is a decrease in the number of species normally found in warm seas, while boreal forms appear in the Red Crag and increase as a percentage of the fauna in successively younger beds. The Chillesford Beds, resting on Red Crag and Norwich Crag, extend from Chillesford (Suffolk) to near Wroxham (Norfolk) and are thin micaceous sands and clays with a boreal molluscan fauna. Of restricted distribution, they may have been deposited in an estuary.

The Weybourne Crag of northern Norfolk with a yet more arctic fauna was the last marine deposit to be deposited in Britain, apart from marginal raised beach material.

The Pleistocene ice age

The Pleistocene saw a major ice age which affected areas as far apart as New Zealand and North America. The effects were most pronounced in the Northern Hemisphere and greatly affected the British Isles. A major glacial epoch of this kind is an infrequent occurrence and the previous one took place in Permian times affecting countries in the Southern Hemisphere and

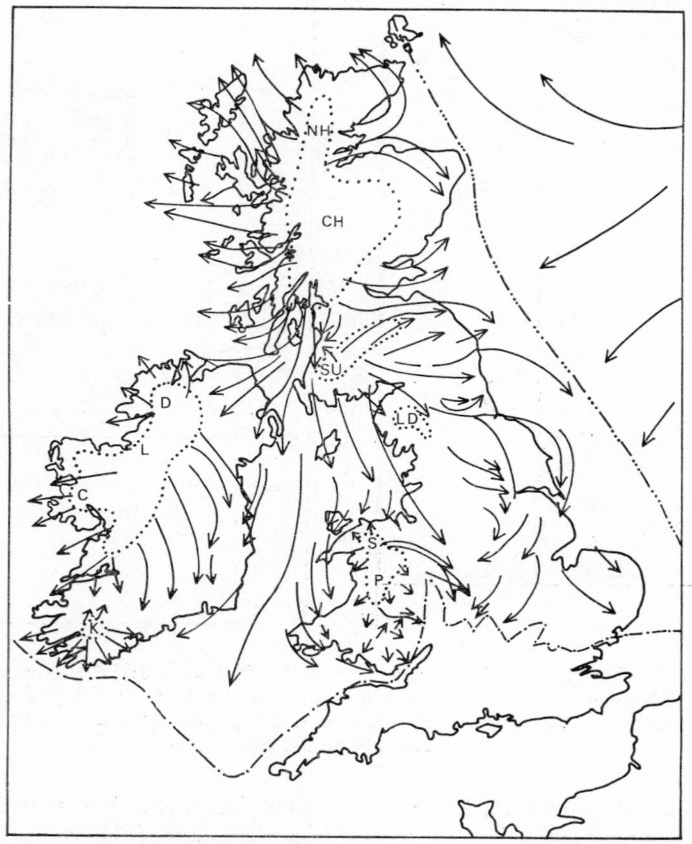

Fig. 16.1 Centres of ice accumulation and principal directions of ice flow during the maximum glaciation of Britain. (From L. J. WILLS, 1951, *A Palaeogeographical Atlas*, Plate XXI. Blackie, London.) C—Connemara, CH—Central Highlands, D—Donegal, K—Kerry, L—Leitrim, LD—Lake District, NH—Northern Highlands, P—Plynlimmon, S—Snowdon, SU—Southern Uplands.

India. No glaciations other than those of the Pleistocene had affected northern latitudes since Precambrian times.

The snow-line, the line above which snow accumulates since precipitation exceeds wastage due to ablation, melting, evaporation, etc., is at the present day of the order of 5,000 feet above sea level in the latitude of northern Britain. As a result, the highest mountain being 4,406 feet, no permanent snow-fields are found. During the Pleistocene on several occasions the snow-

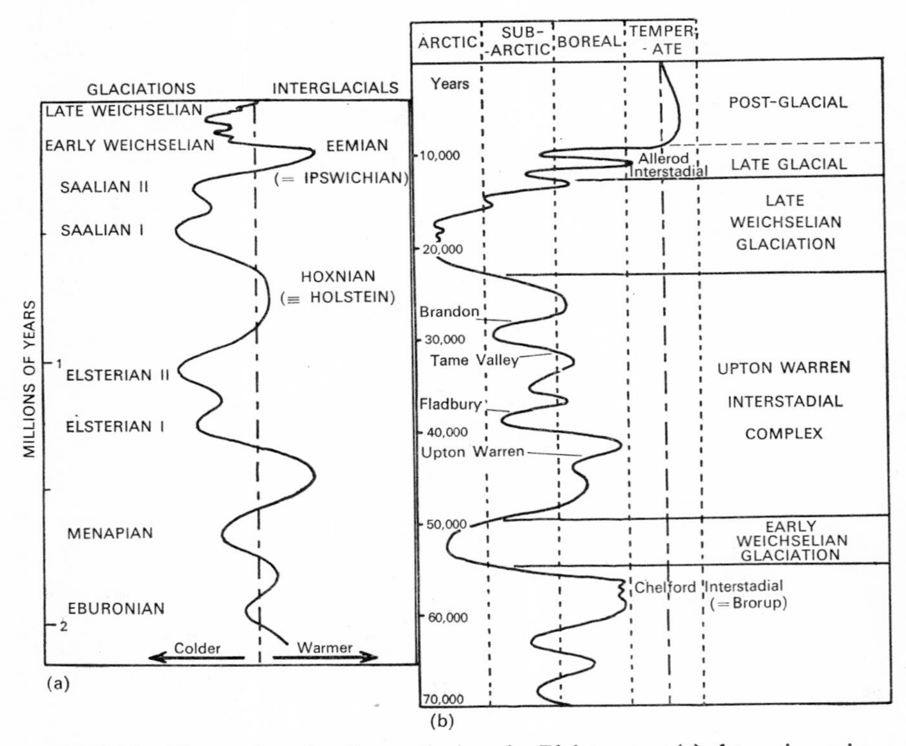

Fig. 16.2 Fluctuations in climate during the Pleistocene: (*a*) the main variations; (*b*) details of climatic variation during the last 70,000 years. (From F. W. SHOTTON, 1964, *Proc. XII Int. Congr. Ent., London*; and G. R. COOPE and C. H. S. SANDS, 1966, *Proc. R. Soc., B*, **165**, 389.)

line descended to a much lower level, to 2,000 feet or even less, so that all the main mountain areas became centres of accumulation (Fig. 16.1). The climate must have been colder by at least 4°C, on the average, throughout the year. Ice flowed outwards from centres of accumulation, guided by the form of the pre-glacial topography, coalescing to form ice sheets, so that much of Britain was covered by an ice cap similar to those still present in Greenland and Antarctica. At the maximum extent of the ice sheet (or sheets) the whole of Ireland and Britain as far south as the Bristol Channel and Essex were covered.

While the term 'Ice Age' seems imprecise, it conveniently covers the several periods of glaciation which affected the British Isles during the Middle and Upper Pleistocene. Although there is evidence from the faunas of the Crag deposits of a gradual deterioration in climate during the Lower Pleistocene, it was not until Elsterian times that ice from centres spread outwards into the lowlands of Scotland, northern England, the Midlands and East Anglia, or that the Irish Sea and North Sea were occupied by ice. However, the mountains of Britain may have had valley glaciers as early as the Eburonian (Fig. 16.2). It is most probable that some areas became glaciated before others and that the lowest glacial deposits are not everywhere contemporaneous. For this reason they no longer are taken as the base of the Pleistocene and, in fact, early Pleistocene deposits are not widespread in Britain. The climate of the Pleistocene was far from being uniformly severe, and cold periods (glacial periods) alternated with periods of milder climate known as interstadials and longer periods known as interglacials. During some of these intervals the climate was considerably warmer than at the present day. As a result of successive glacial periods, each giving rise to drift deposits, the stratigraphy of the Pleistocene is highly complex and is discussed later in this chapter.

Pleistocene sedimentation

During any glacial period Britain can be considered to be divided into three types of area. These are the upland areas of accumulation, the glaciated lowlands and the periglacial areas not covered by outflowing ice sheets but affected by permafrost. The upland areas were greatly eroded and typical glacial landforms developed—the formation of corries where snow-fields accumulated, the occurrence of pointed horn peaks (though not so well developed as in the Alps) and arrêtes which separate adjacent corries. Valleys which 'steered' the outflowing ice became over-deepened U-valleys and have the characteristic stepped longitudinal profiles. The chief effect in the uplands was erosive and material was removed to leave bare rock pavements, often striated by the passage of ice, and ice-scooped hollows. Glacial deposits found in these areas remain, chiefly, from the retreat stages of the glacial periods when valleys were occupied by valley glaciers. These deposited morainic material. The moraines often pass laterally into raised beach deposits in the sea lochs of western Scotland. The glaciated lowlands were the recipients of ice-borne material known collectively as drift. This is extensive and was sometimes deposited to a thickness of over 100 feet. The most characteristic deposit is boulder clay—more often called till—and the name reveals the chief characteristic of glacially transported material, a complete lack of size sorting, material ranging in size from fine ground material (rock flour) to large boulders. In the Forth-Clyde lowlands, in Lancashire and in Ireland around

Belfast and extending southwestwards, the till has been sculptured into drumlins. These are undoubtedly related to areas where the ice sheet was thick (probably sufficiently thick for the resulting pressure to cause basal melting), for further south the drift is formless. As well as till, clays, silts, sands and gravels occur. These were deposited in a number of ways, in ice-dammed lakes, by subglacial drainage and by ice. Deposits from meltwaters distributed beyond the limits of the ice are called outwash deposits. Recent work on the mechanism of ice flow shows that shear planes occur and that these carry material from the sole of the ice sheet to the surface, whereas in general debris which falls on to ice and is carried along as moraine gradually finds its way downwards to the base of the ice. This discovery helps to explain the complexity of glacial deposits and suggests that rapid alternations of till and gravel do not necessarily infer rapid advances and retreats of the ice. Similarly the study of deposition from subglacial drainage shows that some sediments formerly ascribed to glacial (ice-dammed) lakes were not so formed. Associated water-cut channels (called overflow channels) from ice-dammed lakes generally thought to result from water flowing marginally along ice sheets may, in many cases, also be a product of subglacial drainage. Glacial deposits have been highly contorted and, in some cases described in recent years, folding and thrusting is present.

In the glaciated lowlands during interstadials the ice sheets withdrew but the climate remained cold so that the deposits suffered the effects of permafrost. The features which occurred in periglacial areas south of the ice sheets can also be seen, in some cases, in the glaciated areas affecting the interstadial deposits and soils. The most widespread features resulting from frost are the formation of stone polygons or, on sloping ground (due to soil movement), stone stripes. These are now seldom easily seen as a result of several thousand years of post-glacial weathering and, of course, the effects of agriculture. However, large-scale patterned ground has been described from Warwickshire. More commonly sections show the effects of frost action; festoons and cryoturbation structures are widespread, while larger structures due to frost, called pingos, can be found (for example, affecting the Bovey Tracey Beds).

Important deposits in areas to the south of the glaciated areas are river terraces, remnants of former flood plains elevated and dissected as a result of uplift. In the Midlands the earlier extensive glaciation deposited till, then lake sediments were laid down. This area lay to the south of the ice sheets of later glaciations and river terraces formed. Also found in areas bordering ice sheets are wind-blown deposits of loess. Extensive in Europe where they are important stratigraphically, in Britain they occur in the south but are not well known.

The wide variety of deposition in the Pleistocene of both glacial and other sediments makes correlation difficult. The nature of the evidence from glaciated areas is so different from that derived from extra-glaciated areas that it is not always possible to relate the sediments and the geological history they represent of the diverse areas.

Divisions of the Pleistocene

The Pleistocene is divided into Lower, Middle and Upper, and further divided into 'stages'. The latter are of relatively short duration since the whole of the Pleistocene Period is only about 2 m.y. The stages really correspond to major climatic changes, for the basis of stratigraphical division is into tills and the interglacial deposits (Table 16.2).

Naturally, climatic changes which caused glacial periods in Britain affected the whole of northern Europe. When Britain was glaciated Scandinavia was the site of a much larger ice cap which at its maximum encroached the British area and on several occasions extended into the north German and Polish plain, where glacial deposits are well known. The Alps, where valley glaciers still remain as remnants of the much more extensive glaciations, were too far south for their ice sheets to coalesce with those of northern Europe. They have been subject to intensive study over a long period, and the glacial deposits of this region and of northern Europe present problems beyond the scope of this book. Even the problems of correlation within the British area are too complex to deal with except in outline.

TABLE 16.2

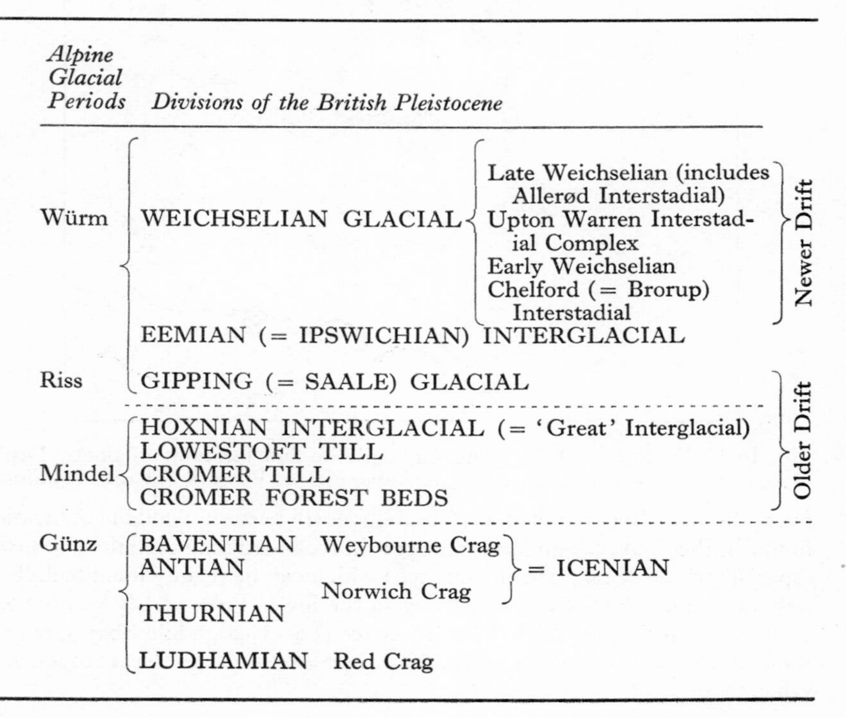

Alpine Glacial Periods	*Divisions of the British Pleistocene*		
Würm	WEICHSELIAN GLACIAL	Late Weichselian (includes Allerød Interstadial)	Newer Drift
		Upton Warren Interstadial Complex	
		Early Weichselian	
		Chelford (= Brorup) Interstadial	
	EEMIAN (= IPSWICHIAN) INTERGLACIAL		
Riss	GIPPING (= SAALE) GLACIAL		Older Drift
Mindel	HOXNIAN INTERGLACIAL (= 'Great' Interglacial)		
	LOWESTOFT TILL		
	CROMER TILL		
	CROMER FOREST BEDS		
Günz	BAVENTIAN	Weybourne Crag	= ICENIAN
	ANTIAN	Norwich Crag	
	THURNIAN		
	LUDHAMIAN	Red Crag	

Provenance of drift

Till deposits are useful indicators of the direction of ice flow, both by the provenance of the till material and its erratics, and the fabric of the till. Shelly drift dredged by ice from the floor of the Irish Sea and the Moray Firth provides vital evidence of flow direction as do certain characteristically coloured tills derived from the Chalk (widespread in eastern England) and

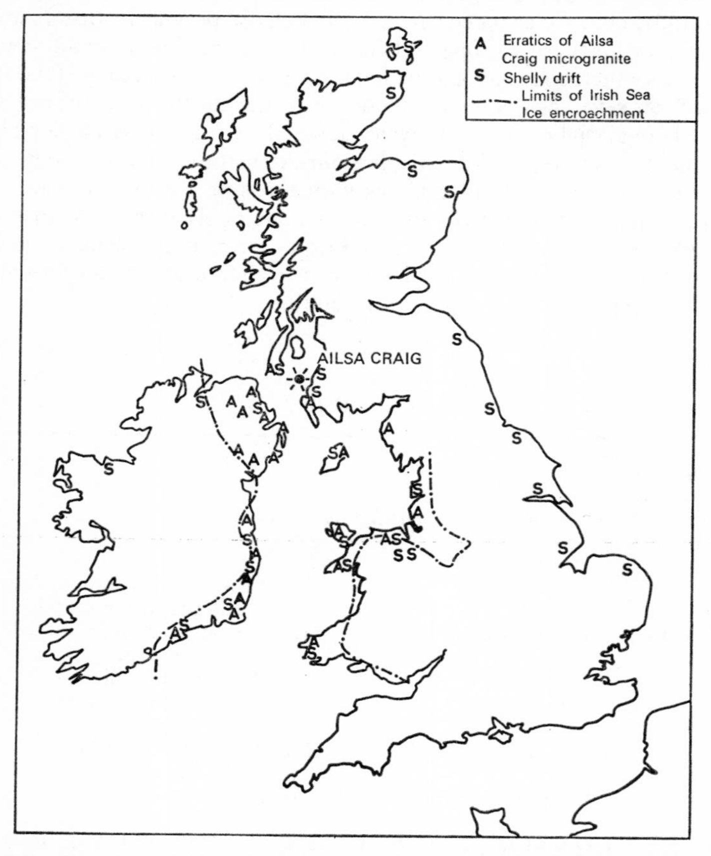

Fig. 16.3 Erratics of Ailsa Craig microgranite and locations of Shelly Drift. (From J. K. CHARLESWORTH, *The Quaternary Era*, Edward Arnold, London.)

from the Old Red Sandstone (found in north-eastern Scotland). Erratics found in the drift play an important part in reconstructing directions of flow, especially those of rare petrological type which can be readily identified. The reibeckite microgranite of Ailsa Craig in the lower Firth of Clyde provided erratics which fix the limits of Irish Sea Ice (Fig. 16.3), while Shap Granite, some of the Borrowdale volcanic rocks and many others are of less spectacular value.

Palaeontological evidence of climate

Tills are unfossiliferous (except for derived shelly material and fossiliferous boulders) and give little information about the climate except that it was glacial. The deposits of the periglacial areas and deposits formed during interstadials and interglacials yield both flora and fauna.

The Pleistocene saw in Britain the abundance of a wide variety of animals including many present British species, many species which were formerly of much wider geographical range—which then included Britain but no longer does so—and some extinct forms. The larger animals are less sensitive to climatic change than many invertebrates and it is a fact that the presence of either the woolly mammoth or the reindeer does not necessarily indicate an arctic climate, since they are found as fossils with a variety of animals typical of temperate climate, and individual animals range far from their usual habitat. Similarly, the presence of Hippopotamus in Pleistocene sediments as far north as the English Midlands does not confirm a tropical climate there, although this animal is now confined to the tropics and subtropics. Nevertheless, the doctrine of uniformitarianism works overall despite the formerly greater climatic tolerance, implicit in their greater geographical range, of some species.

The most significant indicators of climate have proved to be the flora, as revealed by fossil pollen, and insect populations, especially beetles. Beetle faunas, along with all invertebrates, show a history of rapid shifts of population (and some extinction) in response to the varying climate. Thus it is possible to find in the insect faunas of Pleistocene lake deposits, chiefly peats, and in silts of river terraces, forms not now found in Britain—although less than 10% are extinct. Beetles have exoskeletons very resistant to decay and, as fossils, are almost incredibly abundant. Many species are now seen to be restricted to northern latitudes in Fennoscandia and to high altitudes further south, inferring a subarctic climate of continental aspect for Britain at that time (about 32,000 years ago) (Fig. 16.4). The most systematic method of using beetles as climatic indicators is to plot as a histogram the ground-beetles (carabids) present, grouping them according to their present ecological distribution in Fennoscandia.

It is not surprising that the flora of Pleistocene times reflected changes in climate. Palynology, the study of pollen (and spores) reveals the climatic changes. Flowering plants all produce pollen, from 0·1 to 0·01 mm in size, in immense quantities and this is readily dispersed by the wind. That which settles in lakes or bogs becomes incorporated in sediments or in peat and can be retrieved thousands of years later, since pollen is resistant to chemical and physical change. Pollen is identified and counted and a pollen spectrum prepared showing the percentage of important types. Tree pollen and non-tree pollen are normally dealt with separately. Among the more important tree pollens as indicators of climate are those of the beech (Fagus), oak (Quercus), alder (Alnus), birch (Betula), willow (Salix) and pine (Pinus). Zones of vegetation are recognized ranging from the coldest, Alpine (divided into Upper,

Middle and Lower), through zones of Betula and Conifer to the warmest, Quercus. The climatic zone to which a particular deposit belongs can be seen from its pollen spectrum.

Successive deposits provide a sequence of pollen spectra which combined form a pollen diagram (Fig. 16.5). This reveals changes in climate with time

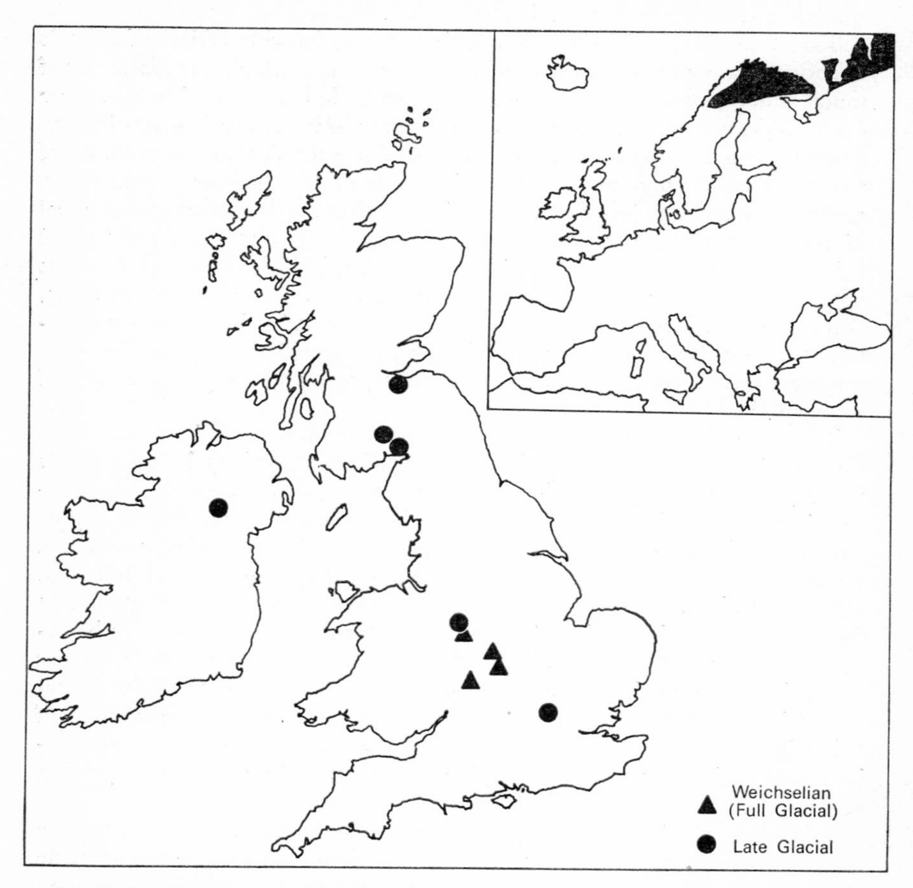

Fig. 16.4 The present distribution in northern Europe of the carnivorous ground-beetle *Diachila arctica* and known occurrences in glacial sites in Britain. (From G. R. COOPE, 1967, *I.N.Q.A. Congress.*)

and so can be used as a stratigraphic tool. Areas not widely separated must have experienced the same changes in climate: if two pollen diagrams from adjacent areas show the same sequence of fluctuations in climate, then the sediments from which the pollen samples were obtained can be correlated.

The changes in climate during the later part of the Pleistocene have been worked out in very great detail from the kind of evidence just described. The main changes in climate were as follows. The Eemian faunas from Thames terrace deposits at Isleworth and Trafalgar Square suggest a temperature a few degrees above that of the present day, at least in summer. The organic deposits from Chelford in Cheshire belonging to the Chelford interstadial indicate a rather cool climate with temperate pine forests flourishing. After the early Weichselian glaciation the climate varied considerably and evidence from sites such as Upton Warren (Worcestershire) and Tame—not of course contemporaneous—indicate a boreal climate, while others such as Fladbury indicate colder, subarctic conditions. It must be mentioned that the evidence of pollen analysis does not agree with that of the insect fauna in the Upton Warren deposits. After the late Weichselian glaciation and the retreat of ice from all but the highest ground in Britain, the spread of species of animals and plants of northern affinity over much of Britain has been demonstrated.

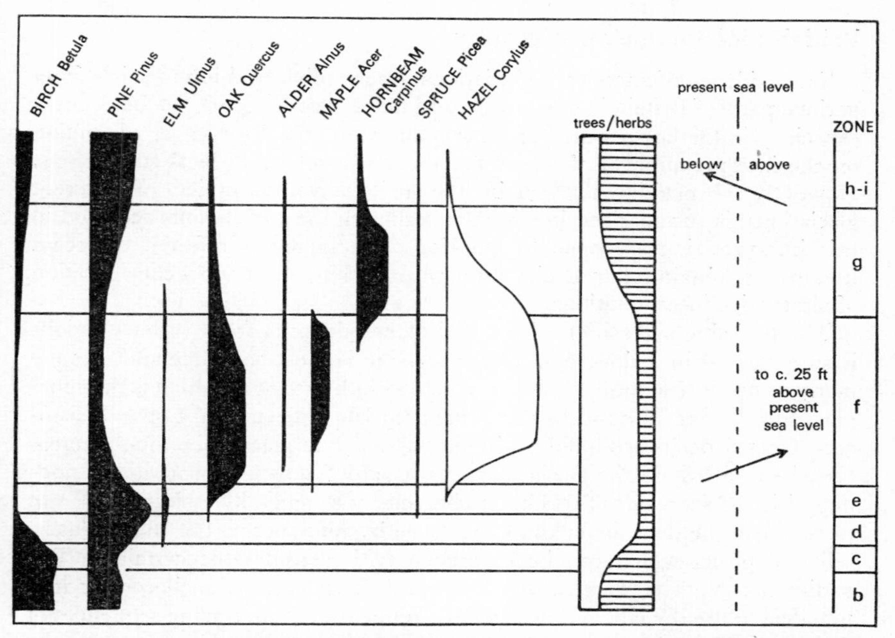

Fig. 16.5 A typical pollen diagram: pollen diagram through the Last (Ipswichian) Interglacial. (From R. G. WEST, 1960, *Advmt. Sci.*, **16,** 428.)

Dating Pleistocene deposits

The chronology of the Pleistocene can be worked out in considerable detail from the relative age of numerous events. Climatic evidence helps to relatively date deposits. In lake deposits counting of seasonally banded deposits, varves (or rhythmites), has given exact dating in some cases and can be used for dating such sediments up to 10,000 years old.

The most important method of dating Pleistocene sediments is by means of radiocarbon dating. This enables beds up to at least 30,000 years to be dated. The isotope of carbon, C_{14}, is created by cosmic ray bombardment and occurs in part of the atmospheric carbon dioxide. This is taken up by all living organisms, either directly (by plants) or indirectly (by animals eating plants). C_{14} disintegrates with a half-life of 45,000 years and the total amount remaining in organic material is a measure of the elapsed time since the animal or plant died and metabolism ceased. As a result of already great numbers of datings, the absolute chronology of the later glacial periods and interglacials is well known. Earlier Pleistocene deposits can only be dated, as yet, by the use of the potassium-argon method (and others) referred to in Chapters 1 and 3 which, because of the long half-life of potassium, is not well suited to dating such relatively recent events. It does not give an adequate degree of accuracy.

British Pleistocene stratigraphy

The evidence of successive glaciations, interstadials and interglacials from various parts of Britain is well known and much detailed work has been done, especially in the last decade. There are, however, still divergences of opinion on the interpretation of observed facts, even in relatively local successions. However, the position reached by the major advances of ice, of different glaciations, is summarized in Fig. 16.6 although the correlations accepted in this interpretation are open to question. The reader is referred to recent literature to obtain fuller details of the problems involved and a consideration of alternative interpretations.

The probable correlation of the Pleistocene deposits from various regions is summarized in Table 16.3. The succession is most complete and, despite many complexities, more readily understood in East Anglia which is therefore placed in the first column. In East Anglia the late Pliocene and Lower Pleistocene Crag deposits are followed by glacial and interglacial deposits, whereas elsewhere in Britain the glacial deposits usually follow a long period of non-deposition. As a result of this latter character it is generally easiest to attempt to correlate the deposits of the Pleistocene by reference to the latest Weichselian deposits which can be correlated with considerable certainty. The earlier the deposits, in general the more difficult correlation becomes: it is practical to use the late Weichselian as a reference datum tracing sequences of deposits back in time, attempting to maintain correlation of sequences as the history of different areas is pursued.

The history of events during Middle and Upper Pleistocene times, deduced largely from East Anglia, is as follows: the Cromer Forest Bed Series, freshwater and estuarine beds of late Lower Pleistocene age, were followed by the deposition of the Arctic Freshwater Bed. The earliest glacial deposits are the Cromer Till. The Corton Beds, shelly sands and clays, were probably deposited during an interstadial before the deposition of the Lowestoft Till (Great Eastern Glaciation), also known as the Lower Chalky Boulder Clay. This highly contorted till includes huge blocks or erratics of chalk, some over 100 yards in length. The Hoxnian interglacial lake silts of Suffolk are not found further north where the Gipping Till (Little Eastern Glaciation), also known as the Upper Chalky Boulder Clay, succeeds the Lowestoft Till. The material of these tills is essentially similar, being derived from Cretaceous and Jurassic

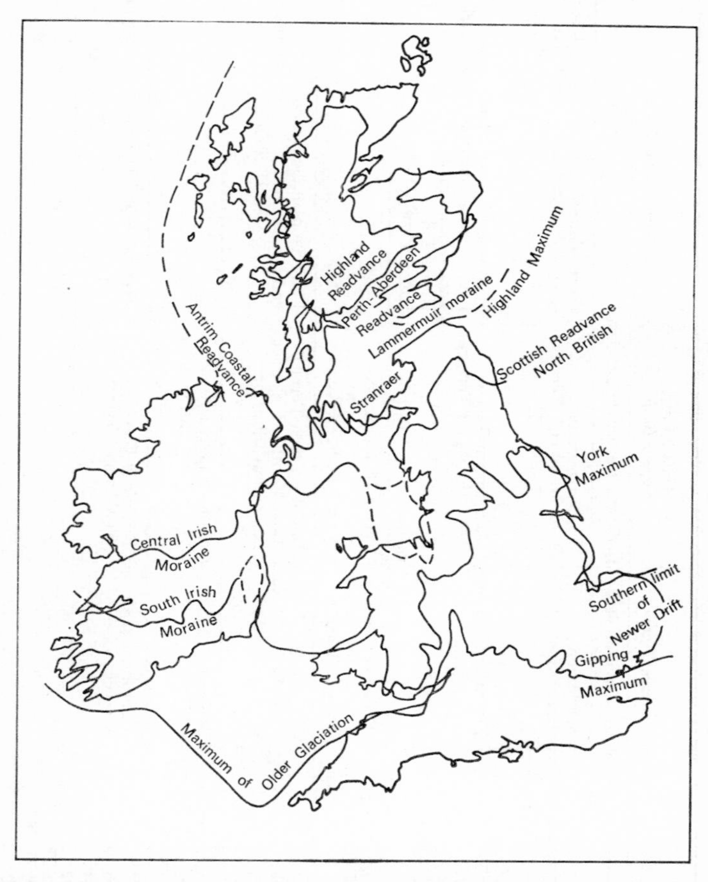

Fig. 16.6 Limits of glaciations in Britain. (Based on R. G. WEST, 1963, and other sources.) For alternative interpretations and correlations consult listed references.

beds over which the ice had moved from a northerly or north-westerly direction. The directions of ice movement and the extent of the ice sheets is known in detail. The youngest till, the Hunstanton Till, the equivalent of the Hessle Till of Lincolnshire and Yorkshire, is brown and contains little Cretaceous material having been derived from a more northerly source. It is found only in the northernmost part of East Anglia and is the only till of the Newer

TABLE 16.3

Stage	*East Anglia*	*Lower Thames*	*Midlands and Severn Basin*	*Northern England*	*Scotland*
FLANDRIAN	Peat, alluvium, late raised beaches, estuarine clays				
WEICHSELIAN	Solifluxion Patterned ground	Solifluxion Buried channel	Worcester Terrace Avon Number 1 Terrace	Corrie glaciers Allerød interstadial	Highland Readv. 50 ft Raised Beach Zone III Allerød interstadial Zone II
	Hunstanton Till (= Hessle Till)	Floodplain Terrace	Main Severn Terrace Avon Number 2 Terrace	Scottish Readvance, Main Lake District Glaciation York Glaciation Hessle Till Upper Purple Till	Perth–Aberdeen Re-advance 100 ft Raised Beach Zone I Moray Firth–Strathmore Glaciation
IPSWICHIAN	Interglacial deposits at Ipswich and Hinton, Cambridge	Trafalgar Square interglacial Taplow Terrace	Severn Kidderminster Terrace Avon Number 4 Terrace		
GIPPINGIAN	Gipping Till	Coombe Rock, etc.	Bushley Green Terrace Avon Number 5 Terrace Bagington-Lillington Gravel	Early Scottish Glaciation and Lower Drift of Holderness	Greater Highland Glaciation
HOXNIAN	Interglacial deposits at Hoxne and Clacton Nar Valley marine beds	Boyne Hill Terrace	Nechells deposit, Birmingham		
LOWESTOFT-IAN	Lowestoft Till Corton Beds Cromer Till	Coombe deposits Gravels, Loess	Bubenhall Clay Wooldridge Terrace	Scandinavian Glaciation	? Scandinavian Glaciation
CROMERIAN	Arctic Freshwater Bed Cromer Forest Bed Series				

Based chiefly on West with data from Flint, Shotton and other sources.

Drift age, although there are widespread gravels, sands and loams—deeply weathered and iron-stained and including glacial erratics—of late glacial age.

South of ice sheets meltwater often became dammed up to form lakes. These have been described from the Vale of Pickering between the Yorkshire Moors and Wolds, the Welsh Borders and in the Midlands south-east of

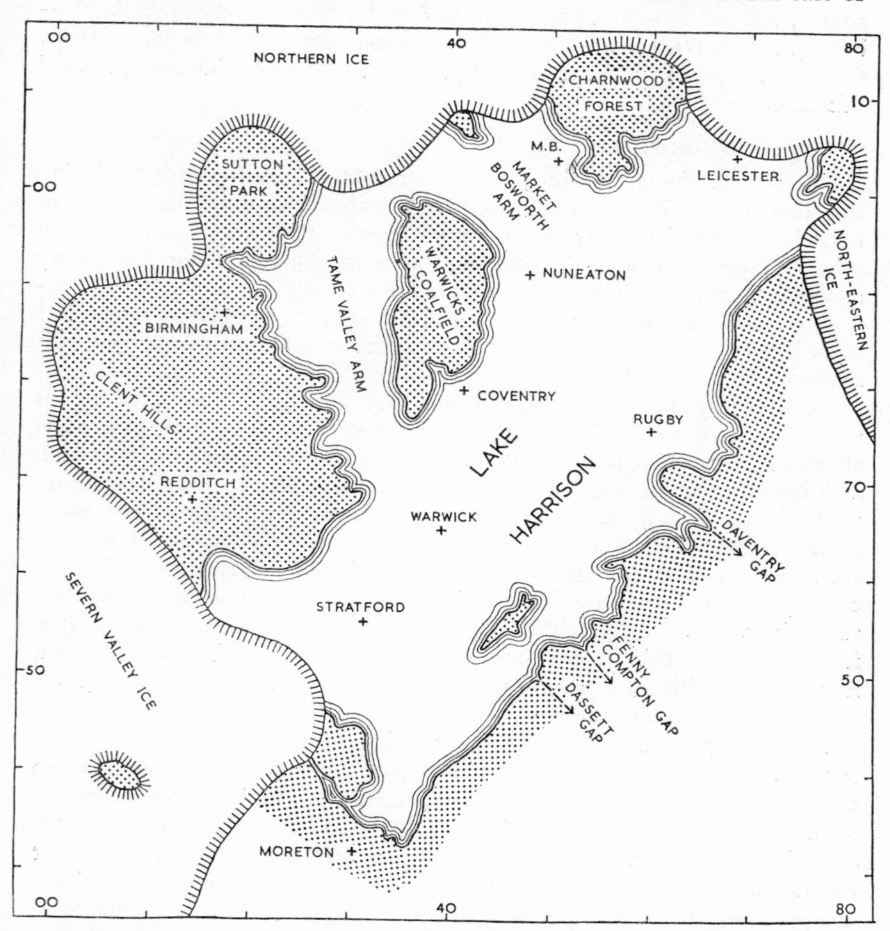

Fig. 16.7 Lake Harrison, an ice-dammed lake in the Midlands, at its maximum extent. (From F. W. SHOTTON, 1953, *Phil. Trans. R. Soc. B*, **237**, 209.)

Birmingham. The deposits of the latter, called Lake Harrison (Fig. 16.7), were the subject of a paper which established the modern approach to Pleistocene stratigraphy and are therefore briefly described here. The form of Lake Harrison bears little relationship to the present topography since it was impounded on an early glacial surface now covered to varying depths by drift.

The succession of clays, gravels and sands laid down in the lake has been established. They are collectively referred to the Older Drift. They now occupy high ground since there has been uplift (discussed in the next section) and the younger Pleistocene deposits or Newer Drift comprise river terraces laid down in the subsequently formed Avon Valley. The highest terrace is, of course, the earliest—with later terraces found at successively lower levels—but the Newer Drift as a whole is topographically lower than the Older Drift. Clearly, the law of the order of superposition is not applicable to deposits of this kind.

Changes in sea level

There is considerable evidence of changes in sea level in Britain during late Tertiary and Pleistocene times. Crustal subsidence under the weight of ice is believed to be partly elastic but partly plastic, flowage taking place in subcrustal layers. As a result, when ice sheets melt there is a negative gravity anomaly. Doming, or upwarping, of the crust takes place as the flowage reverses and the land is ultimately restored almost to its pre-glacial height, but there is some delay in the restoration of isostatic equilibrium during which sea and lake beaches are cut.

Undoubtedly the later changes in elevation of Britain are due to crustal upwarping as a result of the removal of the ice sheets. To some extent the effects of this isostatic uplift would be offset by the rise in sea level due to the addition of water previously locked up as ice. The evidence of uplift is found in the form of marine-cut erosion platforms, or raised beaches where these erosion surfaces have a cover of beach material, and river terraces. The latter are remnants of former floodplains dissected by the rejuvenation of rivers consequent upon uplift (lowering of base level). In the West of Scotland the raised beaches can be related to morainic deposits. On the other hand, it is frequently not possible to correlate river terraces with raised beaches, although it is now possible to tentatively correlate the river terraces of some river systems with terraces of others.

The most pronounced post-glacial uplift has occurred where ice sheets attained the greatest thickness—in Scotland. This region continues to rise more rapidly than other areas and is estimated to still be rising 4 mm per annum, for there is a time lag of about 15,000 years between the disappearance of an ice sheet and completion of uplift. It is still not possible to construct accurate isobases (lines joining points of equal uplift) for Britain, as has been done in the case of Scandinavia. However, the accurate measurements of heights of raised beaches are growing in number.

If Britain had been continuously emerging, then the raised beaches could be arranged in chronological order according to their height above sea level, with the earliest formed beach at the greatest elevation. This is substantially true but there has been, as well as uplift, relatively late subsidence which is demonstrable around many parts of the coast by the presence of submerged forests. These flourished formerly on land but are now revealed only at low tide or are found in harbour excavations. From the Forth to the Thames there is also evidence of sea level standing at a lower level than at present from the presence of buried river channels known to be as much as 100 feet or more below present sea level. Many of these are known in great detail, some from colliery workings (for example the Team 'Wash' of Co. Durham and the Forth buried channel).

The chief raised beaches in northern Britain occur at approximate heights of 100 feet, 50 feet and 25 feet above sea level. There is evidence of erosion surfaces as high as 1,200 and even 3,000 feet. These higher older erosion surfaces are undoubtedly pre-glacial and, though they cannot be dated accurately, they are Tertiary in age but later than Tertiary igneous activity. The 1,200-foot platform may be of Pliocene age. The 100-foot raised beach is of late glacial age, for the beach deposits include an arctic fauna, and the raised beach

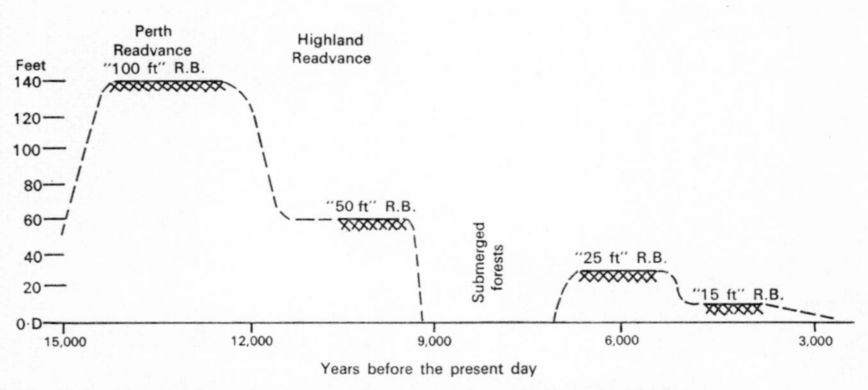

Fig. 16.8 Changes in sea level in late and post-glacial times.

is not found in the higher reaches of the sea lochs which were undoubtedly at that time occupied by glaciers. Beaches occur at heights between 65 and 45 feet, the '50-foot Raised Beach', and represent a temporary halt in uplift. All raised beaches vary in height, in general the highest raised beaches show the greatest variation, this tilting of strand lines being due to greater uplift in the north where the former ice sheets were thickest. Uplift continued until sea level was lower than it is at present; this was the time of the formation of buried channels and the submerged forests. This emergence was followed by submergence during which the 25-foot Raised Beach was cut (and which has subsequently been elevated to that height). It was formed in Neolithic times and is the only raised beach to yield artifacts. It is well developed around the Scottish coasts and those of Northern England and Northern Ireland. It is frequently backed by old sea cliffs and its height above sea level varies from 35 feet to nil. The chronology of the later beaches is given in Fig. 16.8.

In Southern England changes in sea level, resulting from glaciations and isostatic uplift following deglaciation, find their expression in raised beaches and river terraces. These changes were superimposed upon a gradual epeirogenic fall in sea level, for in Pliocene times sea level stood as much as 600 feet above its present level. Deposits fall into three groups: head deposits, river deposits and beach deposits (tills are absent from these southern areas). The clay-with-flints, coombe deposits and brickearths are of late Pleistocene age and show evidence of solifluxion. There is also evidence that the climate was wetter than at present. The brickearths may, however, have been not solely laid down in expanses of water but be in part wind-borne. It seems likely that they are equivalent to some of the loess deposits of Northern Europe which they lithologically resemble.

This chapter, of necessity an incomplete record of the later geological history of Britain, aims to provide no more than a framework for further reading of the proliferating literature.

Further Reading

Chapters 1 and 2

Principles and Palaeontology

AGER, D. V. (1963) *Principles of Palaeoecology*. McGraw-Hill, New York and Maidenhead.

AMERICAN CODE OF STRATIGRAPHICAL NOMENCLATURE (1961) *Bull. Am. Ass. Petrol. Geol.*, **45**, 645.

BENNISON, G. M. (1962) Palaeontological and physical evidence of the palaeoecology of some early species of non-marine lamellibranchs. *Lpool Manchr geol. J.*, **3**, 41.

BOUCOTT, A. J. (1953) Life and death assemblages among fossils. *Am. J. Sci.*, **251**, 25.

BROADHURST, F. M. (1959) *Anthraconaia pulchella* sp. nov. and a study of palaeoecology in the Coal Measures of the Oldham area of Lancashire. *Q. Jl geol. Soc. Lond.*, **114**, 523.

CRAIG, G. Y. and HALLAM, A. (1963) Size frequency and growth-ring studies of *Mytilus edulis* and *Cardium edule*, and the palaeoecological significance. *Palaeontology* **6**, 731.

CRAIG, G. Y. and JONES, N. S. (1966) Marine Benthos, Substrate and Palaeoecology. *Palaeontology*, **9**, 30.

DONOVAN, D. T. (1966) *Stratigraphy: An Introduction to Principles*. Murby, London.

DUNBAR, C. O. and RODGERS, J. (1957) *Principles of Stratigraphy*. Wiley, New York and Chichester.

EAGAR, R. M. C. (1960) A summary of the results of recent work on the palaeoecology of Carboniferous non-marine lamellibranchs. *4eme Congrès pour l'avancement des études de Stratigraphie et de Géologie du Carbonifère, Heerlen*, p. 137.

EAGAR, R. M. C. (1962) Boron content in relation to organic carbon in certain sediments of the British Coal Measures. *Nature, Lond.*, **196**, 428.

FALCON, N. L. and KENT, P. E. (1960) Geological results of petroleum exploration in Britain 1945–1957. *Mem. geol. Soc. Lond.*, **No. 2.**

GEORGE, T. N. (Chairman) (1967) Report of the Stratigraphical Code Subcommittee. *Proc. geol. Soc. Lond.*, **1638,** 75.

HALLAM, A. (1965) Environmental causes of stunting in living and fossil marine benthonic invertebrates. *Palaeontology*, **8,** 132.

HALLAM, A. *et al.* (1967) Depth indicators in marine sedimentary environments. *Marine Geology* (*Special issue*), **5,** 329.

HARLAND, W. B., SMITH, A. GILBERT and WILCOCK, B. eds. (1964) The Phanerozoic time-scale. *Geol. Soc. Lond.*, **120 S.**

HEDBERG, H. D. L. (1965) Chronostratigraphy and Biostratigraphy. *Geol. Mag.*, **102,** 451.

HOLLAND, C. H. (1964) Stratigraphical Classification. *Sci. Progr. Lond.*, **52,** 439.

HOLMES, A. (1965) Principles of Physical Geology, 2nd edn. *Nelson, London.*

MILLER, T. G. (1965) Time in Stratigraphy. *Palaeontology*, **8,** 113.

SHAW, A. B. (1964) *Time in Stratigraphy*. McGraw-Hill, New York and Maidenhead.

SUTTON, J. (1967) The Extension of the Geological Record into the Pre-Cambrian. *Proc. Geol. Ass.*, **78,** 493.

Chapter 3

Precambrian

The first four references give complete bibliographies on the areas mentioned; some more important and more recent papers are given beneath.

ANDERSON, J. G. C. (1965) The Precambrian of the British Isles. In *The Geologic Systems: The Precambrian*, vol. 2. ed. K. RANKAMA. Wiley, New York and Chichester.

CHARLESWORTH, J. K. (1963) *Historical Geology of Ireland.* Oliver and Boyd, Edinburgh and London.

CRAIG, G. Y., ed. (1965) *The Geology of Scotland.* Oliver and Boyd, Edinburgh and London.

WRIGHT, A. E. In Press. The Precambrian of England, Wales and South-East Ireland. Proc. Conf. on Continental Drift, Gander, 1967. *Mem. Am. Ass. Petrol. Geol.*

BOWES, D. R. In Press. Lewisian of Scotland. Proc. Conf. on Continental Drift, Gander, 1967. *Mem. Am. Ass. Petrol. Geol.*

DEARNLEY, R. and DUNNING, F. W. (1968) Metamorphosed and deformed pegmatites and basic dykes in the Lewisian complex of the Outer Hebrides and their geological significance. *Q. Jl geol. Soc. Lond.*, **123,** 335.

EVANS, C. R. (1965) Geochronology of the Lewisian basement near Lochinver, Sutherland. *Nature, Lond.*, **207,** 54.

FLINN, D. (1967) The metamorphic rocks of the Southern Part of the Mainland of Shetland. *Geol. J.*, **5,** 251.

JOHNSON, M. R. W. and STEWART, F. H., eds., (1963) *The British Caledonides.* Oliver and Boyd, Edinburgh and London.

KHOURY, S. G. (1968) The structural geometry and geological history of the Lewisian rocks between Kylesku and Geisgeil, Sutherland, Scotland. *Krystalinikum*, **5,** (in press).

KILBURN, C., PITCHER, W. S. and SHACKLETON, R. M. (1965) The stratigraphy and origin of the Portaskaig Boulder Bed Series (Dalradian). *Geol. J.*, **4,** 343.

MILLER, J. A. and BROWN, P. E. (1965) Potassium-Argon Age Studies in Scotland. *Geol. Mag.*, **102,** 106.

PITCHER, W. S., ELWELL, R. W. D., TOZER, C. F. and CAMBRAY, F. W. (1964) The Leannan Fault. *Q. Jl geol. Soc. Lond.*, **120,** 241.

PITCHER, W. S. and SHACKLETON, R. M. (1966) On the correlation of certain Lower Dalradian successions in Northwest Donegal. *Geol. J.*, **5,** 149.

RAMSAY, J. and SPRING, J. (1962) Moine stratigraphy in the Western Highlands of Scotland. *Proc. Geol. Ass.*, **73,** 295.

ROBERTS, J. L. (1966) Sedimentary affiliations and stratigraphic correlation of the Dalradian rocks in the South-west Highlands of Scotland. *Scott. J. Geol.*, **2,** 200.

STEWART, A. D. In Press. The Torridonian Sediments of North-West Scotland. Proc. Conf. on Continental Drift, Gander, 1967. *Mem. Am. Ass. Petrol. Geol.*

TAYLOR, J. H. (1958) Pre-Cambrian sedimentation in England and Wales. *Eclog. geol. Helv.*, **5,** 1078.

TARNEY, J. (1963) Assynt Dykes and their Metamorphism. *Nature, Lond.*, **199,** 672.

WILLIAMS, G. E. (1966) Palaeogeography of the Torridonian Applecross Group. *Nature, Lond.*, **209,** 1303.

WILLIAMS, G. E. (1968) Torridonian weathering, and its bearing on Torridonian palaeoclimate and source. *Scott. J. geol.*, **4,** 164.

Chapters 4, 5 and 6

Lower Palaeozoic

ANDERSON, T. B. (1965) The Evidence for the Southern Upland Fault in North-East Ireland, *Geol. Mag.*, **102,** 383.

BAILEY, R. J. (1964) A Ludlovian facies boundary in south Central Wales. *Geol. J.*, **4,** 1.

BASSETT, D. A. (1963) The Welsh Palaeozoic Geosyncline: a review of recent work on stratigraphy and sedimentation. In *The British Caledonides*, eds. M. R. W. JOHNSON and F. H. STEWART. Oliver & Boyd, Edinburgh and London.

BASSETT, D. A. and WALTON, E. K. (1960) The Hell's Mouth Grits: Cambrian greywackes in St. Tudwal's peninsula, North Wales. *Q. Jl geol. Soc. Lond.*, **116,** 85.

BEAVON, R. V., FITCH, F. J. and RAST, N. (1961) Nomenclature and diagnostic characters of ignimbrites with reference to Snowdonia. *Lpool Manchr geol. J.*, **2,** 601.

BOSWELL, P. G. H. (1961) The case against a Lower Palaeozoic geosyncline in Wales. *Lpool Manchr geol. J.*, **2,** 612.

BRENCHLEY, P. J. (1964) Ordovician ignimbrites in the Berwyn Hills, North Wales. *Geol. J.*, **4,** 43.

CRAIG, G. Y. and WALTON, E. K. (1959) Sequence and structure in the Silurian rocks of Kirkcudbrightshire. *Geol. Mag.*, **96,** 209.

CRIMES, T. P. and OLDERSHAW, M. A. (1967) Palaeocurrent determinations by magnetic fabric measurements on the Cambrian rocks of St. Tudwal's Peninsula, North Wales. *Geol. J.*, **5,** 217.

CUMMINS, W. A. (1957) The Denbigh Grits; Wenlock Greywackes in Wales. *Geol. Mag.*, **94,** 433.

CUMMINS, W. A. (1962) The greywacke problem. *Geol. J.*, **3,** 51.

DEAN, W. T. (1958) The faunal succession in the Caradoc Series of South Shropshire. *Bull. Br. Mus. nat. Hist.* (*Geology*), **3,** 191.

DEAN, W. T. (1959) The Stratigraphy of the Caradoc Series in the Cross Fell Inlier. *Proc. Yorks. geol. Soc.*, **32,** 185.

DEAN, W. T. (1960) The Ordovician Rocks of the Chatwall District, Shropshire. *Geol. Mag.*, **97,** 163.

DEAN, W. T. (1963) The Stile End Beds and Drygill Shales in the east and north of the English Lake District. *Bull. Br. Mus. Nat. Hist.* (*Geology*), **9,** 47.

DEAN, W. T. (1964) The geology of the Ordovician and adjacent strata in the southern Caradoc district of Shropshire. *Bull. Br. Mus. Nat. Hist.*, (*Geology*), **9,** 257.

DEAN, W. T. and DINELEY, D. L. (1961) The Ordovician and Associated Pre-Cambrian Rocks of the Pontesford District, Shropshire. *Geol. Mag.*, **98,** 367.

DEWEY, J. F. (1963) The Lower Palaeozoic stratigraphy of central Murrisk, County Mayo, Ireland, and the evolution of the South Mayo trough. *Q. Jl geol. Soc. Lond.*, **119,** 313.

FURNESS, R. R. (1965) The petrography and provenance of the Coniston Grits, east of the Lune Valley, Westmorland. *Geol. Mag.*, **102,** 252.

GEORGE, T. N. (1963) Palaeozoic Growth of the British Caledonides. In *The British Caledonides*, eds. M. R. W. JOHNSON and F. H. STEWART. Oliver & Boyd, Edinburgh and London.

GRIFFITHS, D. H., KING, R. F. and WILSON, C. D. V. (1961) Geophysical investigations in Tremadoc Bay, North Wales. *Q. Jl Geol. Soc. Lond.*, **117,** 171.

HARPER, J. C. (1948) The Ordovician and Silurian Rocks of Ireland. *Proc. Lpool geol. Soc.*, **20,** 48.

HOLLAND, C. H. (1958) The Ludlovian and Downtonian rocks of the Knighton district, Radnorshire. *Q. Jl geol. Soc. Lond.*, **114,** 449.

HOLLAND, C. H. (1965) The Siluro-Devonian boundary. *Geol. Mag.*, **102,** 213.

HOLLAND, C. H., LAWSON, J. D. and WALMSLEY, V. G. (1962) Ludlovian Classification—A Reply. *Geol. Mag.*, **99,** 393.

INGHAM, J. K. (1966) The Ordovician Rocks in the Cautley and Dent districts of Westmorland and Yorkshire. *Proc. Yorks. geol. Soc.*, **35,** 455.

JACKSON, D. E. (1961) Stratigraphy of the Skiddaw Group between Buttermere and Mungrisdale, Cumberland. *Geol. Mag.*, **98,** 515.

JONES, O. T. (1956) The geological evolution of Wales and the adjacent regions. *Q. Jl geol. Soc. Lond.*, **111,** 323.

KELLING, G. (1961) The stratigraphy and structure of the Ordovician rocks of the Rhinns of Galloway. *Q. Jl geol. Soc. Lond.*, **117,** 37.

LAWSON, J. D. (1954) The Silurian Succession at Gorsley (Herefordshire). *Geol. Mag.*, **91,** 227.

LAWSON, J. D. (1955) The Geology of the May Hill Inlier. *Q. Jl Geol. Soc. Lond.*, **111,** 85.

MATLEY, C. A. and STACEY WILSON, T. (1946) The Harlech Dome, north of the Barmouth estuary. *Q. Jl geol. Soc. Lond.*, **102,** 1.

MITCHELL, G. H. (1956) The geological history of the Lake District. *Proc. Yorks. geol. Soc.*, **30,** 407.

MOSELEY, F. (1964) The succession and structure of the Borrowdale Volcanic rocks north-west of Ullswater. *Geol. J.*, **4,** 127.

PHIPPS, C. B. (1962). The revised Ludlovian stratigraphy of the type area—a discussion. *Geol. Mag.*, **99,** 385.

PHIPPS, C. B. and REEVE, F. A. E. (1967) Stratigraphy and Geological History of the Malvern, Abberly and Ledbury Hills. *Geol. J.*, **5,** 339.

POTTER, J. F. and PRICE, J. H. (1965) Comparative sections through rocks of Ludlovian-Downtonian age in the Llandovery and Llandeilo Districts. *Proc. Geol. Ass.*, **76,** 379.

READING, H. G. and POOLE, A. B. (1961) A Llandovery shoreline from the southern Malverns. *Geol. Mag.*, **98,** 295.

RICKARDS, R. B. (1964) The graptolitic mudstone and associated facies in the Silurian strata of the Howgill Fells. *Geol. Mag.*, **101,** 435.

RUST, B. R. (1965) The stratigraphy and structure of the Whithorn Area of Wigtownshire, Scotland. *Scott. J. Geol.*, **1,** 101.

SIMPSON, A. (1963) The stratigraphy and tectonics of the Manx Slate Series, Isle of Man. *Q. Jl geol. Soc. Lond.*, **119,** 367.

WALLACE, P. (1968) The sub-Mesozoic palaeogeology and palaeogeography of north eastern France and the Straits of Dover. *Palaeogeog., Palaeoecol., Palaeoclim.*, **4,** 241.

WALMSLEY, V. G. (1958) The geology of the Usk Inlier (Monmouthshire). *Q. Jl geol. Soc. Lond.*, **114,** 483.

WALTON, E. K. (1963) Sedimentation and Structure in the Southern Uplands. In *The British Caledonides*, eds. M. R. W. JOHNSON and F. H. STEWART. Oliver & Boyd, Edinburgh and London.

WHITAKER, J. H. MCD. (1962) The geology of the area around Leintwardine, Herefordshire. *Q. Jl geol. Soc. Lond.*, **118,** 319.

WHITTARD, W. F. (1952) A geology of South Shropshire. *Proc. Geol. Ass.*, **63,** 143.

WILLIAMS, A. (1962) The Barr and Lower Ardmillan Series (Caradoc) of the Girvan district, south-west Ayrshire, with a description of the brachiopods. *Mem. geol. Soc. Lond.*, **No. 3.**

WOOD, A. and SMITH, A. J. (1958) The sedimentation and sedimentary history of the Aberystwyth Grits (Upper Llandoverian). *Q. Jl Geol. Soc. Lond.*, **114,** 163.

ZIEGLER, A. M. (1965) Silurian marine communities and their environmental significance. *Nature, Lond.*, **207,** 270.

Chapter 7

Caledonian

More detail of the events described in this chapter may be obtained from Charlesworth, 1963; Craig, 1965, Chapters 3, 4, 6 and 7; and Johnson and Stewart, 1963, Chapters 2, 6 and 7; while more up-to-date reviews are contained in the papers by Brindley, Brown, Dewey, Harris, Johnson, Leggo and Phillips in the Proceedings of the Gander Conference, Newfoundland, 1967, *Mem. Am. Assoc. Petrol. Geol.*, in press.

ANDERSON, T. B. (1965) The Evidence for the Southern Upland Fault in North-East Ireland. *Geol. Mag.*, **102,** 383.

BARBER, A. J. (1965) The History of the Moine Thrust Zone, Lochcarron and Lochalsh, Scotland. *Proc. Geol. Ass.*, **76,** 215.

BROWN, P. E., MILLER, J. A., SOPER, N. J. and YORK, D. (1965) Potassium-argon age pattern of the British Caledonides. *Proc. Yorks. geol. Soc.*, **35,**103.

CHARLESWORTH, J. K. (1963) *Historical Geology of Ireland.* Oliver & Boyd, Edinburgh and London.

CRAIG, G. Y., ed. (1965) *The Geology of Scotland.* Oliver & Boyd, Edinburgh and London.

JOHNSON, M. R. W. and HARRIS, A. L. (1967) Dalradian-?Arenig relations in parts of the Highland Border, Scotland, and their significance in the chronology of the Dalradian orogeny. *Scott. J. Geol.*, **3,** 1.

JOHNSON, M. R. W. and STEWART, F. H. (1963) *The British Caledonides.* Oliver & Boyd, Edinburgh and London.

KENNEDY, W. Q. (1946) The Great Glen Fault. *Q. Jl geol. Soc. Lond.*, **102,** 41.

MCKERROW, W. S. (1962) The chronology of Caledonian folding in the British Isles. *Proc. natn. Acad. Sci. U.S.A.*, **48,** 1905.

MOSELEY, F. (1968) Joints and other structures in the Silurian rocks of the southern Shap Fells, Westmorland. *Geol. J.*, **6,** 79.

READ, H. H. (1961) Aspects of Caledonian magmatism in Britain. *Lpool Manchr geol. J.*, **2,** 653.

SHACKLETON, R. M. (1953). The structural evolution of North Wales. *Lpool Manchr geol. J.*, **1,** 261.

SIMPSON, A. (1968) The Caledonian history of the north-eastern Irish Sea region and its relation to surrounding areas. *Scott. J. geol.*, **4,** 135.

TREMLETT, W. E. (1965) Caledonian orogeny of the Central Irish Sea Region. *Nature, Lond.*, **207,** 154.

Chapter 8

Devonian

ALLEN, J. R. L. (1960) Cornstone. *Geol. Mag.*, **97,** 43.

ALLEN, J. R. L. (1962) The Petrology, origin and deposition of the highest Lower Old Red Sandstone of Shropshire, England. *J. sedim. Petrol.*, **32,** 657.

ALLEN, J. R. L. (1963) Depositional Features of Dittonian Rocks: Pembrokeshire compared with the Welsh Borderland. *Geol. Mag.*, **100,** 385.

ALLEN, J. R. L. (1965) The sedimentation and Palaeogeography of the Old Red Sandstone of Anglesey, North Wales. *Proc. Yorks. geol. Soc.*, **35,** 139.

ALLEN, J. R. L. and TARLO, L. B. (1963) The Downtonian and Dittonian Facies of the Welsh Borderland. *Geol. Mag.*, **100,** 129.
BALL, H. W. (1951) The Silurian and Devonian Rocks of Turner's Hill and Gornal, South Staffordshire. *Proc. Geol. Ass.*, **62,** 225.
BALL, H. W. and DINELEY, D. L. (1952) Notes on the Old Red Sandstone of the Clee Hills. *Proc. Geol. Ass.*, **63,** 207.
BARTON, R. M. (1964) *Geology of Cornwall.* Barton Ltd., Truro.
CAPEWELL, J. G. (1957) The stratigraphy and sedimentation of the Old Red Sandstone of the Comeragh Mountains and adjacent areas, Co. Waterford, Ireland. *Q. Jl geol. Soc. Lond.*, **112,** 393.
DEWEY, H. (1961) *British Regional Geology. South-west England.* H.M.S.O., London.
DINELEY, D. L. (1961) The Devonian System in South Devon. *Field Studies,* **1,** 121.
ERBEN, H. K. (1964) Facies Development in the marine Devonian of the Old World. *Proc. Ussher Soc.*, **1,** 92.
FRIEND, P. F. and HOUSE, M. R. (1964) Devonian Period. *Phanerozoic time scale, Geol. Soc. Lond.* **120 S,** 223.
GALLOIS, R. W. (1965) *British Regional Geology. The Wealden District.* H.M.S.O., London.
HOLLAND, C. H. (1965) The Siluro-Devonian Boundary. *Geol. Mag.*, **102,** 213.
HOUSE, M. R. (1963) Devonian ammonoid successions and facies in Devon and Cornwall. *Q. Jl geol. Soc. Lond.*, **119,** 1.
HOUSE, M. R. and SELWOOD, E. B. (1964) Palaeozoic palaeontology in Devon and Cornwall. *Present Views on Some Aspects of the Geology of Cornwall and Devon. 150th Anniversary of Roy. geol. Soc. Corn.*, 45.
LAMBERT, J. L. M. (1965) A Reinterpretation of the Breccias in the Meneage crush zone of the Lizard Boundary, South-west England. *Q. Jl geol. Soc. Lond.*, **121,** 339.
MIDDLETON, G. V. (1960) Spilitic Rocks in South-east Devonshire. *Geol. Mag.*, **97,** 192.
RAYNER, D. H. (1963) The Achanarras Limestone of the Middle Old Red Sandstone, Caithness, Scotland. *Proc. Yorks. geol. Soc.*, **34,** 117.
SELWOOD, E. B. (1961) The Upper Devonian and Lower Carboniferous Stratigraphy of Bocastle and Tintagel, Cornwall. *Geol. Mag.*, **98,** 161.
SIMPSON, S. (1959) *Lexique Stratigraphique International,* **3a** *Angleterre, Pay de Galles, Écosse.* Devonian.
TARLO, L. B. (1965) Siluro-Devonian Boundary. *Geol. Mag.*, **102,** 349.
WEBBY, B. D. (1965) The Middle Devonian Marine Transgression in North Devon and West Somerset. *Geol. Mag.*, **102,** 478.

Chapter 9

Carboniferous

BISSON, G., LAMB, R. K. and CALVER, M. A. (1967) Boreholes in the concealed Kent Coalfield. *Bull. geol. Surv. Gt Br.*, **26,** 99.
BLACK, W. W. (1958) The Structure of the Burnsall-Cracoe district and its bearing on the origin of the Cracoe Knoll-Reefs. *Proc. Yorks. geol. Soc.*, **31,** 391.

CRAIG, G. Y. and NAIRN, A. E. M. (1956) The Lower Carboniferous Outliers of the Colvend and Rerrick Shores, Kirkcudbrightshire. *Geol. Mag.*, **93,** 249.

DEARMAN, W. R. (1959) The Structure of the Culm Measures at Meldon, near Okehampton, North Devon. *Q. Jl geol. Soc. Lond.*, **115,** 65.

DELÉPINE, G. (1951) Studies of the Devonian and Carboniferous of Western Europe and North Africa. *Proc. Geol. Ass.*, **62,** 140.

DUNHAM, K. C. (1948) Lower Carboniferous Sedimentation in the Northern Pennines (England). *Rep. Int. geol. Congress.*, 18th Sess., Pt IV, 46.

EDEN, R. A., ORME, G. R., MITCHELL, M. and SHIRLEY, J. (1964) A study of part of the margin of the Carboniferous Limestone 'massif' in the Pindale area of Derbyshire. *Bull. Geol. Surv. Gt Br.*, **21,** 73.

GEORGE, T. N. (1948) Tournaisian Facies in Britain. *Rep. Int. geol. Congress.*, 18th Sess., Pt X, 34.

GEORGE, T. N. (1955a) British Carboniferous Stratigraphy. *Sci. Progr. Lond.*, **43,** 87.

GEORGE, T. N. (1955b) Carboniferous Main Limestone of the East Crop in South Wales. *Q. Jl geol. Soc. Lond.*, **111,** 309.

GEORGE, T. N. (1956) The Namurian Usk Anticline. *Proc. Geol. Ass.*, **66,** 297.

GEORGE, T. N. (1957) Limestones and Dolomites. *Sci. Progr., Lond.*, **45,** 95.

GEORGE, T. N. (1958) Lower Carboniferous Palaeogeography of the British Isles. *Proc. Yorks. geol. Soc.*, **31,** 227.

GEORGE, T. N. and OSWALD, D. H. (1957) The Carboniferous rocks of the Donegal Syncline. *Q. Jl geol. Soc. Lond.*, **113,** 137.

GILL, W. D. and KUENEN, P. H. (1957) Sand volcanoes on slumps in the Carboniferous of County Clare, Ireland. *Q. Jl geol. Soc. Lond.*, **113,** 441.

HODSON, F. and LEWARNE, G. C. (1961) A mid-Carboniferous (Namurian) Basin in parts of the Counties of Limerick and Clare, Ireland. *Q. Jl geol. Soc. Lond.*, **117,** 30.

HUDSON, R. G. S. (1945) The goniatite zones of the Namurian. *Geol. Mag.*, **82,** 1.

HULL, J. H. (1968) The Namurian stages of north-eastern England. *Proc. Yorks geol. Soc.*, **36,** 297.

JOHNSON, G. A. L., HODGE, B. L. and FAIRBAIRN, R. A. (1962) The Base of the Namurian and of the Millstone Grit in north-eastern England. *Proc. Yorks. geol. Soc.*, **33,** 341.

JONES, D. G. and OWEN, T. R. (1961) The age and relationships of the Cornbrook Sandstone. *Geol. Mag.*, **98,** 285.

KENT, P. E. (1966) The Structure of the Concealed Carboniferous Rocks of North-Eastern England. *Proc. Yorks. geol. Soc.*, **35,** 323.

LEITCH, D., OWEN, T. R., and JONES, D. G. (1958) The basal Coal Measures of the South Wales Coalfield from Llandybie to Brynmawr. *Q. Jl geol. Soc. Lond.*, **113,** 461.

MOORE, D. (1957) The Yoredale series of Upper Wensleydale and adjacent parts of north-west Yorkshire. *Proc. Yorks. geol. Soc.*, **31,** 91.

MOSELEY, F. (1954) The Namurian of the Lancaster Fells. *Q. Jl geol. Soc. Lond.*, **109,** 423.

MYKURA, W. (1967) The Upper Carboniferous rocks of south-west Ayrshire. *Bull. geol. Surv. Gt Br.*, **26,** 23.

NEVES, R., READ, W. A. and WILSON, R. B. (1965) Notes on recent spore and goniatite evidence from the Passage Group of the Scottish Upper Carboniferous succession. *Scott. J. Geol.*, **1,** 185.

OWEN, T. R. (1964) The Tectonic framework of Carboniferous sedimentation in South Wales. *Devs Sediment.*, **1,** 301.

OWEN, T. R. and JONES, D. G. (1955) On the presence of the Upper Dibunophyllum Zone (D_3) near Glynneath, S. Wales. *Geol. Mag.*, **92,** 457.

OWEN, T. R. and JONES, D. G. (1961) The nature of the Millstone Grit–Carboniferous Limestone junction of a part of the North Crop of the S. Wales Coalfield. *Proc. Geol. Ass.*, **72,** 239.

PRENTICE, J. E. (1959) Dinantian, Namurian and Westphalian rocks of the district south-west of Barnstaple, North Devon. *Q. Jl geol. Soc. Lond.*, **115,** 261.

PRENTICE, J. E. (1960) The Stratigraphy of the Upper Carboniferous rocks of the Bideford region, North Devon. *Q. Jl geol. Soc. Lond.*, **116,** 397.

READING, H. G. (1964) A review of the factors affecting the sedimentation of the Millstone Grit (Namurian) in the Central Pennines. *Devs Sediment.*, **1,** 340.

READING, H. G. (1965) Recent finds in the Upper Carboniferous of Southwest England and their significance. *Nature, Lond.*, **208,** 745.

ROWELL, A. J. and SCANLON, J. E. (1957a) The Namurian of the north-west quarter of the Askrigg Block. *Proc. Yorks. geol. Soc.*, **31,** 1.

ROWELL, A. J. and SCANLON, J. E. (1957b) The Relation between the Yoredale Series and the Millstone Grit on the Askrigg Block. *Proc. Yorks. geol. Soc.*, **31,** 79.

SHIRLEY, J. (1958) The Carboniferous Limestone of the Monyash-Wirksworth area, Derbyshire. *Q. Jl geol. Soc. Lond.*, **114,** 411.

SIMPSON, S. (1959) Culm Stratigraphy and the Age of the Main Orogenic Phase in Devon and Cornwall. *Geol. Mag.*, **96,** 201.

STUBBLEFIELD, C. J. and TROTTER, F. M. (1957) Divisions of the Coal Measures on Geological maps of England and Wales. *Bull. geol. Surv. Gt. Br.*, **13,** 1.

SULLIVAN, H. J. (1964) Miospores from the Drybrook Sandstone and associated measures in the Forest of Dean Basin, Gloucestershire. *Palaeontology*, **7,** 351.

TROTTER, F. M. (1951) Sedimentation of the Namurian of North-west England and adjoining areas. *Lpool Manchr geol. J.*, **1,** 77.

TRUEMAN, A. E. (1954) *The Coalfields of Great Britain*. Edward Arnold, London.

TURNER, J. S. (1951) The Lower Carboniferous Rocks of Ireland. *Lpool Manchr geol. J.*, **1,** 113.

WESTOLL, T. S., ROBSON, D. A. and GREEN, R. (1955) A Guide to the Geology of the District around Alnwick, Northumberland. *Proc. Yorks. geol. Soc.*, **30,** 61.

WILSON, A. A. (1960a) The Carboniferous Rocks of Coverdale and Adjacent Valleys in the Yorkshire Pennines. *Proc. Yorks. geol. Soc.*, **32,** 285.

WILSON, A. A. (1960b) The Millstone Grit Series of Colsterdale and Neighbourhood, Yorkshire. *Proc. Yorks. geol. Soc.*, **32**, 429.

WOODLAND, A. W. *et al.* (1957) Classification of the Coal Measures. *Bull. geol. Surv. Gt. Br.*, **13,** 6.

Chapter 10

Variscan Orogeny

COE, K., ed. (1962) *Some Aspects of the Variscan Fold Belt*. 9th Inter-Univ. Cong., Manchester Univ. Press, Manchester.
DEARMAN, W. R. (1959) The Structure of the Culm Measures at Meldon, Nr. Okehampton, N. Devon. *Q. Jl geol. Soc. Lond.*, **115**, 65.
DEARMAN, W. R. (1963) Wrench Faulting in Cornwall and Devon. *Proc. Geol. Ass.*, **74**, 265.
DUNHAM, K. C., DUNHAM, A. C., HODGE, B. L. and JOHNSTONE, G. A. L. (1965) Granite beneath Viséan sediments with mineralization at Rookhope. *Q. Jl geol. Soc. Lond.*, **121**, 283.
FITCH, F. J. and MILLER, J. A. (1967) The age of the Whin Sill. *Geol. J.*, **5**, 233.
FYSON, W. K. (1962) Tectonic Structures in the Devonian Rocks near Plymouth, Devon. *Geol. Mag.*, **99**, 208.
GEORGE, T. N. (1962) Tectonics and Palaeogeography in Southern England. *Sci. Progr., Lond.*, **50**, 192.
GEORGE, T. N. (1963) Tectonics and Palaeogeography in Northern England. *Sci. Progr., Lond.*, **51**, 32.
HENDRIKS, E. M. L. (1959) A Summary of Present Views on the Structure of Cornwall and Devon. *Geol. Mag.*, **96**, 253.
JENKINS, T. B. H. (1962) The Sequence and Correlation of the Coal Measures of Pembrokeshire. *Q. Jl geol. Soc. Lond.*, **118**, 65.
KENNEDY, W. Q. (1958) The Tectonic Evolution of the Midland Valley of Scotland. *Trans. geol. Soc. Glasg.*, **23**, 106.
KENT, P. E. (1966) The Structure of the concealed Carboniferous rocks of NE. England. *Proc. Yorks. geol. Soc.*, **35**, 323.
MOSELEY, F. and AHMED, S. M. (1967) Carboniferous joints in the north of England and their relation to earlier and later structures. *Proc. Yorks. geol. Soc.*, **36**, 61.
SIMPSON, S. (1959) Culm Stratigraphy and the age of the Main Orogenic Phase in Devon and Cornwall. *Geol. Mag.*, **96**, 201.
WALMSLEY, V. G. (1958) The Geology of the Usk Inlier. *Q. Jl geol. Soc. Lond.*, **114**, 483.

Chapter 11

Permian

BURGESS, I. C. (1965) The Permo-Triassic Rocks around Kirkby Stephen, Westmorland. *Proc. Yorks. geol. Soc.*, **35**, 91.
CURRY, D. *et al.* (1962) Geology of the Western Approaches of the English Channel. I: Chalky Rocks . . .. *Phil. Trans. R. Soc. B*, **245**, 267.
DAVIES, J. (1961) Two Derived Fossils from the Bunter Pebble Beds of Little Hilbre Island, Cheshire. *Lpool Manchr geol. J.*, **2**, 626.
DAY, A. A. (1958) The Pre-Tertiary Geology of the Western Approaches to the English Channel. *Geol. Mag.*, **95**, 137.
EDWARDS, W. (1951) *The Concealed Coalfield of Yorkshire and Nottinghamshire*, 3rd edit. Mem. Geol. Surv. Gt. Brit.
FOWLER, A. (1955) The Permian Rocks of Grange, Co. Tyrone. *Bull. geol. Surv. Gt Br.*, **8**, 44.

HOLLINGWORTH, S. E. (1942) The Correlation of Gypsum-Anhydrite deposits and the Associated Strata in the North of England. *Proc. Geol. Ass.*, **53,** 141.

HUTCHINS, P. F. (1963) The Lower New Red Sandstone of the Crediton Valley. *Geol. Mag.*, **100,** 107.

KING, W. R. B. (1954) The Geological History of the English Channel. *Q. Jl geol. Soc. Lond.*, **110,** 77.

LAMING, D. J. C. (1965) Age of the New Red Sandstone in South Devonshire. *Nature, Lond.*, **207,** 624.

MEYER, H. G. A. (1965) Stratigraphy of the Permian Evaporites, NW. England. *Proc. Yorks. geol. Soc.*, **35,** 71.

MYKURA, W. (1965) The Age of the Lower Part of the New Red Sandstone of SW. Scotland. *Scott. J. Geol.*, **1,** 9.

SHOTTON, F. W. (1956) Some aspects of the New Red Desert in Britain. *Lpool Manchr geol. J.*, **1,** 450.

SMITH, D. B. (1958) Observations on the Magnesian Limestone Reefs of North-Eastern Durhan. *Bull. geol. Surv. Gt. Br.*, **15,** 71.

SMITH, D. B. (1964) The Permian period. *Phanerozoic Time Scale. Geol. Soc. Lond.* **120 S,** 211.

STEWART, F. H. (1953) Permian Evaporites and Associated Rocks in Texas and New Mexico compared with those of Northern England. *Proc. Yorks. geol. Soc.*, **29,** 185.

STEWART, F. H. (1963a) The Permian Lower Evaporites of Fordon in Yorkshire. *Proc. Yorks geol. Soc.*, **34,** 1.

STEWART, F. H. (1963b) Data of Geochemistry, Chapter Y. Marine Evaporites. *U.S. Geol. Survey Prof. Papers*, **440-Y,** 1.

STONLEY, H. M. M. (1958). The Upper Permian Flora of England. *Bull. Br. Mus. nat. Hist.* **3,** 289.

WAGNER, R. H. (1966) On the presence of probable Upper Stephanian Beds in Ayrshire. *Scott. J. Geol.*, **2,** 122.

WILLS, L. J. (1956) *Concealed Coalfields.* Blackie, London.

Chapter 12

Triassic

CUMMINS, W. A. (1958) Some Sedimentary Structures from the Lower Keuper Sandstones. *Lpool Manchr geol. J.*, **2,** 37.

ELLIOT, G. F. (1953). The Conditions of Formation of the Lower Rhaetic at Blue Anchor, Somerset, and of the Western European Rhaetic generally. *Proc. geol. Soc. Lond.*, 1494. xxxii.

ELLIOT, R. E. (1961) The Stratigraphy of the Keuper Series in Southern Notts. *Proc. Yorks. geol. Soc.*, **33,** 197.

FITCH, F. J., MILLER, J. A. and THOMPSON, D. B. (1966) The palaeogeographic significance of isotopic age determinations on detrital micas from the Triassic of the Stockport-Macclesfield District, Cheshire, England. *Palaeogeogr., Palaeoclim., Palaeoecol.*, **2,** 525.

FRANCIS, E. H. (1959) Rhaetic of Bridgend District, Glamorgan. *Proc. Geol. Ass.*, **70,** 158.

GARRETT, P. A. (1960) Nomenclature of the Keuper. *Nature, Lond.*, **187,** 868.

HALLAM, A. (1960) The White Lias of the Devon Coast. *Proc. Geol. Ass.*, **71,** 47.
KENT, P. E. (1967) Outline geology of the southern North Sea Basin. *Proc. Yorks. geol. Soc.*, **36,** 1.
KLEIN, G. DE V. (1962) Sedimentary structures in the Keuper Marl (Upper Triassic). *Geol. Mag.*, **99,** 137.
RAYMOND, L. R. (1955) Rhaetic Beds and Tea Green Marl of North Yorkshire. *Proc. Yorks. geol. Soc.*, **30,** 5.
ROSE, G. N. and KENT, P. E. (1955) A *Lingula* bed in the Keuper of Nottinghamshire. *Geol. Mag.*, **92,** 476.
SHOTTON, F. W. (1937) Lower Bunter Sandstones of N. Worcs. and E. Shropshire. *Geol. Mag.*, **74,** 534.
SHOTTON, F. W. (1952) Underground Water Supply of Midland Breweries. *J. Inst. Brew.*, **58,** 449.
SMITH, W. CAMPBELL (1963) Description of the igneous rocks represented among pebbles from the Bunter Pebble Beds of the Midlands of England. *Bull. Br. Mus. nat. Hist.* (*Mineralogy*), **2,** 1.
WILLS, L. J. (1956) *Concealed Coalfields*. Blackie, London.

Chapter 13

Jurassic

AGER, D. V. (1956) The Geographical Distribution of Brachiopods in the British Middle Lias. *Q. Jl geol. Soc. Lond.*, **112,** 157.
ARKELL, W. J. (1933) *The Jurassic System in Great Britain*. Oxford Univ. Press, London.
ARKELL, W. S. and DONOVAN, D. T. (1952) The Fuller's Earth of the Cotswolds and its relation to the Great Oolite. *Q. Jl geol. Soc. Lond.*, **107,** 227.
BATE, R. H. (1959) The Yons Nab Beds of the Middle Jurassic of the Yorkshire Coast. *Proc. Yorks. geol. Soc.*, **32,** 153.
BATE, R. H. (1967) Stratigraphy and palaeogeography of the Yorkshire Oolites and their relationship with the Lincolnshire Limestone. *Bull. Br. Mus. nat. Hist.*, (*Geology*), **14,** 111.
BITTERLI, P. (1963) Aspects of the genesis of Bituminous rock sequences. *Geologie Mijnb.*, **42,** 183.
BROWN, P. R. (1963) Algal limestone and associated sediments in the Basal Purbeck of Dorset. *Geol. Mag.*, **100**, 565.
CROWELL, J. C. (1961) Depositional structures from Jurassic Boulder Beds, East Sutherland. *Trans. Edinb. geol. Soc.*, **18,** 202.
DEAN, W. T., DONOVAN, D. T. and HOWARTH, M. K. (1961) The Liassic Ammonite Zones and Subzones of the North-west European Province. *Bull. Br. Mus. nat. Hist.*, (*Geology*), **4,** 437.
DONOVAN, D. T. (1956) The zonal stratigraphy of the Blue Lias around Keynsham, Somerset. *Proc. Geol. Ass.*, **66,** 182.
HALLAM, A. (1959) Stratigraphy of the Broadford Beds of Skye, Raasay and Applecross. *Proc. Yorks. geol. Soc.*, **32,** 165.
HALLAM, A. (1960) A Sedimentary and faunal Study of the Blue Lias of Dorset and Glamorgan. *Phil. Trans. R. Soc. B*, **243,** 1.
HALLAM, A. (1961) Cyclothems, Transgressions and Faunal Changes in the Lias of North-west Europe. *Trans. Edinb. geol. Soc.*, **18,** 124.

HALLAM, A. (1963) Eustatic Control of Major Cyclic Changes in Jurassic Sedimentation. *Geol. Mag.*, **100**, 444.

HALLAM, A. (1964) Liassic Sedimentary Cycles in Western Europe and their relationships to changes in sea-level. *Devs. Sediment.*, **1**, 157.

HALLAM, A. (1966) Depositional environment of British Liassic Ironstones considered in the context of their facies relationships. *Nature, Lond.*, **209**, 1306.

HOWARTH, M. K. (1955) The Domerian of the Yorkshire Coast. *Proc. Yorks. geol. Soc.*, **30**, 147.

HOWARTH, M. K. (1957) The Middle Lias of the Dorset Coast. *Q. Jl geol. Soc. Lond.* **113**, 185.

HOWARTH, M. K. (1962) The Jet Rock Series and the Alum Shale of the Yorkshire Coast. *Proc. Yorks. geol. Soc.*, **33**, 381.

HOWARTH, M. K. (1964) The Jurassic period *Phanerozoic Time Scale. Geol. Soc. Lond.* **120 S**, 203.

HOWITT, F. (1964) Stratigraphy and structure of the Purbeck inliers of Sussex. *Q. Jl geol. Soc. Lond.*, **120**, 77.

HUDSON, J. D. (1962) The Stratigraphy of the Great Estuarine Series (Middle Jurassic) of the Inner Hebrides. *Trans. Edinb. geol. Soc.*, **19**, 139.

HUDSON, J. D. (1963) The Recognition of Salinity-controlled Mollusc assemblages in the Great Estuarine Series (Middle Jurassic) of the Inner Hebrides. *Palaeontology*, **6**, 318.

HUDSON, J. D. (1964) The petrology of the Sandstones of the Great Estuarine Series and the Jurassic Palaeogeography of Scotland. *Proc. Geol. Ass.*, **75**, 499.

KENT, P. E. (1955) The Market Weighton Structure. *Proc. Yorks. Geol. Soc.*, **30**, 197.

LLOYD, A. S. (1964) The Luxembourg Colloquium and the Revision of the Stages of the Jurassic System. *Geol. Mag.*, **101**, 249.

MARTIN, A. J. (1967) Bathonian sedimentation in Southern England. *Proc. Geol. Ass.*, **78**, 473.

MCKERROW, W. S. (1953) Variation in the Terebratulacea of the Fuller's Earth Rock. *Q. Jl geol. Soc. Lond.*, **109**, 97.

MORTON, N. (1965) The Stratigraphy of the Bearreraig Sandstone Series of Skye and Raasay. *Scott. J. Geol.*, **1**, 189.

MORTON, N. and HUDSON, J. D. (1964) The Stratigraphical Nomenclature of the Lower and Middle Jurassic Rocks of the Hebrides. *Geol. Mag.*, **101**, 531.

SMITHSON, F. (1942) The Middle Jurassic Rocks of Yorkshire: a petrological and palaeogeographical study. *Q. Jl geol. Soc. Lond.*, **98**, 27.

TORRENS, H. S. (1967) The Great Oolite Limestone of the Midlands. *Trans. Leics. Lit. Phil. Soc.*, **61**, 65.

WEST, I. M. (1964) Evaporite Diagenesis in the Lower Purbeck Beds of Dorset. *Proc. Yorks. geol. Soc.*, **34**, 315.

WILSON, V. (1948) Mesozoic Rocks East of the Trent. Ch. 7, in *Guide to the Geology of the East Midlands*, ed. C. E. Marshall. Nott'm. Univ. Press, Nottingham.

WOBBER, F. J. (1966) A study of the depositional area of the Glamorgan Lias. *Proc. Geol. Ass.*, **77**, 127.

Chapter 14

Cretaceous

ALLEN, P. (1954) Geology and Geography of the London-North Sea Uplands in Wealden Times. *Geol. Mag.*, **91,** 498.

ALLEN, P. (1959) The Wealden Environment: Anglo-Paris Basin. *Phil. Trans. R. Soc. B*, **242,** 283.

ALLEN, P. (1967) Origin of the Hastings Facies in North-western Europe. *Proc. Geol. Ass.*, **78,** 27.

ANDERSON, F. W. (1962) Correlation of the Upper Purbeck Beds of England with the German Wealden. *Geol. J.*, **1,** 21.

DE BOER, G., NEALE, J. W. and PENNY, L. F. (1958) A guide to the geology of the area between Market Weighton and the Humber. *Proc. Yorks. geol. Soc.*, **31,** 157.

CASEY, R. (1961) The Stratigraphical Palaeontology of the Lower Greensand. *Palaeontology*, **3,** 487.

CASEY, R. (1963) The dawn of the Cretaceous Period in Britain. *Bull. S. East. Un. scient. Socs*, **117,** 1.

CASEY, R. (1964) The Cretaceous Period, *Phanerozoic Time Scale. Geol. Soc. London*, **120, S**, 193.

CASEY, R. and BRISTOW, C. R. (1964) Notes on some ferruginous strata in Buckinghamshire and Wiltshire. *Geol. Mag.*, **101,** 116.

HANCOCK, J. M. (1961) The Cretaceous System in Northern Ireland. *Q. Jl geol. Soc.*, **117,** 11.

HANCOCK, J. M. *et al.* (1965) The Gault of the Weald: a symposium. *Proc. Geol. Ass.*, **76,** 243.

HAWKES, L. (1951) The Erratics of the English Chalk. *Proc. Geol. Ass.*, **62,** 257.

HUGHES, N. F. (1958) Palaeontological Evidence for the Age of the English Wealden. *Geol. Mag.*, **95,** 41.

HUMPHRIES, D. W. (1964) The stratigraphy of the Lower Greensand of the south-west Weald. *Proc. Geol. Ass.*, **75,** 39.

KAYE, P. (1964) Observations on the Speeton Clay (Lower Cretaceous). *Geol. Mag.*, **101,** 340.

KAYE, P. (1966) Lower Cretaceous Palaeogeography of North-west Europe. *Geol. Mag.*, **103,** 257.

KIRKALDY, J. F. (1963) The Wealden and Marine Lower Cretaceous Beds of England. *Proc. Geol. Ass.*, **74,** 127.

MACDOUGALL, J. D. S. and PRENTICE, J. E. (1964) Sedimentary environments of the Weald Clay of south-eastern England. *Devs Sediment.*, **1,** 257.

NEALE, J. W. (1960) The subdivision of the Upper D beds of the Speeton Clay of Speeton, E. Yorkshire. *Geol. Mag.*, **97,** 353.

NEALE, J. W. and SARJEANT, W. A. S. (1962) Microplankton from the Speeton Clay of Yorkshire. *Geol. Mag.*, **99,** 439.

NEALE, J. W. (1968) Biofacies and lithofacies of the Speeton Clay D Beds, E. Yorkshire. *Proc. Yorks. geol. Soc.*, **36,** 309.

OWEN, E. G. and THURRELL, R. G. (1968) British Neocomian rhynchonelloid brachiopods. *Bull. Br. Mus. Nat. Hist.*, (*Geology*), **16,** 101.

SMITH, W. E. (1957; 1961) The Cenomanian Limestone of the Beer District, Devon. *Proc. Geol. Ass.*, **68,** 115; also *Proc. Geol. Ass.*, **72,** 91.

TAYLOR, J. H. (1963) Sedimentary features of an ancient deltaic complex; the Wealden rocks of south-eastern England. *Sedimentology*, **2,** 2.
TRESISE, G. R. (1960) Aspects of the Lithology of the Wessex Upper Greensand. *Proc. Geol. Ass.*, **71,** 316.
WALSH, P. T. (1966) Cretaceous outliers in south-west Ireland and their implications for Cretaceous palaeogeography. *Q. Jl geol. Soc. Lond.*, **122,** 63.

Chapter 15

Tertiary

BADEN-POWELL, D. F. W. (1960) On the nature of the Coralline Crag. *Geol. Mag.*, **97,** 123.
BUTCHER, N. E. (1963) Age of the Alpine Folds of Southern England. *Geol. Mag.*, **100,** 468.
CHARLESWORTH, J. K. *et al.* (1960) The Geology of North-east Ireland. *Proc. Geol. Ass.*, **70,** 429.
CURRY, D. (1965) The Palaeogene Beds of South-east England. *Proc. Geol. Ass.*, **76,** 151.
DAVIES, A. G. and ELLIOTT, G. F. (1958) The palaeogeography of the London Clay Sea. *Proc. Geol. Ass.*, **68,** 255.
GEORGE, T. N. (1967) Landforms and structure in Ulster. *Scott. J. geol.*, **3,** 413.
HESTER, S. W. (1965) Stratigraphy and Palaeontology of the Woolwich and Reading Beds. *Bull. geol. Surv. Gt Br.*, **23,** 117.
HOUSE, M. R. (1961) The Structure of the Weymouth Anticline. *Proc. Geol. Ass.*, **72,** 221.
MITCHELL, G. F. (1965) The St. Erth Beds, an alternative Explanation. *Proc. Geol. Ass.*, **76,** 345.
PHILLIPS, W. J. (1964) The Structures in the Jurassic and Cretaceous Rocks on the Dorset Coast Between White Nothe and Mupe Bay. *Proc. Geol. Ass.*, **75,** 373.
WEST, I. M. (1964) (Corresp.) Age of the Alpine Folds of Southern England. *Geol. Mag.*, **101,** 190.
WOOLDRIDGE, S. W. (1926) The Structural Evolution of the London Basin. *Proc. Geol. Ass.*, **37,** 162.
WOOLDRIDGE, S. W. and LINTON, D. L. (1938) Some episodes in the structural evolution of SE. England . . .. *Proc. Geol. Ass.*, **49,** 264.

Chapter 16

Pleistocene

BOULTON, G. S. and WORSLEY, P. (1965) Late Weichselian Glaciation in the Cheshire-Shropshire Basin. *Nature, Lond.*, **207,** 704.
CATT, J. A. and PENNY, L. F. (1966) The Pleistocene deposits of Holderness, E. Yorks. *Proc. Yorks. geol. Soc.*, **35,** 375.
CHARLESWORTH, J. K. (1957) *The Quaternary Era.* Edward Arnold, London.
COOPE, G. R. (1964) The Response of the British Insect fauna to late Quaternary climatic oscillations. *Proc. XII Int. Congr. Ent., London.*

COOPE, G. R. and SANDS, C. H. S. (1966) Insect faunas of the last glaciation from the Tame Valley, Warwickshire. *Proc. R. Soc., B.* **165,** 389.

COOPE, G. R., SHOTTON, F. W. and STRACHAN, I. (1961) *Phil. Trans. R. Soc. B.,* **244,** 379.

FLINT, R. F. (1957) *Glacial and Pleistocene Geology.* Wiley, New York and Chichester.

GODWIN, H. (1956) *History of the British Flora.* University Press, Cambridge.

KING, C. A. M. and WHEELER, P. T. (1965) The Raised Beaches of the north coast of Sutherland, Scotland. *Geol. Mag.,* **100,** 299.

PENNY, L. F. (1964) A review of the Last Glaciation in Great Britain. *Proc. Yorks. geol. Soc.,* **34,** 387.

SHOTTON, F. W. (1953) The Pleistocene Deposits of the area between Coventry, Rugby and Leamington and their bearing upon the topographic development of the Midlands. *Phil. Trans. R. Soc. B.,* **237,** 209.

SHOTTON, F. W. (1960) Large scale patterned ground in the valley of the Worcestershire Avon. *Geol. Mag.,* **97,** 404.

SHOTTON, F. W. (1964) The geological background to European Pleistocene entomology. *Proc. XII Int. Congr., Ent., London.*

SHOTTON, F. W. (1965) Normal Faulting in the British Pleistocene. *Q. Jl geol. Soc. Lond.,* **121,** 419.

SHOTTON, F. W. and STRACHAN, I. (1959) An investigation of a Peat Moor at Rodbaston, Penkridge, Staffs. *Q. Jl geol. Soc. Lond.,* **115,** 1.

SIMPSON, I. M. and WEST, R. G. (1958) On the stratigraphy and palaeobotany of a Late Pleistocene organic deposit at Chelford, Cheshire. *New Phytol.,* **57,** 239.

SISSONS, J. B. (1967) Glacial stages and radiocarbon dates in Scotland. *Scott. J. geol.,* **3,** 375.

SYNGE, F. M. (1964) Some problems concerned with the glacial succession in south-east Ireland. *Irish Geogr.,* **5,** 73.

SYNGE, F. M. and STEPHENS, N. 1960. The Quaternary Period in Ireland—an assessment, 1960. *Irish Geogr.,* **4,** 122.

TOMLINSON, M. E. (1963) The Pleistocene Chronology of the Midlands. *Proc. Geol. Ass.,* **74,** 187.

WEST, R. G. (1960) The Ice Age. *Advmt Sci.,* **16,** 428.

WEST, R. G. (1963) Problems of the British Quaternary. *Proc. Geol. Ass.,* **74,** 147.

WEST, R. G. and DONNER, J. J. (1956) The Glaciations of East Anglia and the East Midlands—a differentiation based on stone-orientation of the tills. *Q. Jl geol. Soc. Lond.,* **112,** 69.

Index